ADVANCES IN RICE SCIENCE

Botany, Production, and Crop Improvement

ADVANCES IN RICE SCIENCE

Botany, Production, and Crop Improvement

Ratikanta Maiti, PhD
Narayan Chandra Sarkar, PhD
Humberto González Rodríguez, PhD
Ch. Aruna Kumari, PhD
Sameena Begum, MSc
Dasari Rajkumar, MSc

Apple Academic Press Inc.
4164 Lakeshore Road
Burlington ON L7L 1A4, Canada

Apple Academic Press Inc.
1265 Goldenrod Circle NE
Palm Bay, Florida 32905, USA

Exclusive worldwide distribution by CRC Press, a member of Taylor & Francis Group

International Standard Book Number-13: 978-1-77188-854-7 (Hardcover)
International Standard Book Number-13: 978-0-42932-511-3 (eBook)

Library and Archives Canada Cataloguing in Publication

Title: Advances in rice science : botany, production, and crop improvement / Ratikanta Maiti, PhD, Narayan Chandra Sarkar, PhD, Humberto González Rodríguez, PhD, Ch. Aruna Kumari, PhD, Sameena Begum, Dasari Rajkumar.

Names: Maiti, Ratikanta, 1938- author. | Sarkar, Narayan Chandra, 1976- author. | González Rodriguez, Humberto, 1959- author. | Aruna Kumari, Ch., 1972- author. | Begum, Sameena, author. | Rajkumar, Dasari, author.

Description: Includes bibliographical references and index.

Identifiers: Canadiana (print) 20200224018 | Canadiana (ebook) 20200224085 | ISBN 9781771888547 (hardcover) | ISBN 9780429325113 (ebook)

Subjects: LCSH: Rice.

Classification: LCC SB191.R5 M35 2020 | DDC 633.1/8—dc23

Library of Congress Cataloging-in-Publication Data

Names: Maiti, Ratikanta, 1938- author. | Sarkar, Narayan Chandra, 1976- author. | Rodríguez, Humberto González, 1959- author. | Aruna Kumari, Ch., 1972- author. | Begum, Sameena, 1993- author. | Rajkumar, Dasari, author.

Title: Advances in rice science : botany, production, and crop improvement / Ratikanta Maiti, Narayan Chandra Sarkar, Humberto González Rodríguez, Ch. Aruna Kumari, Sameena Begum, Dasari Rajkumar.

Description: Palm Bay, Florida, USA : Apple Academic Press, 2020. | Includes bibliographical references and index. | Summary: "During recent decades, tremendous progress and innovations have been made in rice science with the goal of increasing production of rice to meet the world's growing demands. This new volume, Advances in Rice Science: Botany, Production, and Crop Improvement, provides a concise overview of rice, covering the background and importance of rice; origin, evolution, and domestication of rice; and the world rice production. It goes on to provide new and important recent research advances on many different aspects of rice science and production. The authors look at advances in rice ideotypes, abiotic stress management techniques, biotic stress affecting crop productivity, new methods and technology for cultivation, and new methods and techniques in rice grain quality analysis and processing. It also describes new rice varieties, new hybrid rice technology, and new breeding methods for rice. The overview presented here, along with the informative review of recent advances in rice cultivation, production, and science, makes this volume a comprehensive guide for rice scientists and researchers, agronomists, breeders, and faculty and students of agriculture"-- Provided by publisher.

Identifiers: LCCN 2020016575 (print) | LCCN 2020016576 (ebook) | ISBN 9781771888547 (hardcover) | ISBN 9780429325113 (ebook)

Subjects: LCSH: Rice.

Classification: LCC SB191.R5 M353 2020 (print) | LCC SB191.R5 (ebook) | DDC 633.1/8--dc23

LC record available at https://lccn.loc.gov/2020016575

LC ebook record available at https://lccn.loc.gov/2020016576

Apple Academic Press also publishes its books in a variety of electronic formats. Some content that appears in print may not be available in electronic format. For information about Apple Academic Press products, visit our website at **www.appleacademicpress.com** and the CRC Press website at **www.crcpress.com**

About the Authors

Ratikanta Maiti, PhD
Botanist and Crop Physiologist; Visiting Research Scientist, Forest Science Faculty, Autonomous University of Nuevo León, Mexico

Ratikanta Maiti, PhD, DSc, was a World-renowned botanist and crop physiologist. He worked for nine years on jute and allied fibers at the former Jute Agricultural Research Institute (ICAR), India, and also worked as a plant physiologist on sorghum and pearl millet at ICRISAT (International Crops Research Institute for the Semi-Arid Tropics) for 10 years. For more than 25 years, he was a professor and research scientist at three different universities in Mexico. In addition, he worked for as a Research Advisor at Vibha Seeds, Hyderabad, India, and as a Visiting Research Scientist in the Forest Science Faculty, Autonomous University of Nuevo León, Mexico. As the author of more than 40 books and about 500 research papers, he won several international awards, including an Ethno-Botanist Award (USA) sponsored by Friends University, Wichita, Kansas, the United Nations Development Programme; a senior research scientist award offered by Consejo Nacional de Ciencia y Tecnología (CONACYT), Mexico; and a gold medal from India 2008 offered by ABI. He was Chairman of the Ratikanta Maiti Foundation and chief editor of three international journals. Dr. Maiti died in 2019.

Narayan Chandra Sarkar, PhD
Department of Agronomy, Institute of Agriculture, Visva-Bharati, Sriniketan, Dist- Birbhum, West Bengal, India

Narayan Chandra Sarkar, PhD, is a faculty member in the Department of Agronomy at the Institute of Agriculture, Visva-Bharati University, Sriniketan, West Bengal, India. Several past organizations with which he was affiliated include Vibha Seeds, Syngenta India Ltd, and Nagaland University. He has experience in conducting two international conferences as a convener. He works as a managing editor of an international journal. His main area of research is nutrient management, and his major focus is on the livelihood security of the farming community. Currently, he is deeply engaged with Western Sydney University under Australia India Council funded research projects. Dr. Sarkar has guided sixteen post-graduates students and pursuing 5 research scholars currently. and four PhD students. He has published 32 research papers in national and international journals and has seven books in his credit. He received his MSc in Agronomy from G.B. Pant University of Agriculture & Technology, Pantnagar, India, and his PhD in Agronomy from India's premier institute, Indian Agricultural Research Institute, New Delhi. He received a Junior Research Fellowship during his master's work and an institutional senior research fellowship during his PhD program.

Humberto González Rodríguez, PhD
Professor, Universidad Autonoma de Nuevo Leon, Facultad de Ciencias Forestales (School of Forest Sciences), Linares, Nuevo Leon, Mexico

Humberto González Rodríguez, PhD, is a faculty member at the Autonomous University of Nuevo Leon, Facultad de Ciencias Forestales (School of Forest

Sciences), Nuevo Leon, Mexico. He received his PhD in Plant Physiology from Texas A&M University under the advice of Dr. Wayne R. Jordan and Dr. Malcolm C. Drew. He is currently working on water relations and plant nutrition in native woody trees and shrubs, northeastern Mexico. In addition, his research includes nutrient deposition via litterfall in different forest ecosystems. Dr. Rodríguez teaches chemistry, plant physiology, and statistics.

Ch. Aruna Kumari, PhD

Assistant Professor, Crop Physiology; Professor Jayashanker Telangana State Agricultural University, Agricultural College, Polasa, Jagtial, Telangana, India

Ch. Aruna Kumari, PhD, is an Assistant Professor in the Department of Crop Physiology at Agricultural College, Jagtial, Professor Jaya Shankar Telangana State Agricultural University (PJTSAU), India. She has seven years of teaching experience at PJTSAU and seven years of research experience at varied ICAR institutes and at Vibha Seeds. She was the recipient of a CSIR Fellowship during her doctoral studies and was awarded a Young Scientist Award for Best Thesis Presentation on "In the National Seminar on Plant Physiology." She teaches courses on plant physiology and environmental science for BSc (Ag.) students. She has taught seed physiology and growth and yield and modeling courses to MSc (Ag.) students. She acted as a minor advisor to several MSc (Ag) students and guided them in their research work. She is the author of book chapters in four books. She is also one of editors of the book *Glossary in Plant Physiology* and an editor of six international books, including *Advances in Bio-Resource and Stress Management; Applied Biology of Woody Plants; An Evocative Memoire: Living with Mexican Culture, Spirituality and Religion*; and *Gospel of Forests*. She has published over 50 research articles in national and international journals. Her field of specialization is seed dormancy of rice and sunflower.

Sameena Begum, MSc
Researcher, Professor Jayashanker Telangana State Agricultural University, Agricultural College, Polasa, Jagtial, Telangana, India

Sameena Begum is a researcher at the Professor Jayashanker Telangana State Agricultural University, Agricultural College, Polasa, Jagtial, Telangans, India, where she completed a BSc in Agriculture with distinction and her Master of Science in Genetics and Plant Breeding. During her master's degree program, she conducted research on combining ability, gall midge resistance, and yield and quality traits in hybrid rice (*Oryza sativa* L.) and identified two highly resistant hybrids.

Dasari Rajkumar, MSc
Rice Breeder, Neo Seeds India Private Limited, Hyderabad, India

Dasari Rajkumar is a rice breeder at Neo Seeds India Private Limited, Hyderabad, India. He received his master's degree in botany with specialization in applied stress physiology and molecular biology from the University College of Science at Osmania University, Hyderabad, India. He started his career as a Research Associate and was promoted to Botanist at Vibha Seeds Hyderabad, India. He contributed research on the physiological aspects like drought and salinity tolerance in paddy, cotton, and vegetables. At Bio-Seed Research India, Hyderabad, India, he worked as a senior research associate, where he worked on QTL mapping for drought, salinity tolerance and trait introgression for biotic stress tolerance in rice and actively participated in rice breeding programs. His main area of research is developing high-yielding drought- and salinity-tolerant rice hybrids and varieties of various maturing segments and grain types across India. In addition to this, he is working on biotic stress tolerance in paddy at various hotspots across India and trait introgression program for biotic stress tolerance. Currently, he is working on various aspects of abiotic stress tolerance, such as drought, salinity, heat on field crops (cotton, rice, maize, sunflower), and vegetable crops (tomato, chili, okra). He has more than 15 publications to his credit, including research papers in peer-reviewed national and international journals, and he has authored three books.

Contents

Abbreviations

AAA	active absorption area
AAC	apparent amylose content
ABA	abscisic acid
AC	amylose content
ACC	aminocyclopropane-1-carboxylic acid
ADAR	airborne data acquisition and registration
AF	Australian fragrant
AFLP	amplified fragment length polymorphism
AGB	aboveground biomass
ALPs	amplicon length polymorphisms
AMMI	additive main effects; the multiplicative interactions
ANR	apparent nitrogen recovery
As	arsenic
ASS	alkali-spreading score
ASV	alkali-spreading value
AWD	alternate wetting and drying
BADH2	betaine aldehyde dehydrogenase gene
BEP	Bambusoideae-Ehrhartoideae-Pooideae
BILs	backcross inbred lines
BLB	bacterial blight resistance
BLS	basic linked system
BNF	biological nitrogen fixation
BPH	brown plant hopper
BR	brassinosteroids
Ca	calcium
Cd	cadmium
CDPK13	calcium-dependent protein kinase 13
CDR	characteristic dimension ratio
CERES	crop environment resources synthesis
CF	continuous flooding
CGR	cracked grains ratio
CGR	crop growth rate
CH_4	methane

CM	chemical fertilizers with farmyard manure
CMS	cytoplasmic male sterile
CRTintP1	calreticulin interacting protein 1
CT	cold tolerance
Cu	copper
CWL	continuous waterlogging
CYPs	cyclophilins
DAE	days after emergence
DALYs	disability-adjusted life years
DH	doubled haploid
DMA	dimethylarsinic acid
DOC	dissolved organic carbon
DPPH	2,2-diphenyl-1-picrylhydrazyl
DRO1	deeper rooting 1
DSP	direct-sown plants
DSR	direct seeding of rice
DT	drying time
E	environment
EAR	ERF-associated amphiphilic repression
ECEC	effective cation exchange capacity
El	electrolyte leakage
ELR	elongation ratio
FACE	free air concentration enrichment
FAO	Food and Agriculture Organization
FDR	first derivatives
Fe	iron
FFA	free fatty acids
FFP	farmers' fertilizer practices
FFS	farmer field school
FGP	final germination percentage
FKBPs	FK506 binding proteins
FL	flowering
FRAP	ferric reducing ability power
FS	film solubility
FSV	flour swelling volume
G × E	genotype-by-environment
G	genotype
GA	gibberellic acid

GB	grain breadth
GC	gel consistency
GCMs	general circulation models
GCRPS	ground cover rice production system
GE	genetically engineered
GGE	G and GE interaction
GHGI	greenhouse gas intensity
GI	germination index
GIR	gas-fired infrared
GL	grain length
GLA4	Guang-Lu-Ai 4
GMB	gall midge biotype
GMOs	genetically modified organisms
GR	germination ratio
GR	green revolution
GS	genomic selection
GSTs	glutathione S-transferases
GT	gelatinization temperature
GWAS	genome-wide association studies
GWP	global warming potential
GY	grain yield
H	hardness
H_2O_2	hydrogen peroxide
HI	harvest index
hLF	human lactoferrin
HO	high oleic
HPMC	hydroxy propyl methyl cellulose
HR	husked ratio
HRR	head rice ratio
HRY	head rice yield
HSF	heat stress transcription factor
HSPs	heat shock proteins
HT	heat tolerance
HY3	Hanyou 3
HYVs	high-yielding varieties
IAA	indole-3-acetic acid
ICAP	inductively coupled argon plasma
ICP	plasma emission spectroscopy

ID1	ideotype 1
ID2	ideotype 2
IDD	indeterminate domain
IGP	indo-gangetic plains
ILs	introgression lines
IMM	immunophilin
IPA	isopropanol
IPA1	ideal plant architecture 1
IPM	integrated pest management
IR	infrared
IRC	International Rice Commission
IRRI	International Rice Research Institute
ISSR-PCR	inter-simple sequence repeat polymerase chain reaction
IWR	irrigation water requirement
IWUE	irrigation water use efficiency
K	potassium
KatyRR	Katy red rice
KBNT	Kaybonnet
KDML105	Kao Dawk Mali 105
KO	knockout
LAI	leaf area index
LBR	length to breadth ratio
LCA	life cycle assessment
LCI	life cycle inventory
LD	linkage disequilibrium
LD	long-day
LDPs	long-day plants
LOXs	lipoxygenases
LRI	leaf rolling index
MAEE	microwave-assisted enzymatic extraction
MAS	marker-aided selection
MCLs	maximum concentration levels
MCs	moisture contents
MDA	malondialdehyde
MET	multi-environment trial
Mg	magnesium
miRNA	microRNA
Mn	manganese

MODIS	moderate resolution imaging spectroradiometer
MR	milling recovery
MRPs	*Monascus* rice products
MS	moderately susceptible
MT	moderately tolerant
MV	methyl viologen
MVs	modern varieties
N	nitrogen
N_2O	nitrous oxide
Na	sodium
NARES	National Agricultural Research and Extension Systems
NAS	nicotianamine synthase
NDVI	normalized difference vegetation index
NIL	near-isogenic lines
NIRS	near infrared reflectance spectroscopy
NK	nitrogen and potassium
NO	nitric oxide
NP	nitrogen and phosphorus fertilizer
NPT	new plant type
OMNBR	optimum multiple narrow band reflectance
ORFs	open reading frames
OsSPL14	Souamosa promoter binding protein-like 14
OVB	outer vascular bundles
PACMAD	Panicoideae-Aristidoideae-Chloridoideae-Micrairoideae-Arundinoideae-Danthonioideae
PAR	photosynthetically active radiations
PCA	principal component analysis
PCR	polymerase chain reaction
PEG	polyethylene glycol
PGPR	plant growth promoting rhizobacteria
PGWC	percentage of grain with a white core
PI	panicle initiation
PM	plasma membrane
PMPs	plant-made pharmaceuticals
POD	peroxidase
PPB	participatory plant breeding
P_{PD}	production downturn
PRECIS	providing regional climates for impacts studies

PTDI	provisional tolerable daily intake
PTM	post-translational modification
Put	putrescine
PVS	participatory varietal selection
Q × E	QTL × environment interactions
QTLs	quantitative trait locus
RAPD	random amplification of polymorphic DNA
RBEE	rice bran enzymatic extract
RB-kr	rice bran-super kernel
RBO	rice bran oil
RCS	rice core subset
RGA	rapid generations advance
rhLF	recombinant human lactoferrin
RiFoRG	rice fortification resource group
RILs	recombinant inbred lines
RLD	root length density
RMSD	root mean square deviation
RNAi	RNA interference
ROA	root oxidation ability
ROI	reactive oxygen intermediates
ROL	radial oxygen loss
R-ORS	rice-based ORS
RSP	root surface phosphatase
RUE	radiation use efficiency
RVA	rapid viscosity analyzer
RVI	ratio vegetation index
RW	rice whiteness
RWC	rice whole collection
RWD	root weight density
S	susceptible
SA	salicylic acid
SCT	seedling cold tolerance
SD	short-day
SDPs	short-day plants
SDR	second derivatives
SDSM	statistical downscaling method
SDSs	survival days of seedlings
Se	selenium

SEM	scanning electron microscopy
SeMet	selenomethionine
SF	spikelet fertility
SG	speed of germination
SGR5	shoot gravitropism 5
ShB	sheath blight
Si	silicon
SNAC1	*stress-responsive NAC 1*
SNP	single nucleotide polymorphism
SNPs	silica nanoparticles
SOD	superoxide dismutase
Spd	spermidine
Spm	spermine
SREG	sites regression
SRES	Special Report on Emissions Scenarios
SRI	system of rice intensification
SRS	survival rate of seedlings
SSD	single seed descent
SSH	subtractive suppression hybridization
SSI	stress susceptibility index
SSLP	simple sequence length polymorphism
SSNM	site-specific nutrient management
SSR	simple sequence repeat
Stgstraw	Stuttgart Strawhull
SUB	submergence tolerance
SUB1	*submergence-1*
SWC	square of the white core
T	tolerant
TFs	transcription factors
TGMS	thermosensitive genic male sterility
TI	tolerance index
TLP	thaumatin-like protein
TN1	Taichung native 1
TP	transplanted plants
TPA	instrumental texture profile
TS	tensile strength
T_w	water temperature
USDA	United States Department of Agriculture

VI	vigor index
VIs	vegetation indices
VP	vapor permeability
VWFC	Von Willebrand Factor Type C
W	whiteness
WDR	water-saving and drought-resistance rice
WFCP	wavelet-based filter to determine crop phenology
WHO-ORS	World Health Organization Oral Rehydration Solution
WUE	water use efficiency
YLY6	Yangliangyou 6
Zn	zinc

Acknowledgment

The authors sincerely thank Apple Academic Press for accepting the manuscript and publishing this book within the prescribed timeline.

The authors thank Narayan Chandra Sarkar, PhD and Dasari Rajkumar, MSc for their courtesy in supplying original photographs of the rice plant.

We heartily thank our chief author, Dr. Ratikanta Maiti, an eminent and dedicated scientist for his continuous efforts, motivation, and initiation in writing this book and making this book unique, and for sharing his wisdom as a biologist to pass on to the next generation.

Preface

Rice has been used as a staple food Worldwide since 2000 B.C. It provides food security to the ever-increasing World's human population. Rice has fed more people over a longer period of time than any other crop. As far back as 2500 B.C., rice has been documented in the history books as a source of food for human culture and tradition. Rice is inherently involved in Japanese culture. Besides food security, rice contributes as a base material for several industries, including feed for poultry, livestock, and other commercial purposes. It plays an important role in the economy of many countries, especially in developing and undeveloped countries.

Keeping this in mind, researchers and scientists need to understand the role, nature of the crop, and advancements in research for effective utilization and proper cultivation operations for getting better yield. This will further help to ensure food security for an increasing human population, ultimately increasing the economies of many countries. In the 21st century, significant progress and innovations have been made on rice science, including golden and super rice, and molecular biology. All these attempts are made not only to supply rice in quantity but also quality towards nutritional security. More than 50% of the World population depends on rice as a staple food. Different ideotypes have been put forth to identify rice with higher productivity potential.

From the above point of view, the authors decided to provide a resource of complete information and research literature from several disciplines on rice in the form of the book as a guide for students, researchers, teachers, and others.

The authors have provided information on almost all the aspects of several disciplines of rice and recent literature together under one umbrella, namely *Advances in Rice Science: Botany, Production, and Crop Improvement*. This book covers almost all the aspects of rice, starting from the background of rice crop, to production, origin to domestication, ideotype, botany, physiology of crop growth and productivity, abiotic and biotic factors affecting crop productivity, methods of cultivation, postharvest management, grain quality analysis and to food processing, improvement of rice crop, and research advancements in breeding and biotechnology.

The chapters describe the topics extensively and are enriched with the recent research literature. We need to be concerned that the productivity of rice is affected by several biotic and abiotic stresses, which require concerted interdisciplinary research.

This mode of presentation will help students academicians, and teaching faculty to gain knowledge and understand the crop. Especially this book guides researchers working on rice and rice scientists to understand the relation between several disciplines and implementation of new methods and technologies covered in recent literature for rice crop improvement in order to get higher productivity. A multi-pronged approach needs to be implemented to increase rice productivity to meet the World's demand and hunger. We really need a green revolution in the context of increasing rice productivity.

Most of the books published on rice are on specific aspects; very few books attempted to bring together all disciplines in a concrete form like the present book. We hope this book will satisfy all these requirements. This book attempts to bring together recent advances in different disciplines of rice science.

CHAPTER 1

Background and Importance of Rice

ABSTRACT

This chapter presents the background and importance of rice (including its various industrial products) at the global level and discusses rice-growing environments together with the prospects of rice production. Rice is a staple food and has high demand across the World. Additionally, rice is also a vital source for the supply of essential elements.

1.1 BACKGROUND AND IMPORTANCE OF RICE

Rice is a very important cereal crop. It is under cultivation for more than 7,000 years. At present, it sustains the food necessities of more than half of the World population. Significant research advances have been attained. This involves efficient transformations, development of saturated molecular maps and massive analysis of expressed sequence tags. In this respect, we analyzed the number of complementary DNAs of rice and that of *Arabidopsis* are near equal numbers. Therefore, it is being considered as a model plant of the monocotyledonous group (Takeshi and Shimamoto, 1996).

Besides, its food value for millions of population, various rice products viz. rice brans, husks, etc., are utilized for various purposes.

1.1.1 ORIGIN AND DISPERSAL OF RICE

Khush (1997) discussed the origin, dispersal, cultivation, and variations of rice. He mentioned that the genus of rice *Oryza* contains both two cultivated and 21 wild species. The *O. sativa*, usually referred to as the Asian cultivated rice, all across the global regions is grown in many areas.

The African cultivated rice, *O. glaberrima,* when compared to the Asian cultivated rice its cultivation is confined to small regions of West Africa. This genus *Oryza* was assumed to be originated 130 million years in the past in Gondwanaland. Further, with the shuffling up of Gondwanaland region, different species might have been dispersed into different continents. Most of these cultivated species of rice were opined to have their origin from a common ancestor that had AA genome. The perennial and annual ancestors of *O. sativa* are *O. rufipogon* and *O. nivara*. Likewise, *O. longistaminata* and *O. breviligulata* were the perennial and annual family relatives (ancestors) of *O. glaberrima. O. glaberrima* might have been domesticated probably in Niger River delta. On the basis of genetic affinity backgrounds, the varieties of *O. sativa* are generally categorized into six groups. Among this widely known rice is that of the *indicas* and this matchup to group (I) and *japonicas* to group (VI). The *javanica* rice does also belong to group (VI). These are nominated as tropical *japonicas*, in disparity to temperate *japonicas*, grown largely in temperate climates. The *indica* and *japonica* rice had a polyphyletic derivation (origin). The *indicas* were perhaps mostly cultivated (domesticated) in the Himalayan foothills of Eastern India while the *japonicas* were evidenced to be domesticated in South China. The *indica* rice from India was dispersed all over the regions of tropics and of subtropics; while the *japonica* rice had a move northward from South China. They turned out as temperate ecotypes. The *japonica* rice, moved southward to Southeast Asia and later from there to West Africa and Brazil and became the tropical ecotype. Rice is now grown between 55°N and 36°S latitudes, under dissimilar growing environments such as irrigated, rainfed lowland, rainfed upland, and flood-prone ecosystems. Numerous cultivars have resulted because of their adaptations to varied environments and human selections. Estimations revealed that there are about ~120,000 varieties of rice existing within the World. With the founding of the International Rice Research Institute (IRRI) in 1960, intensification of rice, varietal improvement programs have occurred; several high yielding varieties that were developed were released. Most of these rice varieties are planted across 70% of rice lands in the World. With the extensive taking up of cultivation of these improved varieties of rice, rice production was doubled between 1966 and 1990. In spite of the presence of doubled production for feeding the surplus rice consumers and increased populations, presently; there is a requisite for an increase in rice production by 2025 to an extent of 60%. The novel paraphernalia of molecular and cellular biology viz., anther culture, molecular markers which have aided in the selection and genetic engineering will have additional mounting roles to bring about rice improvement.

Chang (1984) reported about the ethnobotany of rice in Island Southeast Asia with reference to the perspectives of Asia. He mentioned that this genus *Oryza,* containing the rich cultigens *O. sativa* L. and *O. glaberrima* has long antiquity. These had their origin to Gondwanaland and have undergone interspecific differentiation, before the supercontinent were fragmented and drifted apart. This was indicated by the pantropical distribution of the wild species of the genus in a nondisjunct manner across Africa, Oceania, and Latin America. The geographic distribution of the wild species with known genomes (chromosomal complements) reflects a charming picture of a widely dispersion of semiaquatic grass, rice.

The process of near beginning rice cultivation, followed two pathways for its domestication within India and China. Selection of domestication traits found in the near beginning in Yangtze *japonica* and a non-domestication feedback system inferred for 'proto-*indica*.' This domestication process was completed approximately 6,000–6,500 years ago in China and about two millennia, later in India. By this time, there was hybridization in Chinese rice. Later, farming populations increased and then stretched over by migration and introgression of pre-existing populations. These spreading outs can be connected to hypothetical language family dispersal models, which include scattering from China southwards by the Sino-Tibetan and Austronesian groups. In South Asia, much diffusion of rice occurred after Indo-Aryan and Dravidian speakers have taken over rice from the speakers of lost languages of parts of northern India (Fuller, 2011). In this process, the combined efforts of information of modern genetics, ecology, and archaeology have helped in unveiling the process of rice domestication and diversification.

Archeological excavations and environmental archeological investigations conducted for several years in the Shanghai Area have offered immense, valuable, and excellent information for Holocene environmental changes, growth which led to subsequent turndown of human settlements and man-land interactions. Wu et al. (2014) discussed Holocene environmental change and its influences on human settlement in the Shanghai area, East China. Division and circulation of archeological sites between 7,000 and 3,000 cal. yr BP demonstrated a regression process. It advanced southward of the coastline. Temporal and spatial analyses of ^{14}C dates for archeological sites, shell ridges, buried trees, and peat suggested that Holocene environmental changes might have been a major cause of the rise and fall of human settlements and their civilization. A virtual sea-level curve of the Shanghai area which was obtained from dated shell ridges and peat correlated well with the reconstructed sea-level curves of the Yangtze Delta and East China. As this process was going on, the development of human settlements was broken up at least four times in the Shanghai Area, coinciding with four periods of high sea-level, peat accumulation, and amplified in shell ridges. They stated that after this, Neolithic communities migrated to the plain and occupied their lowlands for rice cultivation. In this aspect, the Chenier Ridges played a vital part in giving shelters to the settlers of the Neolithic period. The crumple of Liangzhu Culture about 4,000 cal. yr BP was followed by a lesser amount of developed Maqiao Culture. These studies suggested that the prevalent extreme environmental and hydrological conditions such as terrestrial inundation resulted due to sea-level rise and heavy precipitation led to the end of paddy exploitation and to the social stress that bring about the Liangzhu Culture demise.

Zhu et al. (2012) discussed sequence polymorphisms in wild, weedy, and cultivated rice. The results suggested that seed shattering locus *sh4* had a very negligible function in domestication of Asian rice. The major view on domestication of Asian rice was that the preliminary origin of non-shattering is associated with a single gene of large effect, particularly, the *sh4* locus through the evolutionary substitution of a dominant allele for shattering with a recessive allele for reduced shattering. Sufficient data have been collected to challenge this assumption. In this respect specifically, a few reports have mentioned about the occurrence of occasional seed shattering plants from populations of the untamed progenitor of cultivated rice (*Oryza rufipogon* complex). These plants were homozygous for the putative "nonshattering" *sh4* alleles. The authors tested the *sh4* hypothesis for the domestication of cultivated rice by collecting genotypes and phenotypes from a diverse set of samples of wild, weedy, and cultivated rice accessions. The cultivars were fixed for the putative "nonshattering" allele and nonshattering phenotype, but wild rice accessions were highly polymorphic for the putative "nonshattering" allele (frequency ~26%) having shattering phenotype. All weedy rice accessions are the "nonshattering" genotype at the *sh4* locus but with shattering phenotype. These statistics challenge the widely accepted hypothesis that a single nucleotide mutation ("G"/"T") of the *sh4* locus is the major driving force for rice domestication. Instead, the authors hypothesized that anonymous shattering loci are liable for the early domestication of cultivated rice through abridged seed shattering.

1.2 ROLE OF RICE IN HUMAN CIVILIZATION

Rice plays a key part in human civilization and occupies the space of a vital component of human society. It boosted human civilization, increased the human population and the national economy. There are several evidences that rice is closely associated with human culture and their civilization, and an increase in population.

Rice was basically a tropical semiaquatic swamp grass, before the dawn of agriculture. With the commencement of agriculture, it became a supplementary food to those people, who were much dependent on hunting, fishing, and gathering of other food plants for their subsistence living. Its farming serves to be a subsistence level of livelihood, in several regions of humid tropics where many harsh environments prevail. In areas of proper management practices adopted in the conservation of water, nutrients, or with well soil fertilization, tillage, and weeding practices along with good plant selection are practiced, its yields quite astonishingly increased at a faster pace. The surplus food production that was attained with the adoption of the good management practices helped in meeting the food requirements of the increased population. Henceforth, there was an expansion in rice acreage, an increase in rice yield, and multiple cropping systems. These, in turn, boosted the civilizations flourishing in several countries of Asia. Thereby, this lowly swamp plant during the past two millennia in many Asian countries has provided the impetus for accelerated progress in their national economies and the cultural improvements, even though there were increases in the populations (Chang, 1987).

1.3 RICE PRODUCTS

1.3.1 ENRICHMENT OF SELENIUM CONCENTRATION IN RICE PRODUCTS

In Southern China, Qiuhui et al. (2002) analyzed 30 rice products for selenium content using atomic fluorescence spectrophotometry. The results showed that the selenium concentration in these products ranged from 0.015 to 0.046 μg g^{-1}, which has accounted to only 6–18 μg day^{-1} of daily average dietary intake for an adult by these rice sources. Further, it is presumed that in the inhabitants of southern China, the low Se content of the rice products was primarily accountable for their less Se entire intake. They observed that enrichment in the Se substance of 0.178–0.421 μg g^{-1} in rice products has resulted with foliar application of a Se enhanced fertilizer or by sodium selenite (14–18 g Se ha^{-1}) during the heading stage of rice. They mentioned that Se content in the rice products can be enriched to optimum concentrations by changing the doses of the application of Se enriched fertilizers. Further, they suggested that these enriched rice products of selenium can contribute on an average to an increase in Se intake of 50–100 μg day^{-1}, provided there is a control in the Se concentration in the products in the range of 0.15–0.50 μg.

1.3.2 OCHRATOXIN A IN NON-ORGANIC RICE PRODUCTS

Sixty four rice cultivars grown under non-organic and twenty rice cultivars grown under organic conditions were collected from local and supermarkets to study the presence of ochratoxin A. 7.8% of non-organic rice samples had ochratoxin A ranging from 4.3 to 27.3 μg kg^{-1}. On the other hand, 30% of organic rice samples contained this mycotoxin in the range of 1.0 to 7.1 μg kg^{-1} (Gonzalez et al., 2006).

1.3.3 FERMENTED RICE PRODUCTS FOR HEALTH

Solid-state fermentation is used to prepare certain products from rice which are used as remedial sources for health. They were in use in China for over a period of 1,000 years. *Monascus* rice products (MRPs) are used as health foods in many countries viz., the United States and many Asian countries such as Japan, Taiwan, China, Korea, Thailand, the Philippines, and Indonesia. This particular species produces several viable metabolites, and bioactive compounds having the capacity to act as cholesterol-lowering agents, antibiotics, food colorants, etc. Tseng and Tzann (2007) have discussed about the usefulness of these fermented food products as health products. These exhibit a wide variety of biological activities and pharmacological potentials due to the presence of flavonoids, polyunsaturated fats, pyrrolinic compounds, phytosterols, etc. There are also effective in reducing the triacylglycerol, cholesterol, and lowering of blood sugar. They mentioned that apart from the beneficial effects the fungus involved in the fermentation process for the production of these products also produces some toxins like citrinin, manokolin K that affects the hepatic and renal systems and other metabolic processes. Hence, it is very much essential to keep in view the safety of these products with fewer amounts of toxic metabolites and later be promoted as health foods.

1.3.4 HEAVY METAL CONTENTS IN RICE PRODUCTS

Rice acts as a vital supplier of essential elements. Some toxic elements as arsenic, cadmium, lead, etc., might also be there in the rice grain. Few rice and wheat samples of polished grains, flour, noodles, loaf, etc., collected from over 63 cities of Japan and analyzed for cadmium and lead contents (Shinichiro et al., 2001). The results showed that the polished raw uncooked rice had higher cadmium (50 ng g^{-1}) compared to flour (19 ng g^{-1}), while the content of lead in both rice and flour ranged 2–3 ng g^{-1}. The studies have unveiled that the intake of cadmium was more from the rice foods rather than from wheat, in particular, it was noted that the consumption of rice grown along side of Japan sea had higher cadmium contents and contributed to more cadmium intake through diets.

Pure rice bran is helpful as a supplement of a health food supplement. It is commercially marketed as a superfood and in several of the international programs concerned with malnourishment, it is supplied to malnourished children. These rice bran foods have some soluble which are of main concern. No maximum concentration levels (MCLs) are set for arsenic or its derivative species in foodstuffs. Guo et al. (2008) had mentioned that rice contained elevated arsenic levels than the remaining grain crops, wherein, these levels were more in the whole grain (brown) rice rather than the white rice (polished). Further, rice bran, had a 20 fold high levels of inorganic arsenic class 1 carcinogen, at non-threshold concentrations of (~1 mg kg^{-1} dry weight). Some of the rice bran soluble products of Japan and the United States contained inorganic arsenic (0.61 to 1.9 mg kg^{-1}).

There are a few predominant identified forms of Arsenic in rice products are Arsenite [As (III)], arsenate [As (V)], dimethylarsinic acid

(DMA), and selenomethionine (SeMet). A microwave-assisted enzymatic extraction (MAEE) method was developed by Joerge et al. (2009). This method enables in the extraction of total arsenic (As) and selenium (Se) species simultaneously in these rice products. They devised the optimum extraction conditions (500 mg sample, 50 mg of protease XIV, and 25 mg of α-amylase in the aqueous medium during 40 min at 37°C) for this enzymatic extraction, which enabled in total recovery (100 ± 3) of total arsenic contents. This can be utilized to analyze the arsenic content even in rice products and other rice-based cereals.

Similarly, Bruno et al. (2011) while analyzing variable rice samples (white, parboiled white, brown, parboiled brown, parboiled organic and organic white) from different Brazilian regions, observed that these products had a mean total arsenic content of 222.8 ng g^{-1}. They found that with a daily ingestion of 88 g of rice, there would be a daily intake of inorganic arsenic (the most toxic form) of 10% of the provisional tolerable daily intake (PTDI). Further, these had the predominant arsenic forms viz., inorganic arsenic (As^{3+}, As^{5+}) and DMA at 38.7; 39.7%, respectively. They identified that Brazilians through rice consumption are also having a daily intake of different arsenic species.

1.3.5 IMPORTANCE VALUE OF RICE

Kennedy (2002) explained about the significant value of rice through his research publication. Two sequences for the rice genome: one for each of the two major rice varieties (*japonica* and *indica*) planted throughout the World were published. They provided this data to the public domain, as it is of much importance. It enables the nutritional status and health improvement as hundreds of millions of people are dependent on rice. In comparison to any other cereal, food rice can provide more calories. The efforts made to sequence rice genomes are comparable to other cereals.

1.3.6 RICE FIELDS AN ALTERNATE HABITAT TO WATER BIRDS

Rice fields provide food and shelter for most of the avian species, and act as alternative habitat for water birds also. Toral et al. (2011) mentioned that rice fields were important for waterbirds populations in Europe, where they migrated and brood there during the autumn season. On the basis of the last 23 years counts in natural marshes and areas of rice fields that were conducted in Doñana (SW Spain), an index of rice field use for 76 bird species was constructed. Their study has indicated that these birds utilized the rice fields during their autumn migration season and a positive correlation was found between these bird populations and usage of rice fields. Further, they mentioned that due to some changes that were brought in the Common Agriculture Policy in Europe, there was a reduction in the rice cultivation area which has affected these waterbirds' alternate habitat considerably. Therefore, an environmentally friendly approach has to be towards the restoration of earlier marsh areas and the maintenance of rice cultivation rather than their conversion for other uses.

Different studies have been directed on rice on a molecular level to determine the effectiveness of these researches for rice utility and its importance.

1.3.7 EXPRESSION OF RICE DEFENSE STRESS TRANSCRIPTS

Rice plant during different stages of development often encounters the stress arising from various environmental factors. These might be in the form of osmotic stress, heavy metals, wounding, UV radiations, pollutants, etc. Agrawal (2003) reported that in these diverse environmental cues, there arises a transient regulation of *OsOPR1* transcript of the "octadecanoid pathway." The pathway is mostly concerned with the jasmonic acid synthesis. Jasmonic acid in many plant species was found to confer defense resistance against a wide variety of stresses both biotic and abiotic in nature. This has a key part in rice defense/stress and development. Previously, they stated in rice (*Oryza sativa* L. japonica type cv. Nipponbare) the allene oxide synthase (*OsAOS*) and cyclase (*OsAOC*) genes of the octadecanoid pathway. From the two weeks old rice seedling a 12-oxo-phytodienoic acid reductase gene, called *OsOPR1*, coding the last step of the octadecanoid pathway that results in the biosynthesis of jasmonic acid was isolated and by transcriptional profiling, they predicted its polypeptide sequence and other details. This gene that was isolated had a molecular mass of 42465.02 and p*I* of 5.79, and encoded a polypeptide of 380 amino acids long. Against a wide variety of stresses in the leaves of the rice seedlings, within 30 minutes of the onset of the above-mentioned stress, situations this transcript levels were up-regulated and high levels were detected. Further, they observed that co-application together of signaling molecules like jasmonic acid, ethylene, salicylate astonishingly contributed to a large accumulation of this gene transcript which remained at an elevated level even after 30 minutes. The de novo synthesized negative *trans*-acting factor(s) were found to regulate the transient expression of *OsOPR1* transcript. This was evidenced through the usage of cycloheximide. The endogenous expression of this transcript was more with the advancement in the stage of plant development. Thus, their results have stalwartly suggested a regulatory role for *OsOPR1* in rice plant defense/stress response pathway(s) and reproduction.

Trehalose is a non-reducing disaccharide of glucose. It works as a compatible solute. It is involved in the stabilization of biological structures under abiotic stress in bacteria, fungi, and invertebrates. Garg et al. (2002) reported that accumulation of trehalose in rice plants provides high levels of tolerance to various abiotic stresses. They reported of the regulated overexpression of *Escherichia coli* trehalose biosynthetic genes (*otsA* and *otsB*) as a fusion gene for manipulating abiotic stress tolerance in rice.

1.4 RICE GROWING ENVIRONMENTS

Rice is grown largely in different agroclimatic environments of the World. The large variations in agroclimatic and edaphic conditions influence rice productivity.

1.4.1 AGROCLIMATIC ZONE CLASSIFICATION

Bunded rice was the main food crop in Java, Sulawesi, and Sumatra. On the basis of duration of growing period length for bunded rice and upland crops, the agro-climatic zones of rice-growing environments in Indonesia were classified into twelve zones by Oldeman (1980). They took into consideration the length of the wet period. They distinguished these 12 agro-climatic zones with the usage of monthly rainfall maps and long-term monthly rainfall records.

Similarly, in West Africa, Andriesse and Fresco (1991) on the source basis of ecological and agronomic parameters, have undertaken the characterization of rice-growing environments. They took into consideration the classificatory principles viz., climate, soils, toposequence, land types and rice cropping system. Based on these classificatory principles they distinguished 18 diverse environments and characterized these zones. They have highlighted the constraints and implications of rice research to increase rice production in each of these environmental zones.

1.4.2 LIMITING FACTORS TO RICE PRODUCTION

Rice production in lowlands West Africa is limited because of several limiting factors. An analysis of the soil samples (172) collected from river flood plains and within inland swamps of West Africa (Buri et al., 2000) showed that a large proportion of these soils are deficient in the available zinc and sulfur. The topsoil had moderate levels of Fe, Mn, Cu, and Ni. They observed that in the river flood plains the sulfate-S had a positive correlation with total C, available P, eCEC, and Clay, however, in swamp soils it had a positive association with total C, available P, eCEC, similarly, zinc had with total C, and available P and negative with eCEC. Similar relation as that of zinc was noted with nickle. Though manganese had a positive association with total C and available P, it had a negative association with pH. The agroecology of these regions was found to have a strong effect on the nutrient's availability and distribution.

In West Africa, irrigated systems, covers about 12% of the regional rice-growing area, wherein rice production occurs along the agro-ecological gradient, i.e., from the forest zone to the Sahara desert margins. Becker et al. (2003) in their study identified that along with these ecological gradients rice yield gaps exist and the farmers' yields are between 0.2 and 8.7 Mg ha^{-1}. Simulated experiments have shown that weed control brings an increase in yield by about 1.0 Mg ha^{-1}, while N management by 1–2 Mg ha^{-1}. Proper N management and weed control may enable in overcoming the yield gap by 57–80%.

The identification of the high yielding cultivar for a specific environment on the basis of both genotype (G) and genotype × environment (GE) interactions would be more helpful to breeders and producers. Many yield assessments are derived on the basis of G and environment (E) effects. In most of the cases, these estimates and interactive effects remain insufficient. Stanley et al. (2004) investigated the utility of additive main effects; the multiplicative interactions (AMMI) model analysis. Grain yields (GY) were interpreted based on the G and GE interaction (GGE) biplots, which were acquired from sites regression (SREG) model analysis, in six rice cultivars viz., Cocodrie, Cypress, Jefferson, Lemont, Saber, and Wells grown in Texas. They determined the stability and adaptableness of specific cultivars, in an AMMI biplot, through the plotting of their nominal GYs at specific environments. Their study is more useful for the identification of mega environments (environments with the same highest yielding cultivar. Thus, through the results of GGE biplots of SREG model analysis, they have demonstrated the comparative yield performance of cultivars at a precise environment and identified the highest yielding cultivars at these different mega environments.

1.4.3 CO_2 FERTILIZATION EFFECT ON RICE YIELDS

Xiong (2009) analyzed the impact of greenhouse gas-induced climate change and also the direct CO_2 fertilization effect on rice yields and production in China, by the usage of regional climate model PRECIS (providing regional climates for impacts studies). They simulated the climate change scenarios with the CERES (crop environment resources synthesis) rice crop model of (1961–1990) and future (2011–2100) rice yields and production both under A2 and B2 climate change scenarios. The results demonstrated a reduction in expected rice yields of the future without the CO_2 fertilization cause. These exhibited some augments in a few areas. Yield variability exhibited the largest increase under A2 climate change scenario. It is specified that favorable climatic conditions that exist for single rice cropping may spread out its cultivation to more parts of northern China. Likewise, the double rice cropping systems may exhibit their shift towards the northern portion of the Yangtze River basin. Thus, he mentioned that China's perspective of rice production is expected to be increased in the upcoming years as a result of the CO_2 fertilization effect and also to its shift to rice-producing areas. The results have identified three rice cultivation regions viz., Sichuan basin, Yangtze River basin, and Huang-Huaihai Plain as exceedingly receptive to upcoming climatic changes.

1.4.4 VARIATION IN SOIL NUTRIENT STATUS

The soil nutrient status varies considerably across agro-ecological zones in a country. In many of the low lands, rice cultivation is intensified. Buri et al. (2009) ascertained the soil fertility status of two major agro-ecological zones of Ghana and watershed areas. They found that the soils of JoloK-waha were highly acidic. The Exchangeable cations (K, Ca, Mg, and Na) across areas within Forest agroecology were moderate and with reference to calcium and others, these were relatively low across the Savannah. The total carbon and nitrogen levels were also low; however, these were present to a higher level in the Mankran region. Soils of lowlands within the JoloKwaha watershed were deeper, while those of the Mankran water-shed were shallow. They specified that adoption of effective simple eco-technologies viz., "Sawah" system would be useful in enhancement and maintenance of the nutrient availabilities in these lowlands and enables for the attainment of sustainable rice production.

The soil fertility characteristics and material nature vary across the soils, influencing the potentiality of rice cultivation. Susumu et al. (2010) have made an assessment of these aspects in the soils of West African lowlands as well as soils of tropical Asia, for evaluating their potentiality in rice cultivation. They collected soil samples from major lowland ecosys-tems (185 locations), flood plains (62 locations) and from 13 countries of West Africa. These exhibited small values of pH, total C, and N, available (Bary-2) P, exchangeable Ca and Mg, effective cation exchange capacity (ECEC), and clay content. They had good mineralogical composition. The clay and primary minerals present consisted of kaolinite and quartz predominantly. This has given an indication that most of the lowlands of West Africa have stumpy nutrient-holding capacities and also an inad-equate potential for nutrient supply. It is observed that with increasing rainfall, soil pH, available P, exchangeable bases (Ca, Mg, K, and Na), and ECEC decreased, while total C, total N, and exchange acidity (Al and H) exhibited an increase. This inclination was observed due to the existence of enhanced biomass production and soil weathering sequence patterns which are under the influence of the climate. Thus, their study has indicated that compared to the paddy soils of tropical Asia, these lowland soils of West Africa have lesser values of general fertility parameters and poorer mineralogical characteristics, and deficit levels of S and Zn also prevail, constraining rice production potentiality of West Africa than in Tropical Asia.

In rice-growing environments, in soils across Sub-Saharan Africa, Silicon (Si) is deficient. Critical deficiency of Si arises when its concen-tration levels are below 5%. Si supply is required either from external or internal inputs or from irrigation water. Tsujimoto et al. (2014) chemically analyzed the Si concentration in plant and soil samples which were collected from 99 fields of this region. Their results indicated that in straw, its concentration is in the range of 1.7–8.4%. In 68% of the fields, its concentration levels were much below the critical deficiency level. They observed that water-soluble Si content and straw Si concentrations were much correlated. Multiple regression analysis unveiled that there was 59% variation in straw Si concentration. This variation has arisen due to variability in the plant-available Si in soils, rice-growing environments, N application rates, and mineralizable N in soils. The regression model revealed that an improvement of plant-available Si in soils could likely bring an increase in the Si contents in straw. This increase is by 0.043% per mg kg^{-1}. Further, their results have indicated a decrease in Si content 0.0068% per kg N ha^{-1} input. Their survey indicated the presence of low Si nutrient status in soils, limiting the plant-available Si and its deficiency may increase in many fields with abundant application of N fertilizers. They specified that there is a need for quantification of the association between the blast incidence and Si nutrient status of soils for developing better management practices of rice production.

1.5 PROSPECTS OF RICE PRODUCTION

Chang and Luh (1991) made a summary of the projections of rice production. In the first chapter edition in the 1980 edition of this book, entitled "Rice in its Temporal and Spatial Perspectives" (Lu and Chang, 1980), they explained crop history, World production, and supply and demand up to 1978 in detail. Henceforth, there was a continuous trend in the spread of the high-yielding varieties (HYVs), with an increase in GY, leading to increases in World production, and expansion in consumption and international trade. Here, the authors focus on variations that take place during the past decade. Rice statistical data were based mainly on the Food and Agriculture Organization's World Crop and Livestock Statistics (FAO, 1987), Quarterly Bulletins of Statistics (FAO, 1988, 1989), and the World Rice Statistics 1987 of the (IRRI, 1988).

1.5.1 *CULTURAL ROLE OF RICE*

Few studies have demonstrated the role of rice for various purposes which may be linked to human culture.

The cultural role of rice is well recognized. Bellon et al. (2000) discussed the role of genetic conservation for rice farmers. They mentioned that the existing rice genetic resources were utilized effectively to solve few of the present-day food problems. Most of the landraces of rice, collected over a period of several decades, have been well utilized as 'parents' of the high-yielding, pest-resistant, and well-adapted varieties. These have immensely contributed to enhanced yield increases. With the increase in the gains of its productivity, to most of the rice consumers, the cost of rice availability was also half the cost that prevailed during 1960s. The assortment of diversity in rice was thought to evolve diversity of the rice 1000's of years back. Asian and African peasant farmers, mostly women have selected various types to suit the needs and cultivation practices. Thus, the selection process has resulted in the development of copious rice varieties, having adaptive capacities to a wide range of agroecological conditions, and also having resistance abilities to a various insect pests and diseases. Many varieties of Asian rice, *Oryza sativa*, are impracticable to estimate, even though claims of more than 100,000 have been made (Chang, 1985, 1995). Asian rice varieties illustrate large variations in numerous characters, such as plant height, tillering ability, maturity, and size of panicles, among others. There are also large variations in grain characters like size, shape, and color which were more practically helpful for distinguishing dissimilar varieties. Few wild species appear as weeds in and around rice fields, and even hybridize unsurprisingly with the cultivated forms. This complicated association between cultivated and wild forms has enriched the diversity of the rice crop in conventional agricultural systems, where farmers often grow mixtures of varieties to provide a buffer against the risk of complete loss.

The emission of methane (CH_4) depends mostly on plant-borne material, which might be obtained from rotting tissue or root exudates. The quantity and quality of root exudates is to a large extent under the influence of many factors viz., mechanical impedance, occurrence of toxic elements, and prevalence of nutrient deficiencies. The water status of growing medium and the nitrogenase activity in the rhizosphere also has their role. In general, methane oxidation in rice fields mostly occurs in the rhizosphere zone. In this zone, there exists an overlapping of the concentration gradients of CH_4 and oxygen. The capacity of CH_4 oxidation is thus a function of the downward transport of oxygen through the aerenchyma tissue. This, functions in turn even as a conduit for CH_4 from the soil to the atmosphere (Wassmann and Aulakh, 2000).

Simultaneously, they discussed about the part of rice plants in regulating mechanisms of methane missions (CH_4) in rice fields. They mentioned that there is a requirement of comparative studies of various cultivars of rice to have an understanding about the interactions, mechanisms of CH_4 consumption in the rhizosphere and its transport through the rice plant so as to devise effective strategies of mitigation.

Physiological responses to chilling, together with antioxidative enzyme activity, were studied in rice (San et al., 2003) for identifying chilling tolerance mechanisms. They exposed the plants to 15°C (cold-acclimated) or 25°C (nonacclimated) for 3 d, under 250 $\mu mol\ m^{-2}\ s^{-1}$ photosynthetically active radiations (PAR). Apart from the above, they have exposed all these plants later to chilling temperature at 5°C for 3 d and allowed for their recovery at 25°C for 5 d. Their results have shown that cold-acclimated leaves were less affected and have recovered fast from chilling injury. There was induction of expression and activity of antioxidative enzymes CAT and APX in leaves and SOD, CAT, APX, and GR in roots of these plants. Thus, they deduced that the enzymes CAT and APX have a key important role in cold acclimation and chilling tolerance and an increased activity of these antioxidants in roots are more important for cold tolerance (CT), rather than their increased activities in the shoots. Thus, they mentioned that chilling-sensitive rice plants can be made tolerant by cold acclimatization.

Oxidative damage occurs due to aging, biotic, and abiotic stresses. Salicylic acid (SA), a key endogenous signal involved in mediation of a defense gene expression and disease resistance in many dicotyledonous species. Yang et al. (2004) reported that role of endogenous salicylic acid in rice plants. They have generated SA deficient transgenic rice by expressing the bacterial salicylate hydroxylase that degrades SA. Depletion of high levels of endogenous SA in transgenic rice did not affect defense gene expression. However, it reduced the plants ability in detoxifying the reactive oxygen intermediates (ROI). Further, it contained high levels of superoxide and H_2O_2. These transgenic rice exhibited spontaneous formation of lesions, which varied according to age and light. An exogenous application of analog of SA viz., benzothiadiazole complemented the deficiency of salicylic acid and led to a suppression of ROI levels and lesion formations. Further, these exhibited increased susceptibilities to oxidative stress, on infection with the blast fungus (*Magnaporthe grisea*).

Ishimaru et al. (2006) stated that rice plants absorb iron in the form of Fe^{3+} phytosiderophore and Fe^{2+}. The graminaceous monocots alone have the Strategy II iron (Fe) uptake system. In this system, iron is absorbed as a Fe^{3+} phytosiderophore. They have isolated the *OsIRT2* gene from rice. This gene is highly homologous to *OsIRT1*. An analysis by real-time PCR has revealed that both these viz., *OsIRT1* and *OsIRT2* were expressed principally in roots. These transporters are induced to a large extent by low iron conditions. *OsIRT1* promoter-*GUS* analysis has unveiled that *OsIRT1* is expressed in the epidermis and exodermis of the elongating zone and also in the inner layer of the cortex of the mature zone of Fe deficient roots. Positron emitting tracer imaging system analysis has shown that these rice plants were able to fill up a Fe^{3+} phytosiderophore and Fe^{2+}. This result has indicated that, besides the possession of absorbing a Fe^{3+} phytosiderophore, the rice plant also possesses a fresh Fe uptake system. This enables the direct absorption of Fe^{2+}. This strategy would be more advantageous for effective growth in submerged conditions.

Metals like manganese (Mn) and iron (Fe) are needed for metabolism, whereas cadmium (Cd) is toxic for almost all living organisms. Ishimaru et al. (2012) characterized the role of rice NRAMP5 in Manganese, Iron, and Cadmium transport. The identification of these metal transport systems is essential for breeding better crops. They found that *OsNRAMP5* helps in the transportation of Mn, Fe, and Cd in rice. *OsNRAMP5* expression was limited to roots epidermis, exodermis, and outer layers of the cortex and also in tissues around the xylem. *OsNRAMP5* confined to the plasma membrane (PM). *OsNRAMP5* RNAi (OsNRAMP5i) plants stored less Mn in the roots, less Mn and Fe in shoots, and xylem sap. The *OsNRAMP5* suppression promoted the translocation of Cd to shoots. The study highlights the significance of this gene for utilizing rice in phytoremediation of heavy metals particularly, cadmium. The data also revealed that *OsNRAMP5* also contributes to Mn, Cd, and Fe transport in rice. It has importance in plant growth and development.

1.5.2 *PRICE STABILIZATION*

Price stabilization forms the basis of the food policy. The stabilization of rice prices in developing Asian countries is very important. Many international trade agreements though have pushed the World in the direction of free trade, with reference to price stabilization it remained still inconsistent to have a completely free trade. Therefore, it is very important in developing Asia, which possesses a huge share of rice in economic output. A policy of pure price stabilization for rice, without dependable protection (either subsidization or taxation), is very much essential for the creation of macroeconomic stability for achieving sustained rapid economic growth of the population. Alternatively, it enables the generation of more significant equity gains; it also helps to guard the poor farmers, as well as the consumers from these sharp fluctuations in price. Though many Asian governments were successful in stabilizing rice prices, yet there are no mechanisms of improved stabilizations that can mediate the prevalence of low costs (Dawe, 2001).

1.6 CONCLUSION

Rice, a very important cereal crop originated most likely 130 million years ago in Gondwanaland and spread into different continents. When compared to other cereal food, rice can provide more calories and it is a crucial source of necessary main elements which enables in improvement of the nutritional status and health as hundreds of millions of people are dependent on rice though its production was doubled. To feed the additional rice consumers and increased populations, presently, there is a requirement for an increase in rice production by 2025 by an extent of 60%. The large variations in agroclimatic and edaphic conditions influence rice productivity. The recognition of a maximal yielding cultivar, for a unique environment, on the grounds of both genotype (G) and genotype × environment (GE) interactions would be more helpful to breeders and farmers/producers. Further, the novel means of molecular and cellular biology viz., anther culture, molecular markers which have aided in choosing better ones and genetic engineering will have additional increasing roles to bring about rice improvement.

KEYWORDS

- **enrichment**
- **growing environment**
- **importance**
- **price stabilization**
- **production**
- **rice**

REFERENCES

Agrawal, G. K., Jwa, N. S., Shibato, J., Han, O., Iwahashi, H., & Rakwal, R., (2003). Diverse environmental cues transiently regulate OSORI of the "octadecanoid pathway" revealing its importance in rice defense/stress and development. *Biochemical and Biophysical Research Communications, 310*, 1073–1082.

Alexander, V., (2015). Commemorative publication for Uwe Bläsing on the etymology of middle Korean 'rice.' *Turk Dilleri Araştırmaları, 25*, 229–238.

Andriesse, W., & Fresco, L. O., (1991). A characterization of rice-growing environments in West Africa. *Agriculture, Ecosystems and Environment, 33*, 377–395.

Bellon, M. R., Pham, J. L., & Jackson, M. T., (2000). *In*: Maxted N., Ford-Lloyd B.V., Hawkes J.G. (eds) Plant Genetic Conservation. Springer, Dordrecht. pp. 263–289.

Batista, B. L., Souza, J. M., De Souza, S. S., & Barbosa, F. Jr. (2011). Speciation of arsenic in rice and estimation of daily intake of different arsenic species by Brazilians through rice consumption. *Journal of Hazardous Materials, 191*, 342–348.

Becker, M., Johnson, D. E., Wopereis, C. S., & Sow, A., (2003). Rice yield gaps in irrigated systems along an agro ecological gradient in West Africa. *Journal of Plant Nutrition and Soil Science, 66*, 61–67.

Buri, M. M., Iassaka, R. N., Fujii, H., & Wakatsuki, T., (2009). *Comparison of Soil Nutrient Status of Some Rice Growing Environments in the Major Agro-Ecological Zones of Ghana*. WFL Publisher Science and Technology Meri- Rastilantie 3 B, FI-00980 Helsinki, Finland.

Buri, M. M., Masunaga, T., & Wakatsuki, T., (2000). Sulfur and zinc levels as limiting factors to rice production in West Africa lowlands. *Geoderma, 94*, 23–42.

Chang, T. T., & Luh, B. S., (1991). Overview and prospects of rice production. In: Luh B.S. (eds) Rice. Springer, Boston, MA. pp 1-11.

Chang, T. T., (1984). The ethnobotany of rice in island Southeast Asia. *Asian Perspectives, 26*, 69–76.

Chang, T. T., (1987). The impact of rice on human civilization and population expansion. *Interdisciplinary Science Reviews, 12*, 63–69.

Dawe, D. (2001). How far down the path to free trade? The importance of rice price stabilization in developing Asia. *Food Policy, 26*, 163–175.

Fuller, D. Q., (2011). Pathways to Asian civilizations: Tracing the origins and spread of rice and rice cultures. *Rice, 4*, 78–92.

Garg, A. K., Kim, J. K., Owens, T. G., Ranwala, A. P., Kochian, V., & Wu, R. J., (2002). Trehalose accumulation in rice plants confers high tolerance levels to different a biotic stresses. *PNAS, 99*, 15898–15903.

Gonzalez, L., Juan, C., Soriano, J. M., & Molt, J. C., (2006). Occurrence and daily intake of ochratoxin A of organic and non-organic rice and rice products. *International Journal of Food Microbiology, 107*, 223–227.

Guo-Xin, S., Paul, W., Anne-Marie, C., Yong-Guan, Z., Claire, D., Andrea, R., Joerg, F., Rafiqul, I., & Andrew, M., (2008). Inorganic arsenic in rice bran and its products are an order of magnitude higher than in bulk grain. *Environmental Science and Technology, 42*, 7542–7546.

Ishimaru, Y., Suzuki, M., & Tsukamoto, T., (2006). Rice plants take up iron as a Fe^{3+}-phytosiderophore and as Fe^{2+}. *The Plant Journal, 45*, 335–346.

Ishimaru, Y., Takahashi, R., & Nishizawa, N. K., (2012). Characterizing the role of rice NRAMP5 in manganese, iron and cadmium transport. *Scientific Reports, 2,* 286.

Jorge, L., Guzmán, M., Laura, H. R., Mizanur, R., & Skip, K., (2009). Simultaneous extraction of arsenic and selenium species from rice products by microwave-assisted enzymatic extraction and analysis by ion chromatography-inductively coupled plasma-mass. *Journal of Agricultural and Food Chemistry, 57*, 3005–3013.

Kennedy, D., (2002). The importance of rice. *Science, 296*, 13.

Khush, G. S., (1997). Origin, dispersal, cultivation, and variation of rice. *Plant Molecular Biology, 35*, 25–34.

Oldeman, L. R., (1980). The agro climatic classification of rice-growing environments in Indonesia. *Paper Proceedings of a Symposium on the Agro Meteorology of the Rice Crop*, 47–55.

Qiuhui, H., Licheng, C., Juan, X., Yanling, Z., & & Genxing, P., (2002). Determination of selenium concentration in rice and the effect of foliar application of Se-enriched fertilizer or sodium selenite on the selenium content of rice. *Journal of the Science of Food and Agriculture, 82*, 869–872.

San, J. S., Burgos, N. R., Eak, T. H., Han, O., Ho, B., Jung, S., & Ock, J., (2003). Ant oxidative enzymes offer protection from chilling damage in rice plants. *Crop Science Abstract—Crop Physiology Metabolism, 43*, 2109–2117.

Shinichiro, S., Zuo-Wen, Z., Takao, W., Haruo, N., Matsuda-Inoguchi, N., Kae, H., & Masayuki, I., (2001). Cadmium and lead contents in rice and other cereal products in Japan in 1998–2000. *Science of the Total Environment, 281*, 165–175.

Stanley, O., Samonte, P. B., Wilson, L. T., Clung, A. M., & Medley, J. C., (2004). Targeting cultivars onto rice growing environments using AMMI and SREG GGE biplot analyses. *Crop Breeding Genetics and Cytology, 45*, 2414–2424.

Susumu, S., Moro, B., Issaka, R. N., Paul, K., & Toshiyuki, W., (2010). Soil fertility potential for rice production in West African lowlands. *Japan Agricultural Research Quarterly, 44,* 343–355.

Takeshi, I., & Shimamoto Ko., (1996). Perspectives of becoming a model plant: The importance of rice to plant science. *Science Direct, 1*, 95–99.

Toral, G. M., Aragones, D., Bustamante, J., & Figuerola, J., (2011). Using land stat images to map habitat availability for water birds in rice fields. *Wiley Online Librar., Ibis, 153*(4), 684–694.

Tseng-Hsing, W., & Tzann, F., (2007). In *Monascus* rice products. *Advances in Food and Nutrition Research, 53*, 123–159.

Tsujimoto, Y., Muranaka, S., Saito, M., & Asai, H., (2014). Limited Si-nutrient status of rice plants in relation to plant-available Si of soils, nitrogen fertilizer application, and rice-growing environments across sub-Saharan Africa. *Field Crops Research, 155*, 1–9.

Wassmann, R., & Aulakh, M. S., (2000). The role of rice plants in regulating mechanisms of methane missions. *Biology and Fertility of Soils, 31*, 20–29.

Li Wu., Cheng Zhu., Chaogui Zheng., Feng Li., Xin hao Wang., Lan Li., & Wei Sun., (2014). Holocene environmental change and its impacts on human settlement in the Shanghai Area, East China. *CATENA, 114*, 78–89.

Wei Xiong., Declan Conway., Lin, E., & Ian Holman., (2009). Potential impacts of climate change and climate variability on China's rice yield and production. *Inter-Research Climate Research, 60*, 23–35.

Yang, Y., Qi, M., & Mei, C., (2004). Endogenous salicylic acid protects rice plants from oxidative damage caused by aging as well as biotic and a biotic stress. *The Plant Journal, 40*, 909–919.

Zhu, Y., Ellstrand, N. C., & Rong, B., (2012). Sequence polymorphisms in wild, weedy, and cultivated rice suggest seed shattering locus sh4 played a minor role in Asian rice domestication. *Ecology and Evolution, 2*, 2106–2113.

CHAPTER 2

World Rice Production

ABSTRACT

Rice is grown widely as a staple crop, a dietary source in many countries but the productivity of rice depends on various agro-climatic and other factors. Thus, in this chapter, the factors affecting rice production and trade together with the rice production future requirements were discussed.

2.1 RICE PRODUCTION AND COMMERCE

Generally, a wide range of dissimilarities is observed in the set of varieties which is identified as highly important to crop production and attracted growing attention from research, farmers, and agricultural policymakers. The single variety has a narrow genetic base and cannot resist or tolerate all the possible stresses which lead to yield reduction, and these losses can be lowered by varietal diversity having a broader stress tolerance base. However, in the presence of pest infestations or bad weather varietal diversity may also reduce the variability of yield. In this context, Widawsky et al. (1998) studied Chinese rice production for varietal diversity and yield variability.

During the reform period (1980–1993) in Chinese agriculture particularly in rice production to ascertain the technological change, technical, and allocative efficiency effects on improvement Fan (2000) presented a frontier shadow cost function approach. The results determined that first phase reforms (1979–1984) showed a major impact on technical efficiency of rural which give emphasis to the production system decentralization. However, in the interim of reforms of second phase, where expectations were to be concentrated on rural markets, liberalization of technical efficiency has enhanced scantily and allocative efficiency improved to some extent.

In the Southeastern United States, the effect of different climate scenario change, spatial scales on the assumed yield variations of maize (*Zea mays* L.), winter wheat (*Triticum aestivum* L.) and rice (*Oryza sativa* L.) were considered by Tsvetsinskaya et al. (2003) using CERES family of crop models. By means of control and doubled CO_2 runs of a high-resolution regional climate model and a coarse resolution general circulation model climate change scenarios were made. For each scenario, three diverse cases were considered viz., only climate change, climate change with raised CO_2 and the final with adaptations. In the climate change scenario, corn showed significant yield variations with larger modeled yield reductions or minor rises on the state level. In contrast, for all cases, large reductions in yields were seen in wheat. Mostly owing to low variability in yields rice exhibited significant differences in yield. The results make known that based on the crop and management condition spatial resolution plays an important role in estimations of climate change affects.

From the outlook of grain consumption and world grain commerce and on the fortification of rice production history, Zhang et al. (2005) stated the importance of rice production while discussing food safety and rice production in China. The rice production analysis was done on the basis of rice yield per rice production potential comparison and observed that in China rice yield per hectare still had grade increase space. They point out that rice yield could meet the future demand and the constant improvement of rice yield guarantees the safety of the national grains. Further, rice production improvement strategies to assure China's national food safety was proposed.

In Asia, the influence of climate change on rice production was assessed by Masutomi et al. (2009) in an extensive discussion of the parameter ambiguity in general circulation models (GCMs). In a crop model, on the basis of GCMs projections for three special reports on emissions scenarios (SRES) (18 GCMs for A1B, 14 GCMs for A2, and 17 GCMs for B1) future climate scenarios were recorded. Then for each SRES scenario, the mediocre switch in production (A_{CP}), the standard deviation of the variation in production (SD_{CP}), and the chance of a production downturn (P_{PD}) were calculated by taking the effect of CO_2 fertilization into account. In the 2020s, in almost all climate scenarios due to the greater unfavorable shocks of climate change than the positive chattels of CO_2 fertilization high P_{PD} values for all SRES scenarios were measured indicating the need of immediate adaptive measures irrespective of the emission schemes, in the near forthcoming years. In the 2080s compared with the other scenarios, the A2 scenario with the maximum atmospheric CO_2 congregations bring about a significant decrease in production and had the highest P_{PD} and SD_{CP} among the SRES scenarios. Conversely, the B1 scenario with the lowest atmospheric CO_2 combinations affected a small reduction in production with much lesser SD_{CP} and P_{PD}, than A2. These results of the 2080s proposed that the cutback in CO_2 discharges in long-lasting years has a considerable impact to alleviate reductions in rice production.

In some countries, climate trends arising from technology, CO_2 fertilization, and other factors were large enough to offset a significant part of the increases in average yields. Lobell et al. (2011) reviewed climate trends and global crop production and state that attempts to discern the special effects climate changes to date help to expect how climate change will affect future food availability. They show that except the United States temperature trends for 1980–2008 outpaced one standard deviation of historic year-to-year volatilities in most countries cropping regions and growing seasons. In comparison to a counterfactual deprived of climate trends about 3.8% and 5.5% reduction in overall maize and wheat production was indicated from the models associating four bulkiest commodity crops yields to weather but for soybeans and rice, winners, and losers largely balanced out.

Rice is the main food, partly to half of the World's population. An analysis of Worldwide rice production, supply, trade, and consumption was made by Muthayya et al. (2014). They restated that annually about 480 million metric tons of milled rice is produced of which ~50% of the rice is grown and absorbed in China and India. In Asia, rice is significant for food security because it supplies up to 50% of the dietary calories for millions of persons living in poverty. Since the beginning of the green revolution (GR), there has been an enormous increase in rice production and now it is becoming an important food staple in both Latin America and Africa. Though, the rice is one of the most secured food commodities in World commerce it is deficient in vitamins and minerals which lost during the milling process owing to this population that eats rice face a tremendous risk of vitamin and mineral deficiency. So as to deal with these deficiencies and their associated harmful health effects there is a need to fortify rice by developing advanced technologies. In many countries for the advancement and carrying out of rice fortification multisectoral approaches are required and there are prospects to step up a major part of rice for supply or for adoption in government safety net programs that focus those, who are in most in need, particularly women and children.

Around 18% of the total cultivated field accounts for rice production which is greatly affected by the climate change and water stress. So, the water resources planning and management are essential for conserving the ecosystem purity and safeguard food security. Wang et al. (2014) considered the rice yield, irrigation water requirement (IWR), and water use efficiency (WUE) responses to climate change in China. From 1961 to 2010, the variations of rice yield, water consumption (ET), IWR, WUE and irrigation water use efficiency (IWUE) were studied using rice crop model ORYZA2000 in three typical rice plantation regions of China. The data on future responses to 21st century climate scenarios were obtained from HadCM3 (Hadley Centre Coupled Model version 3) by driving ORYZA2000 and adopting a statistical downscaling method (SDSM). The findings revealed the significant rice yield reduction in all the three regions because of the shortened rice growth duration in the past 50 years. Owing to the significant shortened growth period and depressed WUE and IWUE, the negative impact of climate warm up to rice yield was indicated from the IPCC SRES future projection results. In the meantime, under future climate, the CO_2 concentration increase was found to increase rice yield, assuage crop water requirement and irrigation water consumptions, and increase the water use efficiencies of rice in a convinced scale degree. Further, keeping apprehension on the unpredictable climate change additional research is needed in respect to anthropogenic activities, environmental, and biological factors interactions.

2.2 FUTURE REQUIREMENT OF RICE PRODUCTION

An approach to irrigation water and drainage facilities is essential for accelerated, high-yielding rice production. Though in the large scale rice-producing countries of Asia during the 1970s the irrigation facilities were expanded rapidly the water management was inefficient. Sadiq (1992) discussed water management in rice and stated that as the rice supply, improved investments in new irrigation have reduced and the new water resources development became very costly. In a predictable future, this condition may not be reversed. But enhanced efficiency in water usage is crucial to sustain rice production growth.

More than 75% of all rice production occurs in irrigated areas. Despite the GR focused on both extension of irrigated rice area and improved yield per unit land area the next quantum jump must come about entirely from enhancing yields on existing cropland. Cassman and Pingali (1995) studied the structural changes related with the intensification process and also investigated the main biophysical and economic threats that threaten Asian farmers, should heighten irrigated rice systems further in order to attain more than 60% of yield increase by the year 2025.

From the viewpoints of farm-level, rice-based production systems Tuong and Bhuiyan (1999) reported increasing water-use efficiency in rice production. They found the important causes of water 'loss' from the system are water used during land preparation and seepage and percolation during crop growths. They recommended improving farm-level water-use efficiency, together with the up-scaling of problems from on-farm to system-level water savings.

Increasing future demand for rice will require increased fertilizer N application without causing harm to the environment. In rice systems, Ladha and Reddy (2003) discussed nitrogen fixation and mentioned that the improvement of fertilizer-compassionate varieties in the GR, together with the recognition by farmers the significance of nitrogen, led to the high rates of N fertilizer utilization in rice. To attain food security via sustainable agriculture the fixed nitrogen requirement should be acquired by biological nitrogen fixation (BNF) instead of using nitrogen fixed industrially. Therefore, the improvement of present BNF systems and the development of N_2-fixing non-leguminous crops such as rice are essential. In this context, they reviewed the potentials and limitations of conventional BNF systems in rice agriculture and the probabilities of realizing *in planta* nitrogen fixation in rice.

Usually, rice crop meets its nitrogen requirement from economic available fertilizer urea. But a considerable bulk of the urea-N is lost leading to environmental pollution problems which can be reduce using biological N fixation (BNF) technology. In sustainable rice production, Choudhury and Kennedy (2004) reviewed the prospects and potentials of BNF systems. They mentioned that 30–50% of the required urea-N can be replaced with the aquatic biota Cyanobacteria and *Azolla* which are additive to the plants' N requirements. The other possible substitutes for urea-N are BNF diazotrophic bacteria like *Azotobacter*, *Clostridium*, *Azospirillum*, *Herbaspirillum,* and *Burkholderia.* Further, the rice plant growth physiology or root morphology can be improved with the *Rhizobium* and considerable amounts of atmospheric N can also be fixed by green manure crops among which *Sesbania rostrata* has the maximum atmospheric N_2-fixing ability and it has the possibility to completely substitute the urea-N in rice cultivation.

In the confront of climate change, increasing population, agricultural land loss and competing demands for water to meet growing demand, sustaining, and increasing agriculture output is an important challenge. This desires the need of climate models associated with variations in precipitation and extreme events frequency. In China, Xiong et al. (2009) discussed future cereal production and demonstrate that by the 2040s all the effects of climate change will be comparatively modest using crop and water simulation models and two climate scenarios and socio-economic change (downscaled from IPCC SRES A2 and B2). They found that −18% (A2) and −9% (B2) of total yield decreases owing to the negative interactive effects of other drivers. These findings are highly reliant on climate scenario, socio-economic development pathway and the CO_2 fertilization effects on crop yields which may almost entirely compensate the reductions in production. Further, they find the combined effects of higher crop water requirements (due to climate change) and increasing demand for non-agricultural use of water (due to socio-economic development) caused water availability an important limiting factor in future cereal production. Finally, they concluded that the CO_2 crop yield response is very ambiguous and the effects of the extreme events on crop growth and water availability are expected to be underestimated and in China, per capita, cereal production can be maintained by adopting land and water management and progress strategies in agricultural technology. They are confident because PRECIS simulates much wetter conditions than a multi-model average.

2.3 CONCLUSION

Rice is a staple food and cultivated in different countries for its great demand in the World. China is reported to be the World's leading producer of rice followed by India. Favorable climates have been noted to lead to increased crop productivity and water deficit affects crop production. Rice requires increased fertilizer N application and to achieve food security via sustainable agriculture the fixed nitrogen requirement should be acquired by BNF instead of using nitrogen fixed industrially. Variation of rice yields in different countries has been found to be due to climate change adoption of mitigation methods of greenhouse gas emissions. Various factors such as cultural, technological, climatic, biotic factors affect rice production. Innovative technologies have been developed to increase rice productivity.

KEYWORDS

- **agro-climatic conditions**
- **general circulation models**
- **irrigation**
- **production**
- **rice production**
- **statistical downscaling method**

REFERENCES

Cassman, K. G., & Pingali, P. L., (1995). Intensification of irrigated rice systems: Learning from the past to meet future challenges. GeoJournal, *35*, 299–305.

Choudhury, A. T. M. A., & Kennedy, I. R., (2004). Prospects and potentials for systems of biological nitrogen fixation in sustainable rice production. *Biology and Fertility of Soils, 39*, 219–227.

Fan, S., (2000). Technological change, technical and allocative efficiency in Chinese agriculture: The case of rice production. *Journal of International Development in Jiangsu, 12*, 1–12.

Ladha, J. K., & Reddy, P. M., (2003). Nitrogen fixation in rice systems: State of knowledge and future prospects. *Plant and Soil, 252*, 151–167.

Lobell, D. B., Wolfram, S., & Costa-Roberts, J., (2011). Climate trends and global crop production since 1980. *Science, 333,* 616–620.

Masutomi, Y., Takahashi, K., Harasawa, H., & Matsuoka, Y., (2009). Impact assessment of climate change on rice production in Asia in comprehensive consideration of process/parameter uncertainty in general circulation models. *Agriculture, Ecosystems, and Environment, 131*, 281–291.

Muthayya, S., Sugimoto, J. D., Montgomery, S., & Maberly, G. F., (2014). An overview of global rice production, supply, trade, and consumption. *Annals of New York Academy of Sciences, 1324*, 7–14.

Sadiq, I. B., (1992). *Water Management in Relation to Crop Production: Case Study on Rice First.* Outlook on Agriculture, 21, 4, 293-299. https://doi.org/10.1177/003072709202100408

Tsvetsinskaya, E. A., Mearns, L. O., Mavromatis, T., Gao, W., Daniel, L., & Downton, M. W., (2003). The effect of spatial scale of climatic change scenarios on simulated maize, winter wheat, and rice production in the Southeastern United States. *Climatic Change, 60*, 37–71.

Tuong, T. P., & Bhuiyan, S. I., (1999). Increasing water-use efficiency in rice production: Farm-level perspectives. *Agricultural Water Management, 40*, 117–122.

Wang, W., Zhongbo, Y., Zhang, W., Shao, Q., Zhang, Y., Luo, Y., Jiao, X., & Junzeng, X., (2014). Responses of rice yield, irrigation water requirement, and water use efficiency to climate change in China: Historical simulation and future projections. *Agricultural Water Management, 146*, 249–261.

Widawsky, D., & Rozelle, S., (1998). *Varietal Diversity and Yield Variability in Chinese Rice Production, Farmer's Gene Banks, and Crop Breeding: Economic Analyses of Diversity in Wheat Maize and Rice* (pp. 159–172). Kluwer Academic Publishers.

Xiong, W., Conway, D., Lin, E., Xu, Y., Ju, H., Jiang, J., Holman, I., & Li, Y., (2009). Future cereal production in China: The interaction of climate change, water availability, and socio-economic scenarios. *Global Environmental Change, 19*, 34–44.

Zhang, X., Wang, D., Fang, F., Zhen, Y., & Liao, X., (2005). *Food Safety and Rice Production in China.* Research, Nanjing Agricultural University, Nanjing, China, www.cnki.com.cn (accessed on 15 January 2020).

CHAPTER 3

Origin, Evolution, and Domestication of Rice

ABSTRACT

This chapter discusses research advances on origin, evolution, and domestication of rice. It mentions a regional history of origin, ecotypes, and cultivars. Rice has two groups: *indica* and *japonica*. It has several weed species. The chapter states that significant research advances have been directed on the origin, evolution, and domestication of rice from various angles such as archeological evidences, phylogeny, molecular basis, few mentions single while others multiple origins of rice from wild races.

Rice being a very important staple crop to feed billions of World populations, sufficient research inputs have directed on the origin, evolution, and domestication of rice. Besides these, we discuss regional history, ecotypes, and other basic aspects in the beginning.

3.1 REGIONAL HISTORY OF RICE: THE SEQUENCE OF RESEARCH INPUTS ARE NARRATED HEREIN

Rice has a regional history of origin in different countries, a few of which are cited here. We are discussing the origin of rice in a separate section.

Several shreds of evidence were available from biosystematics, evolutionary biology, biogeography, archaeology, history, anthropology, paleogeology, and paleo-meteorology of rice. On the basis of these evidences that promoted the broad-based cultivation of the Asian cultivated rice (*O. sativa*), and the regionalized cultivation of the African cultigen (*O. glaberrima*) in West Africa sequence of events, Te-Tzu Chang (1976) discussed about the origin, evolution, dissemination, and divergence of both Asian and African rice were discussed by. The rice genus *Oryza* was originated in the Gondwanaland continents and then distributed in the humid tropics of Africa, South America, South, and Southeast Asia, and Oceania. In the distant past, both the Asian and African species of rice had a common ancestor from which the parallel and sovereign evolutionary processes take place in Africa and Asia in the following sequence: wild perennial → wild annual → cultivated annual. The differentiation in the above series has been widely contributed by weed races with the corresponding members, *O. longistaminata, O. barthii*, *O. glaberrima*, and the '*stapfii*' forms of *O. glaberrima* in Africa; *O. rufipogon*, *O. nivara*, *O. sativa,* and the '*spontanea*' forms of *O. sativa* in Asia.

They specified that in South Asia the significant climatic changes next to the preceding glacial age, spread of plants dispersal over latitude or altitude, human selection and cultural manipulation, enhanced the differentiation and diversification of the annuals.

In South and Southeast Asia, rice cultivation was taken up on a large scale of area possibly first in Ancient India. In China, wetland culture was practiced earlier than the dryland culture. On the other hand, dryland cultivation is ancient than the lowland culture in hilly areas of Southeast Asia. In the north and central China, the cultural techniques like puddling and transplanting were developed initially then transferred to Southeast Asia. In permanent fields, the planting method advanced from shifting cultivation to direct sowing, later to transplanting in bunded fields.

The three eco-geographic races of rice viz., *Indica*. Sinica or *Japonica* and *Javanica* and the upland, lowland, and deep water cultural types in monsoon Asia, were formed because of the extensive scattering of the Asian cultigen. Mostly because of cultivator's preferences, socio-religious traditions and population pressure within the period of a millennium, the rice varietal types changed considerably. Corresponding to the ecologic divergence process genetic differentiation developed.

The African cultivation experienced little diversification and developed after the Asian cultivation. Owing to the absence of cultivation pressure or dispersal, South America and Oceania wild races retained their prehistoric characters. At habitations where the wild, weed, and cultivated races co-exist, the African and Asian rice are still experiencing evolutionary changes.

With the improved recovery of archeological spikelet, bases, and rice remnants on primary sites in China, India, and Southeast Asia and with the publication of complete draft genomes of *indica* and *japonica* rice, several advances have been achieved in the understanding of rice in both genetics and archaeology. Based on coherently integrated evidence from these fields, Fuller et al. (2010) built a framework and discussed the consilience of genetics and archaeobotany in the complicated history of rice. They state that the existing archaeobotanical evidence, allows the utilization of rice spikelet bases and grain size variation to demonstrate the gradual evolutionary process of domestication, and based on this independent trends in grain size variation, can be known in India and China. Yangtze basin of China was recognized as the earliest center of rice domestication, however, in the Ganges plains of India largely distinct route into rice cultivation can be marked out. For all through the Asia and Madagascar in which rice extended most of its historical series of significant cultivation by the Iron Age, a modernized synthesis of the various rice varieties spread interwoven patterns can be proposed.

3.2 ECOTYPES AND CULTIVARS

Many rice ecotypes existing in the natural condition simulates the cultivated rice and on different characteristics of these ecotypes, several studies have been undertaken.

Rice varieties from five ecotypes (*aus, aman, boro, bulu,* and *tjereh*) and to two groups of Japanese rice (lowland and upland) were tested for their $KClO_3$ resistance, phenol reaction and apiculus hair length and based on these characters, Ueno et al. (1990) classified Asian rice ecotypes *indica-japonica* and Japanese lowland and upland rice (*Oryza sativa* L.). It was reviewed that the ecotypes *aman, boro,* and *tjereh* were classified as *indica* type while the Japanese lowland rice cultivars were classified as typical *japonica* and both types were cultivated in lowland. Few *aus, bulu,* and Japanese upland rice cultivars have shown the features of upland rice and vary from typical *indica* and *japonica*. Finally, the *aus* type, *bulu* type, and *J.u.r.* type were proposed and stated that the lowland rice cultivars can be clearly categorized into *indica* or *japonica*, but a clear classification cannot be done in upland cultivars.

Under field conditions, the red rice ecotypes comprising 11 straw-hulled, five black-hulled and one brown-hulled type were assessed and plant characteristics were described by Noldin et al. (1999). Most of the ecotypes were uniform and stable with a large amount of genetic variability.

In comparison, Lemont, Mars, and Maybelle rice cultivars red rice plants were taller with lighter green color, pubescent leaves and with several tillers per plant. At harvest, most of the ecotypes had highly dormant seed and at more than 25% of seed moisture, the ecotypes were highly susceptible and shattering was observed at about 14 days after anthesis. In contrast, rice cultivars had more total leaf area per plant with a larger flag leaf at anthesis and led to more seeds per panicle than that of red rice. On the other hand, plant features of red rice ecotypes were strongly associated with cultivated rice, which suggests the existence of natural hybridization with cultivated rice.

Two rice cultivars and two red rice ecotypes plants were planted in 3:0, 2:1, 1:2 and 0:3 (rice-red rice) plants/pot proportions and their growth response was studied by Estorninos et al. (2002). On the basis of shoot dry weight, the relative yield of PI 312777 was found to be comparable with the KatyRR and LA3 while Kaybonnet had lower relative yield than KatyRR or LA3. Compared with KatyRR and LA3 red rice ecotypes, Kaybonnet had less competitive ability than PI 312777 and both KatyRR (suspected rice × red rice cross) and LA3 (tall red rice ecotype) dominated the Kaybonnet in production of tiller, decreasing its leaf area extensively. Whereas, PI 312777 reduced the KatyRR growth and its leaf area was similar to that of LA3. The results put forward that the high tillering ability as exhibited by PI 312777 should be considered while breeding the rice cultivars for their competitive ability against weeds and these agronomic characteristics could assist in the advancement of reduced herbicide rate application program.

Spikelet fertility (SF) (seed-set) is one of the key yield components, affected with high temperature. The high-temperature stress effect on rice species, ecotype, and cultivar variances in SF and harvest index (HI) of rice were studied by Prasad et al. (2006), where they grew fourteen rice cultivars of diverse species (*Oryza sativa* and *Oryza glaberrima*), ecotypes (*indica* and *japonica*) and origin (temperate and tropical) and subjected to ambient and high temperature (ambient + 5°C). Across all cultivars, SF was found to be reduced significantly at high temperature however, the variation was observed among cultivars. At high temperature based on reductions in SF, cultivar N-22 was observed as highly tolerant to high temperature whereas cultivars L-204, M-202, Labelle, Italica Livorna, WAB-12, CG-14, and CG-17 were highly susceptible and cultivars M-103, S-102, Koshihikari, IR-8, and IR-72 were moderately susceptible

(MS) to high temperature. These cultivar differences were mainly due to the decreased pollen production and pollen reception (pollen numbers on stigma). Further, at high temperature, this reduced SF lowers the numbers of filled grains and grain weight per panicle decreasing the HI. In species or within ecotypes of tropical and temperature origin, some cultivars were found to be uniformly sensitive to high temperature and no clear-cut variations were observed among them. Thus, at high temperatures, SF can be effectively utilized while screening the rice cultivars at the reproductive phase for heat tolerance (HT).

In irrigated rice production, red rice is an unmanageable weed that spreads and contaminates the commercial rice seed and machinery but seed dormancy is the main problem in this rice. At two locations, the seed longevity of several red rice ecotypes in the soil was ascertained by Noldin et al. (2006). Near Beaumont, Texas, when buried at 5 cm, only three ecotypes had viable seed (<1%) after five months while after two years, 9 ecotypes possessed viable seed when buried at 25 cm. Only less than 1% of seed was viable after 36 months of burial. Near to College Station, in Texas, the freshly harvested seeds of red rice had longer survival when buried at 12 cm than the seeds which were placed on the soil surface. Black hull type, i.e., Texas 4, had the maximum percentage of viable seeds (2%) after 17 months of burial, whereas seeds of commercial rice cultivar were nonviable after 5 months.

In the United States, all red rice found in commercial rice has been conventionally categorized as *Oryza sativa* ssp. *indica*. Across the southern United States, belt red rice samples were assembled and studied by Vaughan et al. (2001) using 18 simple sequence length polymorphism (SSLP) markers allocated across all 12 chromosomes. Some red rice is closely related to *O. sativa* ssp. *indica* cultivated rice, while other was related to *O. sativa* ssp. *japonica*. From all 18 markers, red rice samples of the three states were found to be similar to the *O. rufipogon* accession IRGC 105491. Red rice samples obtained from Arkansas, Louisiana, Mississippi, and Texas fall in a different group comprising a number of *Oryza nivara* and *Oryza rufipogon* accessions from the National Small Grains Center. These findings clearly indicated that the conventional classification of red rice is not perfect. Though the DNA markers based red rice classification as *O. sativa* ssp. *indica, O. sativa* ssp. *japonica*, or *O. rufipogon* is in agreement with the simple morphological traits based classification alone morphological traits, are not enough to consistently classify red rice. Thus, the red rice exhibited more divergence than expected, therefore, its divergence should be considered to develop red rice management approaches.

Worldwide most of the rice-growing areas were affected with the major weed red rice which has more tillers, early shattering habit, and taller growth than domestic rice. Estorninos et al. (2017) evaluated the Kaybonnet (KBNT) rice cultivar growth response to the three red rice ecotypes viz., Louisiana 3 (LA3), Stuttgart straw-hull (Stgstraw), and Katy red rice (KatyRR). Among three red rice ecotypes, LA3, was observed as the tallest and reduced the KBNT tiller density by 51%, at 91 days after emergence (DAE), above ground biomass (AGB) by 35% and yield 80%. Stgstraw was with medium-height and decreased the tiller density of KBNT by 49%, AGB 26% and yield 61% while the KatyRR red rice was shortest, reduced KBNT tiller density by 30%, AGB 16% and yield 21%. At increased red rice density of 25 to 51 plants m^{-2} rice tiller density was decreased by 20 to 48%, rice biomass at 91 DAE by 9 and 44% and yield was reduced by 60 and 70%. Finally, it was concluded that rice growth and yield were significantly reduced by low population and short stature of red rice types.

3.3 ORIGIN

The origin, dispersal, cultivation, and variations of rice were discussed by Khush (1997). He mentioned that there exists in the genus of rice *Oryza,* two cultivated and twenty-one wild species. The *O. sativa*, usually referred to as the Asian cultivated rice, is cultivated all across the World in many areas. The African cultivated rice, *O. glaberrima,* when compared to the Asian cultivated rice its cultivation is limited to small scale in West Africa. This genus *Oryza* was thought to have originated probably 130 million years ago in Gondwanaland. Further, with the fragmentation of this Gondwanaland region, the different species might have been distributed into different continents. Most of these cultivated species of rice were opined to have their origin from a common ancestor that had AA genome. The perennial and annual ancestors of *O. sativa* are *O. rufipogon* and *O. nivara*. Likewise, *O. longistaminata* and *O. breviligulata* were the perennial and annual ancestors of *O. glaberrima*. *O. glaberrima* might have been domesticated probably in Niger River delta. Based on genetic affinities, the varieties of *O. sativa* are generally categorized into six groups. Among these widely known rice are that of the *indicas* and these relate to group (I) and *japonicas* to group (VI). *Javanica* rice belongs to group (VI). These are tropical *japonicas*, in dissimilarity to temperate *japonicas* grown in a temperate climate. The *indica* and *japonica* rice had a polyphyletic derivation. The *indicas* might have cultivated in the Himalayan foothills of Eastern India, while the *japonicas* were evidenced to be domesticated in South China. The *indica* rice from India was dispersed throughout the tropics and subtropics, while the *japonica* rice moved northward from South China. They developed as temperate ecotype. The *japonica* rice moved southward to Southeast Asia and from there to West Africa and Brazil and became the tropical ecotype. Rice, at present, is grown between 55°N and 36°S latitudes, under dissimilar growing environments such as irrigated, rainfed lowland, rainfed upland and flood prone ecosystems. Numerous cultivars have resulted because of their adaptations to varied environments and human selections. It is assessed that, there are about ~120,000 varieties of rice existing in the World. With the founding of International Rice Research Institute (IRRI) in 1960, intensification of rice varietal improvement programs have occurred and several high yielding varieties that were developed were released. Most of these rice varieties are now planted across 70% of rice lands, with the large scale adoption of cultivation of these improved varieties of rice, rice production was doubled between 1966 and 1990. Though the production of rice was doubled, to feed the surplus rice consumers and increased populations, presently, there is a requirement for an increase in rice production by 2025 by an extent of 60%. The novel tools of molecular and cellular biology viz., anther culture; molecular markers which have assisted in the selection and genetic engineering will have additional increasing roles to bring about rice improvement.

North China was known as one of the important centers of agriculture. Based on new discoveries and perspectives on the beginning of agriculture which enlightened the understanding of the transition to agriculture, Shelach (2000) stated the original neolithic cultures of Northeast China and challenges the general viewpoint of north China as an agricultural development homogeneous area. In northeast China, the Xinglongwa (ca. 8,000–6,800 B.P.) and the Zhaobaogou (ca. 6,800–6,000 B.P.) cultures represented only the initial inactive societies. With the introduction of these societies, attempts were made to progress the useful model like the Chinese Interaction Sphere ahead of generalization and north China one sub-region developments were assessed in detail. Based on this data, significant problems regarding the transition to agriculture along with new prospects for forthcoming research in this aspect were addressed.

Zhao et al. (2000) discussed that during the Late Pleistocene and Holocene periods in the middle Yangtze River Valley, which is the probable home of rice (*Oryza sativa* L.) domestication, significant natural and human-induced vegetational changes were unveiled from Phytolith data of Poyang Lake, southern China. During the Late Pleistocene (from >13,500 to ca. 10,500 yr B.P.) *Oryza* could have been a natural vegetation component, however, it might have not well adapted to the cooler and drier climate prevailing during that era. During early Holocene wild *Oryza,* species may have advanced north than viewed presently because this era might have been wetter and more distinctly seasonal. In the central point of Yangtze River Valley by 4,000 yr B.P., rice agriculture seems to have been well developed. Therefore, in southern China during the Late Pleistocene and early Holocene, *Oryza* human exploitation might have significantly affected with the atmospheric CO_2 contents and the precipitation and temperature seasonality, plus whole cooler and drier climates of Pleistocene.

In Asia, a most important type of cultivated rice viz., glutinous rice has ancient cultural importance. The glutinous phenotype evolutionary and geographical origins were traced by Olsen and Purugganan (2002), for which they studied *Waxy* locus allele genealogy of 105 glutinous and non-glutinous landraces from throughout Asia. They put forth evidence that in Southeast Asia a single evolutionary origin might have originated as a result of the splice donor mutation. Compared with an unlinked locus, *RGRC2* at the *Waxy* locus, reduced genetic variation selection characteristic was found to be related with glutinous rice origin and this was revealed by measuring Nucleotide diversity, confirming that this pattern is explicit to *Waxy.* Further, in Northeast Asia the splice donor site mutation was observed in many non-glutinous varieties; this showed that the fractional suppression of this mutation might have a central role to play in the improvement of these varieties. Finally, they reported that the variation in endosperm starch causes glutinous phenotype which is a result of *Waxy* gene intron 1 splice donor site mutation and set forth growing proof on the significance of modifier loci in the evolution of domestication trait.

With the aim of clarifying the cultivation rice subspecific historical changes from 12 Neolithic sites, 80 samples comprising soils, burned soils and pottery fragments were analyzed by Zheng et al. (2003) for phytoliths. The findings reveal the occurrence of keystone phytoliths recognizable as *Oryza* spp. The phytoliths became enlarged with time following a common trend, suggesting that rice was most probably cultivated 7,000 years ago and that *Japonica* rice was domesticated in China.

In rice highly valued grain quality trait, the fragrance is controlled by a betaine aldehyde dehydrogenase gene (*BADH2*). But, the origin and evolution of this gene were not investigated clearly. Thereby, the origin and evolution of fragrance were explored by Kovach et al. (2009) in rice. Eight putatively nonfunctional alleles having different geographic and genetic origins of the *BADH2* gene were identified. Despite the fragrance trait, have multiple origins all fragrant rice varieties had a single allele, *badh2.1* predominantly which includes broadly known Basmati and Jasmine types of fragrant rice. Within the *Japonica,* varietal group the establishment of the *badh2.1* allele single origin was allowed by haplotype analysis and determines the introgression of this allele from *Japonica* to *Indica.* Irrespective of the fragrance phenotype, the ancestral haplotype of *Japonica* across a 5.3-Mb region flanking *BADH2* was almost identical to the basmati-like accessions revealing the strong evolutionary association between Basmati varieties and the *Japonica* gene pool.

The proof of original diversification in Poaceae was provided from the late Cretaceous period (67–65 Ma) of India which revealed that by the latest Cretaceous era, the Poaceae, possessed modern subclades PACMAD (Panicoideae-Aristidoideae-Chloridoideae-Micrairoideae-Arundinoideae-Danthonioideae) and BEP (Bambusoideae-Ehrhartoideae-Pooideae) members including a taxon with proposed affinities to Ehrhartoideae. Here, Prasad et al. (2011) described the additional fossils and assigned them to the rice tribe, Oryzeae, of grass subfamily Ehrhartoideae on the basis of phylogenetic analyses combining molecular genetic data and epidermal and phytolith features across Poaceae. By the late Cretaceous, considerable diversification was revealed within Ehrhartoideae from these new Oryzeae fossils which pushed back the time Poaceae origin altogether. As a result, for grass evolution and palaeobiogeography re-evaluation of current models is required assuming that the fragrance trait was derived from the *Indica* varietal group.

The origins of agriculture in China, particularly rice agriculture origin, significantly influenced the occurrence of Chinese civilization as well as to the development of World history. In recent times, owing to the remarkable progress of archaeobotanical research in China, the academic community started concentrating on the rice agriculture origin studies. The new data and new issues were provided by Zhao (2010) to this study. Recently, in China, a huge amount of plant remains including those associated with the early rice agriculture study were excavated from archeological sites using the flotation technique. The direct archeological evidence and new issues about the rice agriculture origin in China were put forth from this new data. For instance, irrespective of whether the rice was domesticated or not in the Shangshan site, the rice remains dated to ca. 10,000 cal. B.P. from this site suggest the rice cultivation beginning. From the Jiahu site plant remains dated to ca. 8,000 cal. B.P., were recovered by floatation and their quantitative analysis demonstrated that the fishing/hunting/gathering were primary source of food for Jiahu people whereas rice cultivation and animal husbandry products were supplements to their diet. At the Tian-luoshan site, dated to 6,000 to 7,000 cal. B.P., the continuing excavation, using floatation and water-sieving put forward that although rice farming is important it was only a part of a wider existence pattern of the Hemudu Culture and after 6,500 B.P. culmi, the rice domestication beginning was not known clearly.

In advanced human civilization, long history of selection has occurred which resulted in the domestication of rice but the origins and domestication of rice is always controversial and debatable. From the immediate ancestral progenitor of cultivated rice *Oryza, rufipogon* 446 geographically diverse accessions were used by Xuehui Huang et al. (2012) to generate genome sequences. Afterwards from 1,083 cultivated *indica* and *japonica* varieties variation, a comprehensive rice genome map was built. Consequently, 55 selective sweeps took place during the domestication process were searched for marks of selection. The genome-wide patterns and detailed analysis of these sweeps, suggested that from an *O. rufipogon* specific population the first domesticated rice was *Oryza sativa japonica* cultivated nearby Pearl River in southern China and then by crossing *japonica* rice and local wild rice *Oryza sativa indica* was derived. Further, via high-resolution genetic mapping the domesticated related traits are analyzed in addition an efficient resource and genomics approach for rice domestication and breeding were proposed from this study.

The origin and spreading of rice cultivation was discussed by Charles Higham et al. (2015) which are still controversial and debatable. This question of rice cultivation origin was undraped by the more archeological and archaeobotanical discoveries evidences. The beginning of rice cultivation by 8,500–8,000 years BP in the middle Yangzi Valley and its successive distribution to south China and Southeast Asia, was indicated from the archeological as well as archaeobotanical discoveries.

Broad far-reaching database on archeological proof of rice was obtained from 400 sites of mainland East Asia, Southeast Asia, and South Asia and compiled; using this database modeling approach on the geographical origin of rice cultivation in Asia was adopted by Fabio Silva et al. (2015). The geographical origins of rice cultivation were compared using several models and for the rice origins and successive outward diffusion, the most probable region(s) were inferred. Based on the simple geographical features the least-cost distances were calculated by adopting the Fast Marching method. From the archeologically inferred age, power-law quantile regressions versus the least-cost distance from the putative origin(s) goodness of fit were obtained. The hypothetical geographical origins were obtained from the literature containing genetics, archaeology, and historical linguistics and with these origins; the region that best fits the archaeobotanical data was compared. The dual origin model was the best model that fits all existing archeological evidence with Middle Yangtze and the Lower Yangtze valleys, focused centers for the rice cultivation and dispersal.

3.4 EVOLUTION

Asian rice, *Oryza sativa* is deliberated as the World's ancient and leading crop species. The wild rice *O. rufipogonis* suggested as the common ancestor of both the key species of Asian rice, *indica*, and *japonica* from a single-origin model. On the other hand, the separate domestication of both species from diverse wild species was suggested from multiple independent models on domestication. The high genetic differentiation amongst *indica* and *japonica* and numerous phylogenetic studies supported the second view of rice domestication. The evolutionary past of domesticated rice was reassessed by Molina et al. (2011) on molecular basis for which from a different set of wild and domesticated accessions of rice, 630 gene fragments on chromosomes 8, 10, and 12 were resequenced using SNPs. In cultivated rice, on these chromosomes, 20 putative selective sweeps were detected and a rice single domestication origin was strongly supported by the SNP data based on demographic modeling and diffusion-based approach. In addition to this, a single-origin was also pointed by the previously published phylogenetic sequence datasets and Bayesian phylogenetic analyses employing the multispecies coalescent. Lastly, based on the estimate of molecular clock, the rice domestication origin was dated at ~8,200–13,500 years ago which is in agreement with recognized archeological data and they set forth molecular evidence for a single evolutionary origin of domesticated rice.

3.5 DOMESTICATION

From East China, phytolith evidences were obtained based on this rice domestication and climatic change was studied by Lu et al. (2002). They mention that in the East China Sea, from a late-glacial to epicontinental sediments, Holocene sequence fossil rice phytoliths were detected afterward and they might have transported from the middle and/or lower reaches by the Yangtze River. Among the World's earliest domesticated cereal crops, the rice phytoliths appeared first in the sequence. The 13,000 and 10,000 cal. yr BP period was thought to be extremely colder (Younger Dryas) based on phytoliths, pollen, diatoms, and foraminifera records. But with the advent of old climate conditions, domesticated rice phytoliths were disappeared indicating a high climatic effect on human activities during that period. In the area of rice, domestication might have supported by the warmer and wetter conditions that prevailed during the 13,900 to 13,000 cal. yr BP period and after 10,000 cal. yr BP.

During the Holocene, the agriculture evolution, spatial expansion, and humans' population increase necessitate the domestication of plants and animals which subsequently aid in the advancement of technically innovative societies and further, helped in the establishment of permanent settlements and expansion of urban-based societies by people. Accordingly, the plants and animals domestication transformed the early human's occupation from hunting and gathering to selective hunting, raring, and ultimately established agriculture. During 10,000–7,000 cal years BP fast and extensive domestication of plants and animals was suggested based on the earliest evidences of archaeology from all over the tropical and subtropical areas of southwestern and southern Asia, northern, and central Africa, and Central America. Many paleo records indicated the intense humid phase and favorable climates prevalence during the early Holocen and Gupta (2004), reported that in early Holocene the domestication of plants and animals and successive establishment of agriculture were related to climate enrichment.

Successively, the existence of rice domestication with reduced seed shattering was reported by Li et al. (2006). They mention that in the beginning of crop domestication, the plants without naturally shed ripe fruits or seeds were selected. The cereal domestication was promoted by the reduced grain shattering involving large effect genetic loci. The molecular basis of this crucial domestication transition is still unknown because previously amino acid substitution based human selection in the expected DNA binding domain encoded by a gene, was not known. Later, the gene function essential for the normal development of an abscission layer controlling the separation of a grain from the pedicel was undermined with the substitution of amino acid.

Afterward, in China, the proofs of the early initiation (c. 9,000 cal. BP) of rice domestication were put forth by Liu et al. (2007) and they discussed the domestication process from the collection of wild rice to rice domestication by revising previous publications. Then considering the rice morphology and archeological context, they examined the early rice remnants. Three aspects were emphasized viz., the initial rice domestication timing in the Yangzi River region, the earliest domesticated rice existence in the Lower Yangzi and Huai River regions, and the rice grain size change effects on archeological collections. For differentiating, the domesticated rice from wild rice Fuller et al. used grain size enhancement, complete grain shape, and immature rice remains in the archeological record; the problems associated with the occurrence of these features were also discussed in this paper. On the basis of published data and their research on rice, they revealed that in the north as well as south China Neolithic people might have been harvesting wild rice and initiation of rice cultivation ultimately led to domestication by the early Holocene (9,000 cal. BP).

Concurrently, in the 5th millennium BC, of the Lower Yangtze area; assumed for rice domestication and wild rice cultivation and domestication evidences were described by Fuller et al. (2007) and focused on the question of rice cultivation, however, they confront some of the problems in the identification of wild to domesticated rice transition. This information is encouraged by Jiang and Liu in Antiquity (80, 2006). By means of data from Eastern China, for the Lower Yangtze region, they proposed that around 5,000 BC pre-domestication cultivation started followed by the initiation of rice domestication around 4,000 BC and the sedentary hunter-gatherer-foragers collected rice, along with other subsistence foods such as nuts, acorns, and water chestnuts. Therefore, they discussed agriculture as a long term process and its implications for sedentism.

Simultaneously, the rice domestication genetics and phylogenetic were studied by Sang and Ge (2007). They mention that in the crop species, the genetic and phylogenetic studies on the origin and evolution faced a great challenge owing to their genetically different cultivars and ecologically distinct wild progenitors. Increasing phylogenetic evidences suggested that the *indica* and *japonica* rice cultivars have differed genomic backgrounds and were derived separately from genetically distinct wild populations. But the for the grain shattering reduction a domestication gene, *sh4* may possibly originated only once and later fixed in both cultivars. To verify these data, two models were put forward, first, the early introgression significance between autonomously domesticated cultivars was emphasized in the 'combination model' considering the multiple origins of cultivated rice, whereas the introgression from wild species local populations into an ancestral domesticated population was given importance in 'snowballing model' which consider a single origin of cultivated rice. However, the recent and future introgression of beneficial genes from the wild gene pool via conventional and molecular breeding programs can be seen as the continuance of domestication.

Though the date of rice domestication is controversial and debatable while discussing the intricate past of the rice domestication, Sweeney and Couch (2007) mentioned that rice was domesticated dating to 8,000 BC in archeological sites. The two worldwide domesticated species of rice are *Oryza sativa* (Asian) and *Oryza glaberrima* (African). The wild rice was differentiated from domesticated rice by many traits that include variations in: pericarp color, dormancy, shattering, panicle architecture, tiller number, mating type and number and size of seeds.

They state that many present subpopulations have been categorized in the two key subspecies of rice the *indica* and *japonica* and a deep population structure in these groups was exposed by genetic studies by means of various methodologies. The distant past of the partition is likely at more than 100,000 years ago which is earlier than domestication and supported the separate domestications of *indica* and *japonica* from wild ancestor pre-differentiated pools. The sterility and segregation of domestication traits was observed in crosses between these two subspecies signifying that diverse populations are fixed for various allele networks, involved in these characters. Further, in the crosses between the subspecies and between wild and domesticated accessions of rice many domestication QTLs have been detected. They found that a single gene with pleiotropic effects or that closely related clusters of genes bring about these QTL because many QTLs cluster in the same genomic regions. In recent times, from rice numerous domestication loci have been cloned, comprising the gene regulating pericarp color and two loci for shattering. The distribution and evolutionary history of these genes provide vision into the domestication process and the association amongst the subspecies.

The evidences from current genetic studies provided new understandings in complicated history of rice domestication as mentioned by Kovach et al. (2007). In *Oryza sativa* two main subspecies *indica* and *japonica* with origin from genetically distinct gene pools and with *Oryza rufipogonas* a common wild ancestor were revealed from Genome-wide studies of variation signifying multiple domestications of *O. sativa.* But to the rice domestication, another complex layer was provided from the evolutionary history of newly cloned domestication genes. As predictable from the happening of separate domestications, though some alleles continue only within specific subpopulations other most important domestication alleles are commonly found in all cultivated varieties of *O. sativa*. Finally, they reported that important domestication alleles were transferred between divergent rice gene pools with restricted introgression supporting multiple domestications.

The crop domestication is a compound process involving the selection process in plant evolution and related with variations in the DNA which regulate agronomically important traits. Subsequently, Shomura et al. (2008) reported that yields were improved with the deletion of a grain size-related gene during rice domestication. The fine mapping, complementation testing and correlation analysis showed that the sink size can be increased significantly by deleting the recently identified QTL, *qSW5* (*QTL* for seed width on chromosome 5) due to an increase in cell number in the rice flower outer glume. The sink size might have considered an important trait by ancient humans in selection for increased yield of rice. Further, by means of genome-wide RFLP polymorphisms of different rice landraces, two other defective functional nucleotide polymorphisms of rice domestication-related genes were mapped which revealed the historical significance of *qSW5* deletion in artificial selection, cultivation propagation and natural crossings in rice domestication and the rice genome domestication process were enlightened.

During plant domestication, the nature of selection was described by Sang (2007) and he states that excellent example and enriched model to study evolution in plant domestication. During the Neolithic period, many studies were carried out to detect genes related with domestication and the changing aspects of human cultivation practices were clearly understood by the archeological work. Along with these, a better understanding of the selective pressures accompanying crop domestication was provided. Further, he demonstrated that the understanding of the evolutionary selection nature that supplements domestication can be expanded from the twin vantage points of genetics and archaeology.

About 10,000 years ago, the agriculture has been emerged leading to a notable change in human evolutionary history. Neolithic human populations genetic makeup might get affected with the diet shift in agriculture societies and on this genetic locus Darwinian positive selection, was unveiled from the class I alcohol dehydrogenase sequence polymorphism (ADH1BArg47His) limited enrichment in southern China and the nearby locations, but the driving force for this is still not clear. Across China, Yi Peng et al. (2010) analyzed a total of 38 populations (2,275 individuals) including Han Chinese, Tibetan, and other ethnic populations and in these populations, a clear east-to-west cline was demonstrated from the geographic spreading of the ADH1B*47His allele. In south-eastern populations, allele was dominant while it was rare in Tibetan populations. Further, the molecular dating have shown that ADH1B47His allele might have emerged around 10,000~7,000 years ago this is in consistent with the origin time and extension of Neolithic agriculture in southern China. Finally, with the rice domestication unearthed culture artifact locations in China the ADH1B*47His allele geographic distribution was found to be predominant in East Asia.

3.6 CONCLUSION

The origin, evolution of rice is a complex history and highly interesting. Concerted research activities have been directed to unveil this complex history. It also discusses the presence of rice ecotypes and regional history of origin of rice. This entails distribution of wild rice after Gondawana supercontinents fracture in different countries. There exist two groups' rice, *indica*, and *japonica* and there existed several wild species of *Oryza* among which *Oryza sativa* is cultivated globally but *Oryza glaborimma* is mainly cultivated in Africa. Hybridization among wild species contributed to new species of rice. Intensive studies have been undertaken to unveil complex history of origin, evolution, and domestication of from different angles, archeology, phylogeny, and molecular levels, some suggest single and other multiple origin of modern rice. It has been envisaged that the evolutionary history of rice is more intricate. Recent works cast light on the genetics of the transition from wild (*O. rufipogon* and *O. nivara*) to domesticated (*O. sativa*) rice, for better understanding of ancestors and to utilize their useful characteristic features, in developing and breeding novel rice varieties.

KEYWORDS

- **archeology**
- **cultivars**
- **domestication**
- **evolution**
- **geographical distribution**
- **molecular biology**
- **origin**
- **paleobotany**
- **rice ecotypes**

REFERENCES

Charles, H., and Tracey, L.-D. Lu., (2015). The origins and dispersal of rice cultivation. *Antiquity*, *72*, 867–877. https://doi.org/10.1017/S0003598X00087500 (accessed on 15 January 2020).

Estorninos, L. E., Gealy, D. R., & Talbert, R. E., (2002). Growth response of rice (*Oryza sativa*) and red rice (*O. sativa*) in a replacement series study. *Weed Technology*, *16*, 401–406.

Estorninos, L. E., Gealy, D. R., Gbur, E. E., & Talbert, R. E., (2017). Rice and red rice interference: II. Rice response to population densities of three red rice (*Oryza sativa*) ecotypes. *Weed Science*, *53*, 683–689. https://doi.org/10.1614/WS-04-040R1.1 (accessed on 15 January 2020).

Fabio, S., Stevens, C. J., Alison, W., Cristina, C., Ling, Q., Andrew, B., & Fuller, D. Q., (2015). Modeling the geographical origin of rice cultivation in Asia using the rice archeological database. *PLoS ONE*, *10*, e0137024. https://doi.org/10.1371/journal.pone.0137024 (accessed on 15 January 2020).

Fuller, D. Q., Harvey, E., & Qin, L., (2007). Presumed domestication? Evidence for wild rice cultivation and domestication in the fifth millennium BC of the Lower Yangtze region. *Antiquity*, *81*, 316–331.

Fuller, D. Q., Sato, Y. I., & Castillo, C., (2010). Consilience of genetics and archaeobotany in the entangled history of rice. *Archeological and Anthropological Sciences*, *2*, 115–131.

Gupta, A. K., (2004). Origin of agriculture and domestication of plants and animals linked to early Holocene climate amelioration. *Current Science*, *87*, 54–59.

Oka, H. I., (1988). Origin of Cultivated Rice 1st Edition. Elsevier Science. eBook ISBN: 9780444598523

Khush, G. S., (1997). Origin, dispersal, cultivation, and variation of rice. *Plant Molecular Biology*, *35*, 25–34.

Kovach, J. M., Sweeney, S. T., & McCouch, R., (2007). New insights into the history of rice domestication. *Trends in Genetics*, *23*, 578–58. https://doi.org/10.1016/j.tig.2007.08.012 (accessed on 15 January 2020).

Kovach, M. J., Calingacion, M. N., Fitzgerald, M. A., & Couch, S. R., (2009). The origin and evolution of fragrance in rice (*Oryza sativa* L.). *PNAS*, *106*, 14444–14449.

Li, C., Zhou, A., & Sang, T., (2006). Rice Domestication by Reducing Shattering. *Science*, *311*, 1936–1939.

Liu, L., Lee, G. A., & Jiang, L., (2007). Evidence for the early beginning (c. 9000 cal. BP) of rice domestication in China: A response. *The Holocene*, *17*, 1059–1068.

Lu, H., Liu, Z., & Wu, N., (2002). Rice domestication and climatic change: Phytolith evidence from East China. BOREAS. *An International Journal of Quaternary Research*, *31*, 478–385.

Molina, J., Sikora, M., Garud, N., Flowers, J. M., Rubinstein, S., Reynolds, A., et al., (2011). Molecular evidence for a single evolutionary origin of domesticated rice. *PNAS*, *108*, 8351–8356.

Noldin, J. A., Chandler, J. M., & Cauley, G. N., (1999). Red rice (*Oryza sativa*) biology, characterization of red rice ecotypes. *Weed Technology*, *13*, 12–18.

Noldin, J. A., Chandler, J. M., & McCauley, G. N., (2006). Seed longevity of red rice ecotypes buried in soil. *Planta Daninha*, *24*, 611–620.

Olsen, K. M., & Purugganan, M. D., (2002). Molecular evidence on the origin and evolution of glutinous rice. *Genetics*, *162*, 941–950.

Prasad, P. V. V., Boote, K. J., Allen, L. H., Sheehy, J. E., & Thomas, J. M. G., (2006). Species, ecotype, and cultivar differences in spikelet fertility and harvest index of rice in response to high temperature stress. *Field Crops Research*, *95*, 398–341.

Prasad, V., Strömberg, C. A. E., & Leache, A. D., (2011). Late Cretaceous origin of the rice tribe provides evidence for early diversification in Phocaea. *Nature Communications*, *2*, 480.
Sang, T., & Ge, S., (2007). Genetics and phylogenetics of rice domestication. *Current Opinion in Genetics and Development*, *17*, 533–538.
Sang, T., (2007). The puzzle of rice domestication. *Journal of Integrative Plant Biology*, *49*, 760–768.
Shelach, G., (2000). The earliest Neolithic cultures of Northeast China: Recent discoveries and new perspectives on the beginning of agriculture. *Journal of World Prehistory*, *14*, 363–413.
Shomura, A., Izawa, T., Ebana, K., Ebitani, T., Kanegae, H., Konishi, S., & Yano, M., (2008). Deletion in a gene associated with grain size increased yields during rice domestication. *Nature Genetics*, *40*, 1023–1028.
Sweeney, M., & Couch, S., (2007). The complex history of the domestication of rice. *Annals of Botany*, *100*, 951–957.
Te-Tzu, C., (1976). The origin, evolution, cultivation, dissemination, and diversification of Asian and African rice. *Euphytica*, *25*, 425–441.
Ueno, K., Sato, T., & Takahashi, N., (1990). The *Indica-japonica* classification of Asian rice ecotypes and Japanese lowland and upland rice (*Oryza sativa* L.). *Euphytica*, *46*, 161–164.
Vaughan, L. K., Ottis, B. V., Prazak, H. A. M., & Bormans, C. A., (2001). Is all red rice found in commercial rice really *Oryza sativa*? *Weed Science*, *49*, 468–476.
Xuehui, H., Nori, K., & Bin, H., (2012). A map of rice genome variation reveals the origin of cultivated rice. *Nature*, *490*, 497–501.
Yi, P., Hong, S., Xue-Bin, Q., Chun-Jie, X., Hua, Z., Run-Lin, Z. M., & Bing, S., (2010). The ADH1B Arg47His polymorphism in East Asian populations and expansion of rice domestication in history. *BMC Evolutionary Biology*, *10*, 15. https://doi.org/10.1186/1471-2148-10-15 (accessed on 15 January 2020).
Zhao, Z., & Piperno, D. R., (2000). Late Pleistocene/Holocene environments in the middle Yangtze River valley, China and rice (*Oryza sativa* L.) domestication: The phytolith evidence. *Geoarchaeology*, *15*, 203–222.
Zhao, Z., (2010). New data and new issues for the study of origin of rice agriculture in China. *Archeological and Anthropological Sciences*, *2*, 99–105.
Zheng, Y., Matsui, A., & Fujiwara, H., (2003). Phytoliths of rice detected in the Neolithic sites in the Valley of the Taihu Lake in China. *Environmental Archaeology*, *8*, 177–183.

CHAPTER 4

Rice Ideotype

ABSTRACT

Plant breeders look for plant types with ideal plant characteristics, morphology, plant architecture, leaf types, its orientation, leaf area, and other characteristics that contribute to the high yield of a crop cultivar. This chapter briefly highlights some of the research activities carried out in the improvement of ideotype concepts and characters in rice suitable for different environmental conditions and planting densities.

4.1 ARCHITECTURE OF RICE PLANT

The architecture of a plant is crucial in ascertaining the yield potential of a particular cultivar. It constitutes an important collection of those agronomic characters which determines the yield and the grain production. In rice, key plant architectural traits having an influence on the potentiality of grain production are tillering, plant height, and also the morphology of the panicle that is produced. Wang and Li (2005) have reviewed the progress that was made in the exploration of the different mechanisms involved in controlling the rice plant architecture. Though, significant progress has occurred with relevance to the collection and identification of some rice mutants which were defective in plant architecture, there is still a limitation in the discerning of the mechanisms that have a part to play controlling plant height, its tiller production and number, and the development of the panicles. New innovations of these aspects are still emerging.

The rice plant architecture is under genetic control. During the domestication, there were significant changes in the rice plant architecture in comparison to its wild relatives. The intimately primitive associated wild rice species *Oryza rufipogon* is deliberated as the progenitor of cultivated rice (*Oryza sativa*). An important event that has occurred in rice domestication is the change over from the plant architecture characteristic of this wild progenitor to the present-day cultivated species. The molecular basis behind this transition during the domestication process was explored by Jin et al. (2008). They have shown that *PROG1* gene of wild rice plants was involved in regulating the plant architectural traits as a number of tillers and the tiller angle. This gene encoded a nuclear transcription factor which was to a large extent expressed in the axillary meristems. This gene transcription factor was mapped on chromosome 7. The gene was found to have its dominant expression at the site of tiller bud formation in the axillary meristems of the rice plant. The experiments conducted on rice transformation have revealed that during domestication, there was a substitution of an amino acid in the PROG1 protein, which was responsible for the evolution of the plant architecture from that of wild rice species to the present-day architecture of cultivated domesticated rice.

A preferable characteristic trait of plant architecture involved grain yield (GY) is the erectness in the panicle. This is an important characteristic trait of the *japonica* rice. The molecular mechanism behind this panicle erectness trait that was modified during domestication was investigated by Zhou et al. (2009). They have reported that a deletion that has occurred in a quantitative trait gene qPE9-1 might have been linked with panicle erectness and it has led to an improvement in rice plant architecture during the domestication process. They reported that a main quantitative trait locus (QTL), qPE9-1 has an important indispensable role in the control of plant architecture of rice, particularly its panicle erectness. Map-based cloning of this has shown that this R6547 qPE9-1 gene encodes a 426 amino acid protein. Further, it was found to be homologous to the keratin-associated protein 5–4 family. It has three Von Willebrand Factor Type C (VWFC) domains, a transmembrane domain, and one 4-disulfide-core domains. The phenotypic analogies of near-isogenic lines (NIL) and transgenic lines unveiled that this functional allele (qPE9-1) has contributed to the production of drooping panicles. Further, a loss-of-function mutation (qpe9-1) in this gene has led to the production of more erect panicles in the transgenic lines of rice. Besides, they observed that qPE9-1 locus regulated the panicle and grain length (GL), grain weight and ultimately the GY. Thus, they suggested that the panicle erectness trait has contributed to a natural random loss-of-function mutation for the qPE9-1 gene. It has later become the target of artificial selection during *japonica* rice breeding programs.

Research studies have revealed that the rachis branch and spikelet development in rice is mostly controlled by LAX1 and FRIZZY PANICLE 2 Genes (Komatsu et al., 2001). They evaluated two mutants that exhibited transformed panicle architecture in rice (*Oryza sativa* L.). L*ax1-2,* which was a new and stronger allele than *lax* mutant which was reported earlier, exhibited a severe inhibition in the commencement and/or conservation and upkeep of rachis-branches, lateral spikelets, and terminal spikelets. An *in situ* hybridization analysis carried out by the usage of *OSH1,* a rice *knotted1* (*kn1*) ortholog, has led to the confirmation that in these *lax1-2* panicles there were no lateral meristems and these mutants lacked them. These defects that were exhibited have unveiled that the *LAX1* gene is much obligatory for the initiation and for maintenance of the axillary meristems in the rice panicle. Apart from its specific role in the development of lateral meristems it also works as an identity gene of floral meristems, specifying the terminal spikelet meristem formations. They compared the defects in *lax1-1* and *lax1-2* plants. They found that the receptivity's diminished *LAX1* activity is not the same among contrasting types of meristems. However, they found that the primary key branching pattern of the panicle was distinguishable from that of the wild type in the *fzp2* mutant. In this mutant, there was a blockage in the stipulation of both terminal and lateral spikelet meristems. Consecutive continuous rounds of branching developed. These developed at the initiative point of spikelet meristems in wild type panicle. Finally, it has led to the generation of an excessively ramified rachis branched panicle production. Thus, they mentioned that *lax1-1 fzp2* double mutants exhibit a unique, basically additive, phenotype, where the functions of *LAX1* and *FZP2* are genetically independent pathways that occur in rice governing its architecture.

The architecture of vegetative and even the reproductive parts of a plant are ascertained by the activities of the axillary meristems. In *Arabidopsis thaliana*, it was found that the transcription factor LFY, has conferred to the conversion vegetative meristems to the floral meristems. It has conferred to the floral fate in this eudicot species. New meristems arise from the edge of the reproductive shoot apex. Divergent orthologous *LFY* genes were found to be involved in governing this transition from the vegetative-to-reproductive phase. A homolog to these genes which is important in rice is RFL gene. Nagashree et al. (2008) have described *RFL* by knockdown of its expression and by also it's ectopic overexpression in transgenic rice. They found that a cutback in *RFL* expression has brought a spectacular setback in transition to flowering (FL). There was no FL in the phenotype which had a vast decrease in the expression of the RFL. Conversely, they observed that overexpression of *RFL* has led to the triggering of the precocious FL. Their studies have revealed that *RFL* regulates *OsSOC1* (*OsMADS50*), and activates FL. Its expression status was found to have an influence even in the regulation of tillering apart from its ability in facilitating the conversion of the primary growth axis to an inflorescence meristem. Thus, they specified that *RFL* expression regulated spatially and temporally, in the course of the development of vegetative axillary bud (tiller) primordia and inflorescence branch primordia is very much crucial for the production of required number of tillers and panicle branches. Their results offer means for the alteration of rice FL time and plant architecture by the manipulations of the RFL-mediated pathways and bring about a change in rice plant architecture and yields.

Most of the studies have shown rice architecture, as an essential agronomic attribute affecting GY. Gao et al. (2009) reported an innovative putative esterase gene, Dwarf 88, stirring the rice plant architecture. They identified a tillering dwarf mutant *d88.* This mutant was obtained with the treatment of EMS of Lansheng a cultivar of *Oryza sativa* ssp. *Japonica.* The mutant had numerous curtailed tillers and petite panicles and seeds. They observed that there was a drastic decline in the number and size of parenchyma cells throughout the stem marrow cavity. This was related with a lag lengthening in the elongation of parenchyma cells along with the production of delicate fragile tillers and dwarfism in the *d88* mutant. This gene was found to encrypt a putative esterase. Further, this gene was found to be revealed in high levels in most rice organs, particularly its expression was several folds higher in the vascular tissues. The mutant lugged is a nucleotide substitution in the 1st exon of the gene. This exchange has resulted in the substitution of amino acid arginine for glycine. This might have apparently rattled the functionally conserved *N*-myristoylation domain of the protein. They confirmed *D88* gene function by complementation test and antisense analysis. They mentioned that this *D88*, gene serves as a recent category of genes which are concerned with the regulation of cell growth and organ development and accordingly the plant architecture. They discussed about the plausible association between the tiller formation related genes and *D88*. They specified that future detection of the substrate for *D88* might enable in the description of novel pathways convoluted in the regulation of plant development.

The development of new plant types (NPT) with quintessential plant architecture (IPA) has been proposed as a means for enhancing the rice yield potential over that of existing high-yield varieties.

Jiao et al. (2010) reported that control of *OsSPL14* by *OsmiR156* contributed to representative plant architecture in rice. They reported that the cloning and characterization of *IPA1* (ideal plant architecture 1), a semidominant QTL has changed remarkably rice plant architecture and yield. This locus encodes *OsSPL14* (Souamosa promoter binding protein-like 14), which was found to be controlled by microRNA (miRNA) *OsmiR156 in vivo*. They indicated that a point mutation in *OsSPL14* led to a disturbance in *OsmiR156*-directed regulation of *OsSPL14*. It led to the generation of an 'ideal' rice plant, with decreased tiller number, expanded lodging resistance and augmented GY. Their study implied that *OsSPL14* might be helpful in improving rice GYs and enables in breeding of novel elite rice varieties.

In rice breeding, another significant agronomic attribute is the leaf morphology. Hu et al. (2010) undertook a study to recognize and characterize narrow and rolled leaf 1, a unique gene regulating the leaf morphology and plant architecture in rice. They secluded three allelic mutants of narrow and rolled leaf 1 (*nrl1*). The phenotypes of these mutants had a shortened leaf width and semi-rolled leaves. They also exhibited variable degrees of dwarfism. The microscopic analysis has revealed that the *nrl1-1* mutant possessed scanty longitudinal veins and tinier adaxial bulliform cells. The *NRL1* gene was mapped to the chromosome 12. It was observed that *NRL1* gene encode cellulose synthase-like protein D4 (*OsCslD4*). Sequence analyses revealed the occurrence of single base substitution in these three allelic mutants. Genetic complementation and over-expression of the *OsCslD4* gene proved the identity of *NRL1*. They observed that this gene expression level though was seen in all parts of rice, higher expression levels at heading stage were more in those organs which exhibited vigorous growth viz., roots, sheaths, panicles, etc. They observed that this gene expression was low in the leaves of mutant plants compared to the leaves of wild plants. They concluded that *OsCslD4* encoded by *NRL1* has a key role to play in rice leaf morphogenesis and vegetative development.

Lu et al. (2013) has unveiled that the transcription activator Ideal plant architecture1 is vital for regulation of rice (*Oryza sativa*) plant architecture. It can substantially bring about enhancement in GY also. They elucidated its molecular basis and confirmed that it is a functional transcription activator. They through the assays of chromatin immunoprecipitation sequencing distinguished 1,067 and 2,185 genes that were linked with IPA1 binding sites in shoot apices and young panicles, respectively. The squamosa promoter binding protein-box direct binding core motif GTAC was deeply enhanced in IPA1 binding peaks; interestingly, to uncharacterized this indirect binding motif TGGGCC/T was found to be undoubtedly enhanced through the interaction of IPA1 with proliferating cell nuclear antigen promoter binding factor1 or promoter binding factor2. Genome-wide expression profiling by RNA sequencing indicated that IPA1 has various aspects in varied pathways. Moreover, their findings revealed that IPA1 could directly bind to the promoter of rice *teosinte branched1*, a negative regulator of tiller bud outgrowth, to repress rice tillering. IPA1 directly and positively regulate *dense and erect panicle1*, a key gene regulating panicle architecture and influencing the plant height and panicle length. Thus, they mentioned that explanation of target genes of *IPA1* genome-wide will in a better way contribute to advance an understanding of the molecular mechanisms, underlying plant architecture. It also facilitates to breed elite varieties with ideal plant architecture.

Wu et al. (2013) reported that loose plant architecture 1, an indeterminate domain (IDD) protein is associated with the shoot gravitropism. This protein was found to involve in the control of plant architecture in rice. Tiller angle and leaf angle were two key components of rice plant architecture. They play a central role in ascertaining GY. They reported the cloning and characterization of the Loose Plant Architecture 1 (LPA1) gene in rice, and the functional ortholog of AtIDD15/shoot gravitropism 5 (SGR5) gene. They mentioned that LPA1 regulated tiller angle and leaf angle. It controlled the adaxial growth of tiller node and lamina joint, to control the tiller angle and leaf angle. It also affects the shoot gravitropism. Expression pattern scrutiny has suggested that LPA1 effects plant architecture by disturbing the gravitropism of leaf sheath pulvinus and lamina joint. LPA1 influenced gravity perception or signal transduction in coleoptile gravitropism. It acts as a regulator of the sedimentation rate of amyloplasts. This regulation is different from that of LAZY1. They observed that LPA1 encoded a plant-specific IDD protein. It also has defined a unique subfamily of 28 IDD proteins, which had numerous distinctive conserved features. LPA1 was found to be confined in the nucleus and functioned as an active transcriptional repressor. Its activity was conferred due to the presence of a conserved ERF-associated amphiphilic repression (EAR)-like motif. Additionally, this analysis, has proposed that it also participated in a convoluted transcriptional and protein interaction network. It has also emerged some unique functions which were found to be different to that of SGR5. The study indicated that this information assists in development of elite lines for rice breeding.

Dai et al. (2018) reported that *OsMYB2P-1*, a R2R3 MYB transcription factor, is involved in control of phosphate-starvation responses and root architecture in rice. They found that *OsMYB2P-1* was confined in the nuclei. It also displayed transcriptional activation activity. Overexpression of *OsMYB2P-1* in rice increased tolerance to Pi starvation. Its suppression by RNAi in rice rendered the transgenic rice more sensitive to Pi deficiency. Furthermore, primary roots of *OsMYB2P-1*-overexpressing plants were curtained in growth under Pi-sufficient conditions. The primary roots and adventitious roots of *OsMYB2P-1*-overexpressing plants were elongated compared to wild-type plants under Pi-deficient conditions. The results specified it may also be related with regulation of root system architecture. Overexpression of *OsMYB2P-1* also led to a large expression of phosphate-responsive genes such as *OsSQD*, *OsIPS1*, *OsPAP10*, *OsmiR399a*, and *OsmiR399j*. In contrast, overexpression of *OsMYB2P-1* suppressed the expression of *OsPHO2* under both Pi-sufficient and Pi-deficient conditions. The results put forward that *OsMYB2P-1*, may act as a Pi-dependent regulator in controlling the expression of Pi transporters. These findings demonstrated that *OsMYB2P-1* is a novel R2R3 MYB transcriptional factor which is mainly correlated with Pi starvation signaling in rice.

4.1.1 OTHER FACTORS INFLUENCING YIELD COMPONENTS AND PLANT ARCHITECTURE

Gupta et al. (1995) mentioned that grain size and grain number are two essential key components of GY. Moreover, the grain size influences the end-use quality (e.g., flour yield and protein content) for consumer preference. These two traits were also the constituents of the domestication syndrome of crop plants. Several studies we redirected to figure out the genetic and molecular basis of these two important yield-contributing traits.

Yan et al. (2011) reported that a major QTL, Ghd8, plays pleiotropic roles in regulating grain productivity. Rice yield and heading date were two recognizable traits regulated by quantitative trait loci (QTLs). An understanding of molecular mechanisms of concealed rice yield traits is very much essential to develop high-yielding rice varieties. They reported about the cloning and characterization of *Ghd8*, a major QTL with pleiotropic effects on GY, heading date, and plant height. Two sets of near-isogenic line populations advanced for the cloning of *Ghd8*. *Ghd8* were limited down to a 20-kb region encompassing two putative genes. One gene ciphered the *OsHAP3* subunit of a CCAAT-box binding protein (HAP complex). The gene was deliberated as the *Ghd8* candidate. A complementary test proved the identity and pleiotropic effects of the gene. The genetic effect of *Ghd8* was reliant on its genetic background. Regulation of *Ehd1*, *RFT1*, and *Hd3a*, *Ghd8* delayed FL under long-day conditions, but promoted FL under short-day conditions. *Ghd8* up-regulated *MOC1*, a key gene governing tillering and branching; has elevated the number of tillers, primary, and secondary branches. It had produced 50% more grains per plant. Their findings explained the key roles of *Ghd8* in rice yield formation and FL.

Li et al. (2011) reported that natural diversity in *GS5* has a key role in regulating grain size and yield in rice. They unveiled that the QTL *GS5* in rice controls grain size by improving grain width, filling, and weight. *GS5* encrypts a putative serine carboxypeptidase. It operates as a positive manager of grain size. Greater expression of *GS5* is associated with bigger sized grain. The outcomes proposed that this contributes to a large extent to diversity of grain size in rice.

4.2 RICE IDEOTYPE

The concept of ideotypes in crops was first proposed by Donald (1968). Almost all the plant breeding methods are based on the selection of yield attributing traits in plants and the breeding of crop ideotypes serves as an additional valuable approach because the plants having model characteristics influences the photosynthesis, growth, and grain production (in cereals). For a few crops, the ideotype characteristics were cited and effective use of model qualities of this kind is mentioned.

The successful crop ideotype is assumed as a feeble competitor, proportionate to its mass. Therefore, there will be a less competition between the similar plants in the crop community and two negative associations among genotypes may rise from the exploitation of environment with this association of plant forms.

The crop ideotype has less demand for the dry matter formed per unit resources. Additionally in cereals, each entity of dry matter consists of such a number of florets as to make sure that the ear has adequate ability to take all photosynthates either from its own green surfaces or from other parts of the plant. When the high population density intensify the tension by the community on environmental resources and for high fertility these criteria are to be fulfilled.

An ideotype of wheat has a short, solid stem; few, small, erect leaves; a large ear (this specially means many florets per unit of dry matter of the tops); an erect ear; awns; and a single culm.

In order to feed 5.0 billion rice consumers in 2030, the new possible strategies have to be made which are discussed by Gurdev (2005). He stated that GR made significant advances in rice production. The rice production was raised from 257 million tons in 1966 to 600 million tons in 2000 accounting for 130%, however, the densely populated low-income countries population increased by 90%. The continuously increasing population will demand 40% more rice in 2030 which has to be met from less land, with less water, less labor, and fewer chemicals. With the use of high yielding rice varieties with greater yield stability, this challenge of producing more rice can be overcome. Thereby many strategies were employed to increase the yield potential of rice: (1) conventional hybridization and selection procedures, (2) ideotype breeding, (3) hybrid breeding, (4) wide hybridization, and (5) genetic engineering. Further, to develop durable resistance towards diseases and insect and for tolerance to abiotic stresses different conventional and biotechnology methods are being taking up. Now the rice genome sequence availability permitted the identification of the each of 60,000 rice genes functions via functional genomics. The identification of gene function will help in the development of new rice varieties by introducing new genes by means of traditional breeding together with marker assisted selection or by engineering the genes directly into rice varieties.

Kim (1988) undertook physiological studies on low-tillering rice: an ideotype for expanding GY potential. He compared a low tillering, large panicle IR25588 with a high tillering, small panicle IR58. Both varieties possessed similar plant type and growth duration. They differed in tillering ability and panicle size. High tillering IR58 possessed not only higher tiller number but also earlier, higher rate and longer period of tillering compared with low tillering type IR25588. Percent effective tiller was higher in IR25588 (85%) than IR58 (67%) and reduced with increase in nitrogen level and plant density. The six heaviest panicles in both IR25588 and IR58 were the main culm; the first, second, third, and fourth primary tillers; and the first secondary tiller from the second primary tiller. This suggests a new rice ideotype of six tillers per plant to increase GY potential. The top six tillers were initiated and flowered earlier, had bigger culm and leaves per tiller to support many spikelets and better grain filling, and had more inner (IVB) and outer vascular bundles (OVB) at just below the neck node of the panicle which may give better translocation of assimilates. The top six panicles based on grain weight was primarily due to the greater spikelet number per panicle with meager variations in 1,000-grain weight and percent fertility. Spikelet number per panicle for the top six tillers in low down tillering IR25588 was around 200 to 250 spikelets to compensate for low yield.

Peng et al. (1999) mentioned that after the release of IR8 in 1966, 42 additional *indica* rice (*Oryza sativa* L.) cultivars developed by the International Rice Research Institute (IRRI) for the irrigated and favorable rainfed lowlands have been released in the Philippines. In the past 30 years, the maximum yield of IR8 has declined to about 2 Mg ha^{-1}. In the tropics, rice yield potential has maintained to about 10 Mg ha^{-1} with empirical breeding for population improvement within the *indica* germplasm. To break the yield obstacle, many approaches have been assumed. These contain development of a NPT with less tillering capacity and bulky panicles from tropical *japonica* germplasm and utilization of heterosis through intervarietal and intersubspecific hybrids. Under the tropical conditions, the rice yield potential was improved by about 9% with the introduction of hybrid rice between *indicas*. The higher yield potential of *indica/indica* hybrids compared with *indica* inbred cultivars was accredited to a larger biomass production rather than harvest index (HI). NPT breeding could not increase the yield potential because of reduced grain filling and low biomass production. Factors responsible for the reduced grain filling and low biomass production of the NPT lines have been recognized. Selecting parents with good grain filling traits, introduction of *indica* genes into

NPT's tropical *japonica* background and fine-tuning of the original NPT design is expected to increase the performance of the NPT lines. Further, rise in yield potential may be probable from utilization of intersubspecific heterosis between *indica* and NPT lines.

Peng et al. (2008) discussed advancements in ideotype breeding to boost rice yield prospective. The ideotype method has been in use in breeding programs at the IRRI and in China to advance rice yield potential. The first-generation NPT lines were developed from tropical *japonica* at IRRI. The yield potential of lines was very low. This was mainly due to the production of low biomass and poor grain filling. Second-generation NPT lines were obtained by crossing elite *indica* with improved tropical *japonica* and led to subsequent progress. These second-generation NPT lines have more yield potential than the first-generation NPT lines and *indica* check varieties. China's "super" rice breeding project has developed many F_1 hybrid varieties. These hybrids were developed by combining both ideotype approach and intersubspecific heterosis. The yield potential of hybrid varieties was 12 t ha^{-1} which was 8–15% advanced than the standard hybrid varieties. The accomplishment of China's "super" hybrid rice became unbeaten because of its partial development by bring together the good components of IRRI's NPT design, together with the exploitation of intersubspecific heterosis. For example, both designs have given importance to hefty panicle size, a bridged tillering capacity, and superior lodging resistance. China's "super" hybrid rice has attained improvement, as they have given much emphasis to the top three leaves and panicle position within a canopy for meeting the requirement of weighty panicles for a huge source supply. The accomplishment of "super" hybrid rice breeding in China and advancement in NPT breeding at IRRI suggests that the ideotype advance is effectual to break the yield upper limit of an irrigated rice crop.

Lu et al. (2014) undertook a study on yielding potential and associated characteristics of rice ideotype using six varieties of ideotype, curve-slant, and erect plant types in three regions. At the intermediate level of soil fertility, the ideotype rice has shown the maximum yielding potential (528–677 kg/mu), which was higher than that of the curve-slant plant type (422–508 kg/mu) and the erect plant type rice (370–443 kg/mu). The ideotype rice could incarcerate more light and grow quickly throughout the early stage and the upper leaves grew erect (leaf angle about 15°, curvature 0.01 cm (–1)), the coefficient of light extinction (K) grew in a lower speed

(K = 0.7–0.8 at heading stage) during the middle and late stages. This helped to increase the optimal leaf area index (LAI), which holds more photosynthetic organs and improve lighting situation of the middle and lower leaves, thus leading to a reduction in the leaf senescence rate. From the perspective of rice population, the ideotype rice should have leaves erect on the upper part, bend on the middle and horizontal on the lower part, with higher efficiency in light exploitation. The ideotype rice with small leaves, should keep up leaf area through ever-increasing leaf width (by 15%–35%), the leaves should be slender (SLW = 2.9–3.8 mg cm^{-2}) to increase the development capacity of leaf area for the duration of the early growing period and turn deepen (SLW = 4.4–4.5 mg cm^{-2}) to manage leaf area and amplify the quality of leaf photosynthesis during the middle to late periods. It is obligatory to unearth out a morphology index which could be used to assess leaf morphology in terms of leaf length, width, and curvature. For whichever variety, the optimal LAI could be resulted by the leaf morphology index; ultimately, the yield potential could be calculated by the optimal LAI and leaf photosynthesis ability.

Few studies have been assumed to study molecular basis of rice morphological characteristics. Guo et al. (2011) constructed SSR Linkage Map and Analysis of QTLs for rolled leaf in *Japonica* Rice. Genetic investigation of rolled leaf is considered essential to rice ideotype breeding. To spot loci regulating rolled leaf of *japonica* restorer lines, they investigated SSR marker genotypes and phenotypes of flag leaf rolling index (LRI) in Xiushui 79 (P_1, a *japonica* rice variety), C Bao (P_2, a *japonica* restorer line) and 254 recombinant inbred lines (RILs) resulting from the cross between P_1 and P_2, and in two environments. A genetic map of this cross was built, QTLs for LRI were identified, and their interactions with environments were studied. Among 818 pairs of SSR primers, 90 primers exhibited polymorphism between P_1 and P_2, and 12 markers exhibited extremely noteworthy association with LRI in both environments on the basis of single-marker regression analysis. The genetic map covering 74 information loci has a total distance of 744.6 cM, with an average of 10.1 cM between two adjacent loci. Three QTLs (*qRL-1*, *qRL-7,* and *qRL-8-1*) were identified with two softwares: WinQTLCart 2.5 and QTL Network 2.0. *qRL-8-1* was a new locus, accounting for 15.5% and 12.8% of phenotypic variations in the two environments, correspondingly. The phenotypic variation that was elucidated by additive effect was 6.6%. Interaction was not found between *qRL-8-1* genotype and environments.

Jantaboon et al. (2011) adopted ideotype breeding for submergence tolerance (SUB) and cooking quality by marker-aided selection (MAS) in rice. SUB and jasmine-like cooking quality are enviable for rice varieties grown in rainfed and irrigated lowland ecosystems. The varieties IR57514 and Kao Dawk Mali 105 (KDML105) were hybridized with the objective of making an ideotype that combines SUB and jasmine-like cooking quality. By means of the single seed descent (SSD) technique, a bulky population of RILs was developed. They confirmed the prospective of using marker-assisted selection (MAS) in the detection of the ideotype from the offspring. Four markers, R10783Indel, Waxy, Aromarker, and GT11, were utilized to choose the preferential alleles of the Sub1, *Wx*, *badh2* and *SSIIa* loci, respectively. The ideotype was categorized into two groups: ideotype 1 (ID1), carrying the $Sub1^{IR}$, $badh2^{KD}$, Wx^{KD}, and $SSIIa^{KD}$ alleles and consisting of 66 RILs and ideotype 2 (ID2), carrying the $Sub1^{IR}$, $badh2^{KD}$, Wx^{KD}, and $SSIIa^{IR}$ alleles and consisting of 31 RILs. All of the ID1 lines have displayed SUB and jasmine-like cooking quality and have displayed small amylose content (AC), a fragrance, and a high alkali-spreading value (ASV), whereas the ID2s exhibited the same distinctiveness as ID1, with the exception of a low ASV, which was inherited from IR57514. An extensive array of agronomic characters were found in both of the ID groups, and a little of the IDs were advanced in the yield component, as compared to their parents. This paper provides additional support that the exactitude of markers used in MAS can improve the progress of ideotypes in rice.

4.3 CONCLUSION

Ideotype of a crop should have desirable plant characteristics such as leaf morphology, its orientation, branching patterns, canopy for efficient capture of sunlight leading to the high production of the crops. Various ideotype concepts and ideotype characteristics were set forth by different authors for different rice varieties that enabled in contributing to high yield. In our survey of breeding lines of rice, we have observed that the yielding potential could be increased by maintaining the optimum LAI and leaf photosynthesis ability.

KEYWORDS

- **ideal plant architecture**
- **ideotype**
- **indeterminate domain**
- **leaf rolling index**
- **marker-assisted selection**
- **planting density**

REFERENCES

Dai, X., Wang, Y., Yang, A., & Hao, W., (2018). OsMYB2P-1, a R2R3 MYB transcription factor, is involved in regulation of phosphate-starvation responses and root architecture in rice. *Plant Physiology*, *159*, 169–183.

Donald, C. M., (1968). The breeding of crop ideotypes. *Euphytica*, *17*, 385–403.

Gao, Z., Qian, Q., Liu, X., Yan, M., Feng, Q., Dong, G., Liu, J., & Han, B., (2009). Dwarf 88, a novel putative esterase gene affecting architecture of rice plant. *Plant Molecular Biology*, *71*, 265–276.

Guo, Y., Cheng, B. S., & Hong, D., (2011). Construction of SSR linkage map and analysis of QTLs for rolled leaf in *Japonica* Rice. *Rice Science*, *17*, 28–34.

Gurdev, S. K., (2005). What it will take to Feed 5.0 billion rice consumers in 2030. *Plant Molecular Biology*, *59*, 1–6.

Hu, J., Zhu, L., & Zeng, D., (2010). Identification and characterization of narrow and rolled leaf 1, a novel gene regulating leaf morphology and plant architecture in rice. *Plant Molecular Biology*, *73*, 283–292.

Jantaboon, J., Siangliw, M., Im-Mark, S., Jamboonsri, W., Vanavichit, A., & Toojinda, T., (2011). Ideotype breeding for submergence tolerance and cooking quality by marker-assisted selection in rice. *Field Crops Research*, *123*, 206–213.

Jiao, Y., Wang, Y., & Xue, D., (2010). Regulation of OsSPL14 by OsmiR156 defines ideal plant architecture in rice. *Nature Genetics*, *42*, 541–544.

Jin, J., Huang, W., Gao, J. P., Yang, J., Shi, M., Zhu, M. Z., Luo, D., & Lin, H. X., (2008). Genetic control of rice plant architecture under domestication. *Nature Genetics*, *40*, 1365–1369.

Kim, J., (1988). *Physiological Studies on Low-Tillering Rice: An Ideotype for Increasing Grain Yield Potential*. University Library, University of the Philippines at Los Banos.

Komatsu, M., Maekawa, M., Shimamoto, K., & Kyozuka, J., (2001). The LAX1 and FRIZZY PANICLE 2 genes determine the inflorescence architecture of rice by controlling rachis-branch and spikelet development. *Developmental Biology*, *231*, 364–373.

Li, Y., Fan, C., Xing, Y., Jiang, Y., Luo, L., Sun, L., Shao, D., & Xu, C., (2011). Natural variation in GS5 plays an important role in regulating grain size and yield in rice. *Nature Genetics*, *43*, 1266–1269.

Lu, C., Gu, F. Z., & Jiangshi, L. M., (2014). Studies on yielding potential and related characteristics of rice ideotype. *Scientia Agricultura Sinica*, 1991–05.

Lu, Z., Yu, H., Xiong, G., Wang, J., Jiao, Y., Liu, G., Jing, Y., Meng, X., Hu, X., Qian, Q., Fu, X., Wang, Y., & Li, J., (2013). Genome-wide binding analysis of the transcription activator ideal plant architecture1 reveals a complex network regulating rice plant architecture. *The Plant Cell*, *25*, 3743–3759.

Nagashree, N., Rao, Kalika, P., Puja, R. K., & Usha, V., (2008). Distinct regulatory role for *RFL*, the rice *LFY* homolog, in determining flowering time and plant architecture. *PNAS*, *105*, 3646–3651. https://doi.org/10.1073/pnas.0709059105 (accessed on 15 January 2020).

Peng, S., Cassman, K. G., Virmani, S. S., Sheehy, J., & Khush, G. S., (1999). Yield potential trends of tropical rice since the release of IR8 and the challenge of increasing rice yield potential. *Crop Science, 39*, 1552–1559.

Peng, S., Khush, G. S., Virk, P., Tang, Q., & Zou, Y., (2008). Progress in ideotype breeding to increase rice yield potential. *Field Crops Research*, *108*, 32–38.

Wang, Y., & Li, J., (2005). The plant architecture of rice (*Oryza sativa*). *Plant Molecular Biology*, *59*, 75–84.

Wu, X., Tang, D., Li, M., Wang, K., & Cheng, Z., (2013). Loose plant architecture1, an INDETERMINATE DOMAIN protein involved in shoot gravitropism, regulates plant architecture in rice. *Plant Physiology*, *161*, 317–329. https://doi.org/10.1104/pp.112.20849 (accessed on 29 January 2020).

Yan, W. H., Wang, P., Chen, H. X., Zhou, H. J., Li, Q. P., Wang, C. R., et al., (2011). A major QTL, Ghd8, plays pleiotropic roles in regulating grain productivity, plant height and heading date in rice. *Molecular Plant*, *4*, 319–330.

Zhou, Y., Zhu, J., Li, Z., Yi, C., Liu, J., Zhang, H., Tang, S., Gu, M., & Liang, G., (2009). Deletion in a quantitative trait gene qPE9–1 associated with panicle erectness improves plant architecture during rice domestication. *Genetics, 183(1)*, 315–324.

CHAPTER 5

Rice Botany

ABSTRACT

This chapter deals with different aspects of botany such as taxonomy, morphology, anatomy of roots, stems, their characterization and classifications, and relation to adaptation to abiotic stresses such as drought, flood. These have been verified by molecular biology.

5.1 TAXONOMY AND PHYLOGENY

Phytolith analysis enables in the recognition of opaline silica bodies. It offers an unswerving way of identifying rice, particularly under situations where conservation of charred botanical remains was meager. Zhao et al. (1998) distinguished rice (*Oryza sativa*, Poaceae) from wild *Oryza* species through phytolith investigation of Asian rice. The outcome of this research, which incorporates all Asian wild Oryza species and a diverse array of traditional *Oryza sativa* cultivars, confirms that rice can be identified with a high level of certainty by the size and qualitative features of a distinctive phytolith, the double-peaked glume cell.

Glaszmann (1986) did a varietal categorization of Asia cultivated rice on the basis of isozyme polymorphism of rice genetics collection. He surveyed electrophoretically detected enzyme variation among 1,688 varieties of Asian cultivated rice (*Oryza sativa* L.). The diversity was elucidated by gene polymorphism at 21 accepted loci. The scrutiny of the enzymatic disparity in a representative section of 120 varieties led to the identification of six varietal groups. Groups I, II, III, IV, and V included rice typically classified as *indica* which were different from one another, Group VI included the conventional *japonica* and *javanica* types. This group consisted mostly of the upland rice from Southeast Asia, and most of those rice which were grown at elevated elevations of the Himalayas.

On the Indian subcontinent, particularly along the foothills of the Himalayas, there was high varietal diversity. On the contrary, it was noted that in groups I and IV, there were most varieties from Southeast Asia and East Asia.

Subsequently, Glaszmann (1987) undertook isozymes study and categorization of Asian rice varieties. He detected enzyme dissimilarity by starch gel electrophoresis used with an objective to look into the genetic structure of *Oryza sativa* L. species. Among 1,688, fifteen polymorphic loci coding for eight enzymes in traditional rice from Asia were surveyed. Multivariate analysis of the data revealed six varietal groups, with two major ones, form groups I and VI, two minor ones, groups II and V, and two satellite ones, groups III and IV. Group I is observed all through tropical Asia; it includes most Aman rice in Bangladesh, the Tjereh rice in Indonesia and the Hsien rice in China. Group VI is observed typically in temperate regions and in high elevation areas in the tropics; it includes most upland rice from Southeast Asia, the Bulu rice from Indonesia and the Keng rice from China. Groups II, III, IV, and V reveal regular variances from groups I and VI. This suggests another evolutionary history. Groups II and V are found in the Indian subcontinent from Iran to Burma. Familiar components of these are Aus rice from Bangladesh for group II and Basmati rice from Pakistan and India for group V. Groups III and IV are restricted to some deepwater rice in Bangladesh and Northeast India. Based on the analogy with other categorizations, Group I could be deliberated as the "*Indica*" type and Group VI as the "*Japonica*" type.

Ghareyazie et al. (1995) undertook the classification of 35 Iranian rice varieties together with two typical *Indica* and three typical *Japonica* varieties by means of ALP and PCR-based RFLP. They compared the possibility of employing a PCR-based approach to find DNA polymorphism for rice germplasm categorization with that of Southern-based RFLP analysis. They used thirteen mapped RFLP markers hybridization probes against Southern blots, where 12 of the 13 probes have identified polymorphism in the varieties. In this study, amplicon length polymorphisms (ALPs) were noticed with 6 of the 15 sets of primers. Finally, RFLPs were identified for 11 sets of primers, owing to point mutations and to addition/deletion events that were too minute to be detected as ALPs. As PCR, products are effortlessly produced and may be examined in each aspect using restriction endonucleases that cut rice DNA repeatedly, PCR-based RFLP analysis is a helpful contrivance for the categorization of rice germplasm.

Virk et al. (2000) determined the usefulness of dissimilar classes of molecular marker for classifying and enlightening differences in rice (*Oryza sativa*) germplasm. They used comparable numbers of markers from four molecular marker systems (AFLP, isozymes, ISSR, and RAPD) for enlightening genetic diversity and discerning between infraspecific groups of *Oryza sativa* germplasm. Every marker system classified the germplasm into three chief groups (with isozymes and AFLPs). The germplasms were classified with differences (by ISSR). Partial agreement existed only at highest levels of genetic similarity, as to relationships between individual accessions when different markers were used. When variance was partitioned in the midst of and within the three subspecific groups, it has indicated greater variation amongst than within groups using AFLP and isozymes, with the overturn for RAPD and ISSR. Each marker system has given similar results for measurements of polymorphism, using standard heterozygosity and effectual number of allele. They explained these based on the relation to a range of genetic resources preservation activities, and the suitability of extrapolating to other sets of germplasm mainly of other crop species.

Cheul et al. (2010) undertook categorization of rice (*Oryza sativa* L. *japonica* nipponbare) immunophilins (IMM) (FKBPs, CYPs) and expression patterns under water stress. FK506 binding proteins (FKBPs) and cyclophilins (CYPs) are ubiquitous abundant proteins regulating metabolic processes acting as chaperones or involved in isomerization of proline residues during protein folding. They belong to the peptidyl-prolyl *cis/trans* isomerase (PPIase) superfamily. Both these were jointly designated as IMM. They classified these and provided information about their evolutional or functional importance offering the possibility that exists in rice for the manipulation of its responses towards water stress. It is determined that like other green photosynthetic organisms viz., *Arabidopsis* (23 FKBPs and 29 CYPs) and *Chlamydomonas* (23 FKBs and 26 CYNs); rice also possessed maximum number of IMM genes, many of which were upregulated by water stress.

Yu et al. (2002) by whole-genome shotgun sequencing, investigated draft sequence of the Rice Genome (*Oryza sativa* L. ssp. *indica* for broadly cultivated subspecies in China, *indica*). The genome had 466 megabases in size, with a 46,022 to 55,615 genes. Even though there was 92.0% of functional coverage in the accumulated sequences, there was only 42.2% of the genome that was in exact 20-nucleotide oligomer repeats. In the analysis, they observed that most of the transposons were in the intergenic regions between genes. The findings unveiled that in rice, only 49.4% of expected rice genes had a homolog in A. thaliana. They mentioned that in rice there were a great fraction of rice genes with no decipherable homologs, because of a gradient in the GC content of rice coding sequences and it led to only 49.4% homology in genes.

Parsons et al. (1997) reported divergent genetic diversity associations in rice (*Oryza sativa* L.) using two PCR-based molecular marker systems: RAPD (random amplification of polymorphic DNA) and ISSR-PCR (inter-simple sequence repeat (SSR) polymerase chain reaction). By the use of RAPD, a set of 14 decanucleotides of arbitrary sequence heading for the amplification of 94 reproducible marker bands, 47 (50%) of which were polymorphic. Apart from those, a set of 9 ISSR primers were used to direct amplification of 71 PCR products, 40 (56%) of which were polymorphic. Subtle differences in the relationships indicated between rice groups using the two types of PCR-based marker led to exploration of their map positions using an intraspecific doubled haploid (DH) mapping population.

Li et al. (2002) discussed the phylogeny of the rice tribe Oryzeae (Poaceae) using nucleotide sequences of the chloroplast gene *matK* for 26 species. These species represented eleven genera of the tribe Oryzeae and three outgroup species. The variation in sequenced fragments ranged between 1,522 base pairs (bp) to 1,534 bp in length. There was 15.4% variability; the variability at phylogenitically informative sites is around 7.9%, where there was exclusion of out groups. The sequence data analysis suggested that species of Oryzeae form a powerfully supported monophyletic group. The tribe Oryzeae can be separated into two monophyletic lineages, which correspond to the traditionally recognized subtribes Oryzinae and Zizaniinae. The first subtribe includes *Oryza* and *Leersia*, while the subtribe Zizaniinae includes the remaining genera. The *matK* sequence data did not accept the close affinities of the monoecious genera in Oryzeae. This has indicated that in this tribe there were possibilities of several origins of the floral structures. From the studies, it has been emphasized that *Porteresia coarctata* is very much closely related to *Oryza* species. The results suggested that it should be treated as an associate in the genus *Oryza* rather than a separate monotypic genus formation in rice stems by light microscopy.

5.2 RICE MORPHOLOGY

5.2.1 BOTANICAL CHARACTERISTICS OF RICE

5.2.1.1 LEAF MORPHOLOGY

5.2.1.1.1 Silica Bodies in Leaf Epidermis

Silicon is generally considered as one of the chief trait required for standard normal growth and development of the rice plant. A number of investigations conducted on the physiological functions of the deposited silica in rice epidermal systems, led to the proposal of 'window hypothesis,' which has stated that this deposited silica bodies might act as window and enable the diffusion of light to the mesophyll tissue for effective photosynthesis. Agarie et al. (1996) tested this hypothesis and investigated the role of silica bodies in the epidermal system of rice: They observed that increased supply of silica increases the silica content of leaves and the number of silica bodies per unit leaf area in the epidermal system. The number of silica bodies per unit leaf area correlated with the supply of silica. Though, there were differences in the silica deposition and development of silica bodies between the leaves of silica treated and non-treated, polymerization of silicon was high in Silicon treated leaves particularly within the silica cells and bulliform cells of the epidermis. Even though there was formation of windows in silicon treated leaves, there were no differences with respect to the optical properties of leaf transmittance, reflectance, and absorbance spectra in treated or untreated leaves. Furthermore, they detected that in leaves of silica treated plants there was less light energy use efficiency and quantum yield. Thus, their research findings have specified that in rice these silica bodies do not have a task of window for light transmission (Figure 5.1).

5.2.1.1.2 Indicators of Phase Change

Leaf shape and leaf anatomy act as indicators of phase change in the grasses from vegetative to reproductive phases. Sylvester et al. (2001) studied these variations in maize, rice, and bluegrass. The results revealed that juvenile leaves of maize had higher coating of epicuticular wax and lacked trichomes and bulliform specialized cells, and epidermal cells of these stained uniform purple, while the adult leaves were pubescent, and lacked the coating of epicuticular waxes. Further, the adult leaves had crenulated epidermal cell walls and had stained purple and blue. The rice blades were pubescent, and were also coated with epicuticular waxes. These also showed purple and blue wall staining. They found that there was a steady increase in the leaf blade width at each node. The leaf blade widths achieved a threshold size much before several nodes have acquired the reproductive competence. Blade-to-sheath length exhibited an analogous trend of incessant alteration followed by irregular alteration prior to reproduction. An assessment of leaf development in these three grasses have shown that there is a rapid initiation of maize primordia relative to leaf blade or sheath growth rather than to that seen in rice or blue grass. Thus, they concluded that leaf shape, as defined by leaf blade width and blade-to-sheath ratio is a consistent marker of phase, while anatomy is not a complete marker of phase change in these grasses. They speculated that because of the variances in the timing of embryonic and postembryonic development, within these grasses there were variations even in their growth patterns (Figures 5.2–5.4).

FIGURE 5.1 Typical rice plant showing various plant structures.

FIGURE 5.2 Rice panicle with spikelets.

FIGURE 5.3 Matured rice panicle.

FIGURE 5.4 Threshed paddy grains.

5.3 ANATOMY

Harris and Clark (1981) studied the root anatomy of hydroponically grown rice. Their electron microscopy studies revealed that rice root contained an epidermis, exodermis, and fibrous layer. The exodermis possessed a suberin lamella along its inner tangential wall. The fibrous layer comprised of thick-walled lignified cells with meager pitting. At young stage of development, they observed that the cortical parenchyma was compact which later expanded and separated to form a zone of cell walls and air spaces in a spoked arrangement. Further, they observed that in regions near to the lateral roots in particular had occasional presence of supporting columns of living parenchyma cells. The endodermis is characteristic for grasses with Casparian strips, suberin lamellae, and tertiary state walls with copious pits. The pericycle and pith become sclerified. Protoxylem elements alternate with protophloem in the young root; later, early metaxylem, late metaxylem, and metaphloem multiply. The exodermis, fibrous layer, lacunate cortex, and endodermis emerge to present a dreadful blockade to radial ion movement in the full-grown portions of the root.

Uga et al. (2009) studied dissimilarity in root morphology and anatomy among accessions of Asian cultivated rice with different genetic backgrounds. Asian cultivated rice shows genetic diversity of root characteristics, but this disparity has not been studied entirely with an orientation to the genetic background. An earlier principal-coordinate investigation study using data on 179 restriction fragment length polymorphisms categorized these accessions into three varietal groups: 13 *japonicas*, 21 *indica*-I, and 25 *indica*-II. Principal-components analysis of the six traits, have shown that the *japonica* group exhibited extensive dissimilarity in root anatomy than the two *indica* groups. In particular, *japonica* upland rice was characterized for having a larger stele and xylem structures. On the contrary, the two *indica* groups also exhibited large difference in root morphology compared to *japonica* group. Between the two *indica* groups, on average, the *indica*-I accessions possessed deeper, thicker roots than the *indica*-II accessions. Their results demonstrated that the *japonica* and *indica* groups have dissimilar genetic diversity with respect to their root characters.

Fibrous root system is made up with seminal and nodal roots with many lateral roots. Rice roots have their own characteristics including well-developed air space in the matured cortex. Morita and Nemoto (1995) studied morphology of the fibrous root system. Morphometric analysis on anatomy of matured rice roots including vascularization suggest that with orientation to root function, though many studies were conducted, the aspects of vascularization have been rarely examined viz., relations between vascular systems of stem and nodal roots; vascular system patterns along the root axes; relations between vascular systems of the parental roots and their lateral roots; and relations among various kinds of vascular elements in the same roots. They suggested that to further recognize both structure and function of rice root, studies on morphology and anatomy with orientation to synchronization in organo- and histogenesis should be carried over.

Lux et al. (1999) studied about the endodermal silicification in developing seminal roots of two upland rice cultivars, IRAT 109 and Moroberekan, and one lowland rice cultivar, Koshihikari of rice. The findings unveiled that there was exclusive deposition of silica in the endodermal cells of seminal roots. There was a basipetal raise in the content of silica in the endodermis in both upland and lowland cultivars. There was a higher intensity of silica deposition in the upland rice cultivars, which resulted in higher drought resistance. Further, it was found that though all the epidermal cells of leaves had silica deposits, maximum silica concentration was present in the silica cells. The deposition of silica in leaves was numerous folds more than that of the endodermis of roots.

Champoux et al. (1995) undertook research in 203 recombinant inbred lines (RILs) developed from a cross between *indica* cultivar Co39 and *japonica* cultivar Moroberekan, for identifying and mapping quantitative trait loci (QTLs) connected with five rice root morphology parameters. These lines in greenhouse conditions were drought-stressed at the seedling, early vegetative, and late-vegetative growth stage and they gave a visual rating based on leaf rolling as to their degree of drought avoidance/tolerance. Most of the identified QTLs exhibited linkage with root thickness, root/shoot ratio, and root dry weight per tiller, and only a few with deep root weight. Root thickness and root dry weight per tiller were found to be the less influenced by environmental changes. Twelve of the fourteen chromosomal regions contain putative QTLs linked with field drought avoidance/tolerance also had QTLs related with morphology of root. Thus, they mentioned that selecting for Moroberekan alleles at marker loci related with the putative root QTLs recognized, may be an effectual approach for changing the root phenotype of rice towards that normally linked with drought-resistant cultivars.

5.3.1 ROOT SYSTEM AND ITS ROLE

O´Toole and Soemartono (1981) developed an effortless technique to characterize rice root systems in relation to drought resistance. They mentioned that drought acts as a restraint to rice yields in large areas of rainfed rice in Asia and even though genotypic variation for root system characteristics is well recognized for rice, development of rice for drought resistance has yet to be attained. They reported about the advance and testing of a straightforward device that measures the pulling force that is a requisite to uproot rice seedlings. The force necessary to uproot and heave seedlings from the paddy soil is more related with root weight, root branching, and thick root number. They illustrated the potential to classify rice cultivars by this technique and also root pulling force relationship to other drought screening methods.

Rice root systems have a crucial role in facilitation of uptake of water and nutrients from soil. Changming et al. (2004) in their four-year experimental study on the effects of different nutrient and water regimes on rice root growth found that there was an improvement in rice root characteristics with organic source of fertilizers incorporation in paddy soil. There was an increase in root length density (RLD), and root weight density (RWD) by 30 and 40%, by organic fertilization viz., integrated use of chemical fertilizers and wheat straw (CS), chemical fertilizers with farmyard manure (CM), respectively, in the alternate wetting and drying (AWD) treatment. Continuous waterlogging (CWL) had an undesirable effect on root active absorption area (AAA), root oxidation ability (ROA) of alpha-naphthylamine (α-NA) (ROA), and root surface phosphatase (RSP) of rice plants in presence of integrated application of organic and inorganic fertilizers, while continuous flooding (CF) caused a reduction in RSP, ROA, and AAA. Similarly, AWD with the incorporation of organic manure led to an increment in the uptake rates, allocation, and transfer of N, P, and K, especially P allocation and transfer to rice ears and grains, has resulted in more number of filled grains panicle^{-1}, 1,000-grain weight and grain yield (GY).

5.3.1.1 ALTERATION OF ROOT SYSTEM ARCHITECTURE FOR DROUGHT AVOIDANCE

Uga et al. (2013) demonstrated that modification of root system architecture improves drought avoidance through the cloning and characterization of *deeper rooting 1* (*DRO1*), a rice quantitative trait locus (QTL) controlling root growth angle. *DRO1* is negatively regulated by auxin. It is concerned with the elongation of cells in the root tip. It causes asymmetric root growth and downward bending of the root in response to gravity. Elevated expression of *DRO1* increased the root growth angle due to which there will be more downward direction growth of roots. The introduction of this trait locus into a shallow-rooting rice cultivar, and resulted line obtained through backcrossing exhibited and increase in deep-rooting feature, which enabled its high performance in drought conditions. Thus, the study proposed that control of root system architecture will throw in a way to drought avoidance in crops.

Banda and Lal (1983a) investigated the association between root growth in three upland rice varieties (IB 6, IR 1529-680-3 and IET 1444) and leaf water potential in the field conducted under two soil moisture regimes (i.e., with, and without supplementary irrigation). They observed that in IB 6 the root system penetrated below 20 cm, while in other two varieties viz., IR 1529-680-3 and IET 1444, it was only limited to between 15 and 20 cm. The leaf water potential of IB 6 was high (less negative), and it retained turgidity much before. Root density at a depth of 15–25 cm exhibited positive association with leaf water potential and also with the advance of accumulative internal plant water stress, while the root density at a depth of 0–10 cm had slight results on leaf water status.

Similarly, Banda Mambani and Lal (1983b) in their study in 10 rice varieties on variable soil moisture regimes along a toposequence effects on root development, plant water status, and GY found that soil moisture regime had a trivial effect on total root dry weight, however, they had an influence on depth and angle of root penetration. Roots of tall varieties (63–83, IB 6 and IRAT 13) penetrated much deeper into the soil profile than those of short varieties. The soil moisture depletion pattern was alike to that of root density profile. Relative water stress enhanced exponentially with reduction in root density at a depth of 25 cm. These ten varieties presented a normal distribution in the GY under adequate soil moisture. Even under the unfavorable moisture regime, stable yield were maintained by variety IB 6 and other tall varieties. Further, they observed that root density at a depth of 25 cm is more linearly related to GYs.

Insaludr et al. (2006) studied morphological and physiological responses of rice (*Oryza sativa*) to phosphorous supply. It is known that rain-fed lowland rice commonly is exposed to stresses from changeable water regimes and nutrient deficiency. Under those conditions, roots have to attain both oxygen and nutrients under unfavorable circumstances, while also acclimatized to change in soil-water regime. The outcomes unveiled that after only 1 day in stagnant conditions, there was 90% decline in root radial oxygen loss (ROL) in subapical zones. At the tip, ROL was maintained. After 4 days in stagnant conditions, there was 11% less in maximum root length and after 8 d, shoot growth was 25 % less, compared with plants in aerated solution. The plants in stagnant solution had 19 % more adventitious roots, 24 % more root porosity, and 26 % higher root/shoot ratio. Rice plants grown with low P supply had fewer tillers under both stagnant and aerated conditions. At low P supply, relative uptake of P declined after 1–2 days in a stagnant solution. Aerated roots at low P supply upheld relative P uptake for 4 days, after which uptake declined to the same levels as in stagnant condition. It is concluded that roots responded quickly to oxygen deficiency with reduced ROL in subapical zones within 1–2 days. This indicates stimulation of a barrier to ROL and these changes in ROL take place at least 2 days before any variations in root morphology, porosity, or anatomy were evident. Under oxygen deficiency, also low relative uptake of P was observed, indicating that a sudden decline in root-zone oxygen adversely affects P nutrition of rice.

Colmer (2003) assessed waterlogging tolerance, root porosity, and radial O_2 loss (ROL) from the adventitious roots, of seven upland, three paddy, and two deepwater genotypes of rice. He observed that upland types, excluding one genotype, were tolerant to 30 days of soil waterlogging. All the 12 genotypes except one showed an increment in the adventitious roots number per stem in waterlogged soil. In stagnant deoxygenated nutrient solution, there was a genotypic variation for root porosity and rates of ROL. There was an increase in adventitious root porosity from 20–26% for plants grown in aerated solution to 29–41% for plants grown in stagnant solution. Growth in stagnant solution also induces a 'tight' barrier to ROL in the basal regions of adventitious roots of five of the seven upland types, all three paddy types, and the two deepwater types. The increased porosity gave a low resistance pathway for O_2 movement to the root tip. The barrier to ROL in basal zones improved the longitudinal O_2 diffusion towards the apex, by decreasing losses to the rhizosphere. The plasticity in root physiology, seemingly contributes to the capacity of rice to grow in varied environments.

Henry et al. (2012) reported that root attributes affect water uptake of rice under drought. They stated that lowland rice roots have an inimitable physiological response to drought because of their adjustment to flooded soil. Rice root characteristics that assist growth under flooded conditions may influence rice response to drought, but the relative role of root structural and functional distinctiveness for water uptake under drought in rice are not elucidated. The authors measured morphological, anatomical, biochemical, and molecular characteristics of soil-grown rice roots to explore the genotypic unevenness and genotype-environment interactions of water uptake under uneven soil water regimes. The results revealed that drought-resistant genotypes exhibited the lowest night-time bleeding rates of sap from the root system in the field. Diurnal fluctuation was predominant. Strongest source of variation for bleeding rates in the field and root hydraulic conductivity (Lp_r) was noted in the greenhouse. It was associated with expression trends of various PIP and TIP aquaporins. In drought resistant genotypes, there was a positive response in root anatomy. There was a reduction in suberization and compaction of sclerenchyma layer cells, particularly suberization of endodermis was still enhanced under drought. This suggested the distinct roles of these two cell layers in oxygen retention under flooded conditions (sclerenchyma layer) and water retention under drought (endodermis).

Root anatomical parameters are useful for selection and breeding of salt-tolerant rice, precise perceptive of root anatomy plasticity under salt stress requires the development of competent screening techniques under these stress conditions. Aybeke (2016) determined anatomical plasticity in root characteristics of 31 dissimilar rice cultivars in response to salt stress. It best rice cultivars there was an improvement in protective anatomical characters from non-saline to saline conditions. Anatomical plasticity was found to be related directly to features of apoplastic barrier. High genotypic difference was found in root anatomy in all cultivars; for cell size; Si presence, its buildup shape and distribution, degree of lignifications and suberization of apoplastic barriers, idioblast cells number and presence of phenolic and gummic substances in them, etc. He mentioned that it is better to select those cultivars possessing stable anatomy for breeding salt-resistant rice cultivars.

Price et al. (2000) constructed a combined RFLP and AFLP linkage map of upland rice and identified the QTLs for root-penetration, from an earlier mapped F_2 of a cross between the two drought-resistant upland rice varieties Bala and Azucena. The map contained a 101 RFLP and 34 AFLP markers. These were present on 17 linkage groups, covering 1680 cM. They assessed root penetration by counting the number of root axes that penetrated a 3 mm-thick layer that consisted of 80% wax and 20% white soft paraffin. Good root penetration was expected to bring a raise in drought resistance, where soil strength was high. Single-marker analysis exhibited seven QTLs for the number of roots that penetrated this wax layer. Seven QTLs for the ratio of penetrated to the total number of roots were detected in identical locations. Among these QTLs, four were explained by the transgressive inheritance of positive alleles from Bala. The resemblance of the identified QTLs with the previous reports of QTLs for root morphology suggested that alleles that improve root penetration ability may also make the roots longer or thicker.

Root morphological features were considered as significant traits for drought resistance in rice varieties. Price et al. (2002) used molecular marker technology for the enhancement of drought resistance in rice to identify QTLs associated with root morphology and other drought resistance-related traits. They screened a mapping population of 140 RILs and the parental varieties Bala and Azucena for root growth in thin glass-sided soil-filled chambers. In the early water deficit treatment, they observed a slow decline in the shoot growth, and a maximum increase in root length. This has indicated that major changes in partitioning occurred, which was also evidenced through variations in root mass, and root to shoot ratios. The difference in partitioning between similar treatments was also observed in different years, which in part reflected the plant responses to soil water and nutrient status. They noticed that in both early and late water deficit treatments there was an extraction of water first from the upper sections of the rooting zone which later progressed to deeper layers. A RLD of 0.4 cm cm^{-3} was required for effective water extractions. In the late water deficit, there was an extraction of water extraction at 100 cm depth. It was related to shoot size and stomatal conductance. Azucena had thicker roots and more roots at depth compared to Bala. Bala cultivar slowed its shoot growth much earlier and became less water-stressed. Azucena has root traits that contributed potentially to drought resistance, while Bala has a number of shoot-related mechanisms that make it adapted to drought-prone environments.

5.3.1.2 DROUGHT INDUCED CHANGES IN ROOT PATTERNS

Root characteristics viz., root length, density, root thickness, and rooting depth and root distributions were mostly selected as crucial traits for selection of cultivars for drought resistance. Many studies have shown that deep rooting cultivars were more resistant to drought than shallow-rooted cultivars. Asch et al. (2005) quantified the effects of varied levels of drought on partitioning of dry matter and root development of three rice cultivars viz., CG14 (*Oryza glaberrima*), WAB56-104 (*O. sativa,* tropical *japonica*, improved) and WAB450-24-3-2-P18-HB (CG14 × WAB56-104 hybrid) in west Africa.

They observed that these cultivars of rice responded to drought stress. They exhibited reduction in plant height, leaf area and biomass production, tiller abortion, changes in root dry matter and rooting depth and also a setback in reproductive development. They observed that when plants were gradually stressed, there was no affect on the assimilate partitioning between root and shoot. In no case, surplus biomass was partitioned to the roots; in contrast, dry matter partitioning to the root completely stopped under severe stress. As a result of the irrigation technique used, there was a little variation in the vertical soil moisture distribution and roots grew deeper under drought stress. This was observed more evidently in the upland adapted WAB56-104. They deliberated about the implications for modeling of drought responses in the upland rice systems.

5.3.2 SEEDS

Seeds and coleoptiles of rice varieties vary in their characteristics. Hoagland and Paul (1978) made a comparative scanning electron microscopy (SEM) of the surfaces of seeds and coleoptiles of several conspecific rice (*Oryza sativa* L.) varieties including the weed red rice and several cultivars: Mochi-Gommi and LA-110 (medium grained); Starbonnet, Labelle, and Bluebelle (long grained). There were numerous trichomes at seed apex and other surfaces of seeds of red rice, LA-110 and Mochi-Gommi. A large central awn with spines on its axis and apical end extended from seed apex was observed only in red rice. Though there was no significant difference in seed, surface tubercles or wax content and structure there were similarities in coleoptiles of all varieties. The coleoptiles had parallel rows of arrangement of stomata on both surfaces (adaxial and abaxial), rodlet structure of epicuticular wax, prickle hairs on the adaxial coleoptile surface, more number of spines on the edge of the apical end (older portion), two types of papillae on both the adaxial and abaxial surfaces. Though they could detect the varietal difference among developing coleoptiles on a macroscopic scale, morphological differences at the SEM level could not be discerned at the SEM level.

5.3.2.1 EMBRYO

Jones and Thomas (1992) studied the developmental anatomy and ultrastructure of somatic embryos from rice scutellum epithelial cells. They observed that somatic embryos have derived from the basal part of epithelial cells of scutellum. Between 48 and 60 hours of plating, a periclinal division in scutellar epithelial cells occurred as initial event of somatic embryogenesis and basal and terminal cells produced the suspensor and embryo, respectively. A globular embryo was formed after a series of random divisions in the terminal cell. The scutellum was initiated and covered the shoot apex, which developed in a lateral notch. The coleoptile originated from a rim of tissue contiguous the shoot apex. The root apex differentiated internal to the coleorhiza. At maturity, a rice somatic embryo resembled a zygotic embryo. It had a whole scutellum epithelium, which acts as a basis of secondary somatic embryos on supplementary subculturing.

5.3.2.2 AERENCHYMA IN RICE

The aerenchyma spaces formation is generally considered as an adaptation trait to submergence for facilitation of gaseous exchange. In rice, studies revealed that these develop by cell death and lysis. The aerenchyma formation at cellular level is poorly understood. Steffens et al. (2011) reported aerenchyma development in the rice stem. In deep water, rice variety Pin Gaew 56 there was development of aerenchyma from the internodes, absent from the nodes. A lesser degree of constitutive aerenchyma development was seen in two lowland rice varieties. Developmental gradients in the aerenchyma formation was seen between older (a larger number of aerenchyma) and at top of each internode. Ethephon or submergence treatment led to the promotion of aerenchyma development in all genotypes. The preaerenchymal cells had less starch and were devoid of chloroplasts. Further, the cells of these were thin-walled and produced more levels of O_2^- and H_2O_2.

5.3.2.3 ROLE OF SILICON AGAINST RICE ROOT BORER

Rice varieties exhibit variable responses to infestation of rice borers according to the silica content within them. Djamin and Pathak (1967) reported about the significance of silica in conferring resistance to Asiatic rice borer, *Chilo suppressalis* (Walker), in rice varieties. On the basis of the results got from various tests they classified the varieties into three groups viz., resistant, moderately resistant, and susceptible. Inherent varietal characters offered consistent results of the varietal reactions to infestation of Asiatic rice borer infestation. An extremely considerable negative correlation was evidenced between silica content of the stem and susceptibility to the rice borer. High silica content in the plant caused defacing of the larval mandibles and thus interfered with the feeding and boring behavior of the larvae into rice cultivars. Thus, the results conferred that a practical and efficient approach for reduction of rice borer infestation is to use the cultivars possessing high silica contents rather than its application to soil.

5.3.2.4 SILICON MANAGEMENT AND PRODUCTIVITY OF RICE

Rice is a silicon accumulator. Balanced and integrated nutrient management systems are adopted in crops for attaining increased and sustained yields. Savant et al. (1996) reviewed about the significance of silicon management for sustainable rice production. According to them, adequate attention should be given to the beneficial role of silica and its inclusion in integrated nutrient management. Rice plant absorbs large quantities of silicon in the form of monosilicic acid [H_4SiO_4, or Si $(OH)_4$] and deposits it as amorphous SiO. This deposited form of silica is found to be related with cellulose and hemicellulose in the rice plant tops (leaves and hulls). It is present in the form of *cuticle-Si double layer*. Though the solubility of soil Si is low, its content in soil solution is mostly accredited to its dissolution kinetics, which is further influenced by several soil factors viz., Al, Fe oxides, organic matter, redox potential, moisture, etc. They summarized the past and current literature on Si nutrition and discussed its potential role in increasing and sustaining rice productivity and suggested a few suggestions for future research.

5.3.2.5 NITROGEN AND SILICON EFFECTS ON RICE YIELD

Lee et al. (1990) explored the lodging-related traits in paddy with regard to nitrogen and silicon application. They found that there was an increase in leaf blades in the contents of total N, CaO, and MgO at heading while there was a decline in the contents of K_2O, SiO_2, lignin, and hemicellulose with an increase in N rate. Similarly, with an increase in silica application rates, there was a decline in CaO, SiO_2, lignin, and hemicellulose contents, total N and MgO contents. High N rate of application resulted in elongation of 3rd and 4th internodes and reduced lignin and cellulose contents of cell walls, while these were thickened and shortened, and exhibited a decrease in only hemicellulose content in cells walls at 30d with high Si rates of application. Though, lodging occurred at 300 kg N/ha^{-1}, silica applications reduced its effect. Further, while N applications led to an increase in numbers of panicles/hill and spikelets/panicle, an increase in ripened grain percentage and 1,000-grain wt was observed with Si application.

5.3.2.6 STEM ANATOMY AND METHANE EMISSION

The plant-mediated transport is the chief route of methane (CH_4) emission from paddy fields to the aboveground atmosphere. Methane emissions from fields of ten rice cultivars was assessed by Das and Baruah (2008) taking into consideration their anatomical and morphological characteristics. They observed large variations in the methane fluxes among the cultivars, which were found to be regulated to a large extent by the variations in these cultivars in their anatomical and morphophysiological characteristics. The occurrence of a large-sized medullary cavity was identified by microscopic analysis of stems of high- and medium-CH_4-emitting cultivars. Similarly, high-CH_4-emitting varieties exhibited more leaf area and transpirational rates, with more of stomatal frequencies.

Bernier et al. (2008) mentioned that drought stress is one of the severe abiotic restraints of upland rice productivity. There is an immense requirement to augment yield of rice in the upland ecosystem for meeting the food security needs. Though numerous secondary characters were recommended to help selections by plant breeders, most of them were not utilized because of practical applicability, low heritability, or noncorrelative to GYs. Many studies have confirmed that drought-tolerant upland rice can be bred, by direct selections of yield traits for yield in stress environments, wherein use of molecular markers might be of great help for the efficient selections. Though, several QTLs for drought resistance were recognized in rice, a few have their suitability for use in these marker-assisted selections. Identification of a large number of effective drought resistance QTLs may be much effective for the usage of marker aided selections for drought resistance.

Fukai et al. (1995) studied the physiomorphological traits in rice for developing drought-resistant cultivars; they described the patterns of water-stress development in rice fields. They made a review on the various physio-morphological traits and their genetic variations in conferring drought resistance in rice. They proposed that there is a requisite of integration of the plant breeding objectives and the physiological research. They mentioned that there exists a gap in drought associated research for rice in rainfed lowland conditions, which need to be rectified. They put forward that if the efforts of physiological-genetic research were focused on target environments, it enables in improving the breeding program's efficiency along with the relevance of the stress physiology.

5.4 CONCLUSION

Various researches have been undertaken on different aspects of botany such as taxonomy, morphology, the anatomy of root and stem, which were associated with adaptation to some abiotic stresses such as drought, flooding, etc.

KEYWORDS

- **immunophilin**
- **molecular characterization**
- **quantitative trait loci**
- **radial oxygen loss**
- **root oxidation ability**
- **scanning electron microscopy**

REFERENCES

Agarie, S., Agata, W., Uchida, H., Kubota, F., & Kaufman, P. B., (1996). Function of silica bodies in the epidermal system of rice (*Oryza sativa* L.): Testing the window hypothesis. *Journal of Experimental Botany*, *47*, 655–660.

Asch, F., Dingkuhn, M., Sow, A., & Audebert, A., (2005). Drought-induced changes in rooting patterns and assimilate partitioning between root and shoot in upland rice. *Field Crops Research*, *93*, 223–236.

Aybeke, M., (2016). Root anatomical plasticity in response to salt stress under real and full season filed conditions and determination of new anatomical selection characters for breeding salt resistant rice (*Oryza sativa* L.). *Journal of Natural Sciences*, *17*, 87–104.

Banda, M., & Lal, R., (1983a). Response of upland rice varieties to drought stress: I. Relation between root system development and leaf water potential. *Plant and Soil, 73*, 59–72.

Banda, M., & Lal, R., (1983b). Response of upland rice varieties to drought stress: II. Screening rice varieties by means of variable moisture regimes along a toposequence. *Plant and Soil*, *73*, 73–94.

Bernier, J., Atlin, G. N., Serraj, R., Kumar, A., & Spaner, D., (2008). Breeding upland rice for drought resistance. *Journal of the Science of Food and Agriculture*, *88*, 27–939.

Champoux, M. C., Wang, G., Sarkarung, S., Mackill, D. J., O′Toole, J. C., Huang, N., & McCouch, S. R., (1995). Locating genes associated with root morphology and drought avoidance in rice via linkage to molecular markers. *Theoretical and Applied Genetics*, *90*, 969–981.

Changming, Y., Linzhang, Y., Yongxing, Y., & Zhu, O., (2004). Rice root growth and nutrient uptake as influenced by organic manure in continuously and alternately flooded paddy soils. *Agricultural Water Management, 70*, 67–81.

Cheul, J. A., Kim, D. W., & You, Y. N., (2010). Classification of rice (*Oryza sativa* L. japonica nipponbare) immunophilins (FKBPs, CYPs) and expression patterns under water stress. *BMC Plant Biology, 10*, 1–22.

Colmer, T. D., (2003). Aerenchyma and an inducible barrier to radial oxygen loss facilitate root aeration in upland, paddy, and Dee. *Annals of Botany*, *91*, 301–309.

Das, K., & Baruah, K. K., (2008). Methane emission associated with anatomical and morph physiological characteristics of rice (*Oryza sativa*) plant. *Physiologia Plantarum*, *134*, 303–312.

Djamin, A., & Pathak, M. D., (1967). Role of silica in resistance to Asiatic rice borer, *Chilo suppressalis* (Walker), in rice varieties. *Journal of Economic Entomology, 60*, 347–351.

Fukai, S., & Cooper, M., (1995). Development of drought-resistant cultivars using physiomorphological traits in rice. *Field Crops Research*, *40*, 67–86.

Ghareyazie, B., Huang, N., Second, G., Bennett, J., & Khush, G. S., (1995). Classification of rice germplasm: I. Analysis using ALP and PCR-based RFLP. *Theoretical and Applied Genetics*, *91*, 218–227.

Glaszmann, J. C., (1987). Isozymes and classification of Asian rice varieties. *Theoretical and Applied Genetics*, *74*, 21–30.

Harris, W. H., & Clark, L. H., (1981). Observations on the root anatomy of rice (*Oryza sativa* L.). *American Journal of Botany*, *68*, 154–160.

Henry, A., Cal, A. J., Batoto, T. C., Torres, R. O., & Serraj, R., (2012). Root attributes affecting water uptake of rice (*Oryza sativa*) under drought. *Journal of Experimental Botany*, *63*, 4751–4763. https://doi.org/10.1093/jxb/ers150 (accessed on 29 January 2020).

Hoagland, R. E., & Paul, R. N., (1978). A comparative SEM study of red rice and several commercial rice *(Oryza sativa)* varieties. *Weed Science*, *26*, 619–625.

Insaludr, N., Bell, W., Colmer, T. D., & Rerkasem, B., (2006). Morphological and physiological responses of rice (*Oryza sativa*) to limited phosphorus supply in aerated and stagnant solution culture. *Annals of Botany*, *98*, 995–1004.

Jones, T. J., & Thomas, L., (1992). The developmental anatomy and ultra structure of somatic embryos from rice (*Oryza sativa* L.) scutellum epithelial cells. *International Journal of Plant Sciences*, *150*, 41–49.

Lee, D. B., Kwon, T. O., & Park, K. H., (1990). Influence of nitrogen and silica on the yield and the lodging related traits of paddy rice. *Research Reports of the Rural Development Administration, Soil and Fertilizer*, *32*, 15–23.

Li, A., Lu, B. R., Zhang, S. Z., & Hong, D. Y., (2002). A phylogeny of the rice tribe Oryzeae (Poaceae) on matK sequence data. *American Journal of Botany*, *89*, 1967–1972.

Lux, A., Luxová, M., Morita, S., Abe, J., & Inanaga, S., (1999). Endodermal silicification in developing seminal roots of lowland and upland cultivars of rice (*Oryza sativa* L.). *Canadian Journal of Botany*, *77*, 955–960.

Morita, S., & Nemoto, K., (1995). Morphology and anatomy of rice roots with special reference to coordination in organo- and histogenesis. Structure and function of roots. *Developments in Plant and Soil Sciences, 58*, 75–86.

Parsons, B. J., Newbury, H. J., Jackson, M. T., & Ford, B. V. L., (1997). Contrasting genetic diversity relationships are revealed in rice (*Oryza sativa* L.) using different marker types. *Molecular Breeding*, *3*, 115–125.

Price, A. H., Steele, K. A., Gorham, J., Bridges, J. M., Moore, B. J., Evans, J. L., & Richardson, P., (2002). Upland rice grown in soil-filled chambers and exposed to contrasting water-deficit regimes: I. Root distribution, water use, and plant water status. *Field Crops Research, 76*, 11–24.

Price, A. H., Steele, K. A., Moore, B. J., Barraclough, P. P., & Clark, L. J., (2000). A combined RFLP and AFLP linkage map of upland rice (*Oryza sativa* L.) used to identify QTLs for root-penetration ability. *Theoretical and Applied Genetics*, *100*, 49–56.

Savant, N. K., Snyder, G. H., & Datnoff, L. E., (1996). Silicon management and sustainable rice production. *Advances in Agronomy*, *58*, 151–199.

Steffens, B., Geske, T., & Sauter, M., (2011). Aerenchyma formation in the rice stem and its promotion by H_2O_2. *New Phytologist*, *190*, 369–378.

Sylvester, A. W., Vickie, P. C., & Murray, G. A., (2001). Leaf shape and anatomy as indicators of phase change in the grasses: Comparison of maize, rice, and bluegrass. *American Journal of Botany*, *88*, 2157–2167.

O'Toole, J. C., & Soemartono (1981). Evaluation of a simple technique for characterizing rice root systems in relation to drought resistance. *Euphytica*, *30*, 283–290.

Uga, Y., Sugimoto, K., Ogawa, S., Rane, J., Ishitani, M., et al., (2013). Control of root system architecture by deeper rooting 1 increases rice yield under drought conditions. *Nature Genetics*, *45*, 1097–1102.

Uga, Y., Ebana, K., Abe, J., Morita, S., Okuno, K., & Yano, M., (2009). Variation in root morphology and anatomy among accessions of cultivated rice (*Oryza sativa* L.) with different genetic backgrounds. *Breeding Science (Chinese)*, *59*, 87–93.

Virk, P. S., Zhu, J., Newbury, H. J., Bryan, G. J., Jackson, M. T., & Ford, L. B. V., (2000). Effectiveness of different classes of molecular marker for classifying and revealing variation in rice (*Oryza sativa*) germplasm. *Euphytica*, *112*, 275–284.

Yu, J., Hu, S., & Wang, J., (2002). A draft sequence of the rice genome (*Oryza sativa* L. ssp. *indica*). *Science*, *296*, 79–92.

Zhao, Z., Pearsall, D. M., Benfer, R. A., & Piperno, D. R., (1998). Distinguishing rice (*Oryza sativa* poaceae) from wild Oryza species through phytolith analysis, II Finalized method. *Economic Botany*, *52*, 134–145.

CHAPTER 6

Physiological Basis of Rice Growth and Productivity

ABSTRACT

This chapter discusses significant research advances up to 2018 on the physiological basis of rice growth and productivity. It also discusses different growth stages of rice viz. vegetative stage, panicle development stage, grain filling stage and factors affecting rice growth and productivity. Besides, it discusses the mineral deficiency and toxicity symptoms in rice long with their management strategies.

6.1 GROWTH AND DEVELOPMENT

Rice has various stages of crop growth mentioned below:

- **Growth Stage 1:** Vegetative stage, from germination to vegetative stage.
- **Growth Stage 2:** Panicle development stage, from initiation of panicle to full developed panicle.
- **Growth Stage 3:** Grain filling stage.

A universal scale (to be known as the BBCH scale) and a uniform decimal code for both crop and weed growth stages were proposed by Lancashire et al. (1991). For cereals, the scale and codes are based on the familiar Zadoks code. From the general scales, individual crop-specific scales may be constructed. The scale uniformity is easy to memorize and usage in agricultural practice and very simple for recovery in a computer system. From the specific scales of crops such as rice, maize, oilseed rape, field beans, peas, and sunflower a description of the general scale is provided.

Though many papers have been published on rice growth staging systems, none of them were used broadly to elucidate the rice growth and development. Counce et al. (2000) suggested a constant, adaptive system for expressing rice development. They described three key stages of development in rice: seedling, vegetative, and reproductive. Seedling development stage is comprised of four growth stages: unimbibed seed (S0), emergence of radicle and coleoptile from the seed (S1, S2), and emergence of prophyll from the coleoptile (S3). Vegetative development stage is comprised of few stages, V1, V2, …, V*N*; where N is equal to the final number of leaves with collars on the main stem. Reproductive development includes 10 growth stages on the basis of distinct morphological criteria: panicle initiation (R0), panicle differentiation (R1), flag leaf collar formation (R2), panicle exertion (R3), anthesis (R4), grain length (GL) and width expansion (R5), grain depth expansion (R6), grain dry down (R7), single grain maturity (R8), and complete panicle maturity (R9). They mentioned that the division of rice growth stages on the basis of distinct morphological criteria provides a clear growth-stage determination.

To get better estimations of rice aboveground biomass (AGB) Gnyp et al. (2014) studied hyperspectral canopy sensing. Using fixed band Green Seeker active multispectral canopy sensor, normalized difference vegetation index and ratio vegetation index (GS-NDVI and GS-RVI) were obtained. The GS-NDVI and GS-RVI are general non-destructive methods of crop growth parameters estimation, but the soil and/or water conditions at early crop growth stages and saturation effects at medium to high biomass states affect their performance. In this chapter, optimum measurements were identified with diverse methods: (a) soil adjusted vegetation indices (VIs); (b) optimized narrowband RVI and NDVI; and (c) optimum multiple narrow-band reflectances (OMNBR) models based on raw reflectance, and its first and second derivatives (FDR and SDR). To make different biomass conditions they conducted six rice nitrogen (N) rate experiments in Heilongjiang province of Northeast China. From the experimental and farmers field's hyperspectral field data and AGB, samples were taken from four growth stages of crop. The results unveiled that 21–35% more variability of AGB was explained by six-band OMNBR models comparative to the best performing fixed band RVI or NDVI at various stages of growth. In comparison to raw reflectance-based 6-band OMNBR models, 4%, 6%, and 8% more AGB variability was explained by FDR-based 6-band OMNBR models at stem elongation, booting, and heading stages, respectively. Except for the stem, elongation stage further improvements were not observed in SDR-based 6-band OMNBR models. Compared to the best performing fixed band RVI or NDVI, 18–26% more AGB variability was explained by optimized RVI and NDVI for each growth stage. These outcomes were constant through distinct sites and years. Finally, they determined that the estimation of rice AGB could be significantly improved with the suitable band combinations, optimized narrowband RVI or NDVI, without the requirement of derivative analysis. This estimation can be further improved with the six-band OMNBR models viz. SDR at the stem elongation stage and FDR at other growth stages.

6.1.1 *GERMINATION AND SEEDLING ESTABLISHMENT*

Germination and seedling establishment is a very important phase of a crop for maintaining adequate stands. The following various factors affect this phase.

6.1.1.1 *SALINITY*

In California, the most common method of irrigation for rice is flood irrigation. During the seedling establishment of rice, flood irrigation may cause salinity damage and ultimately reduce yield. Salinity effects on seedling growth and yield traits of rice were studied by Zeng and Shannon (2000). The M-202 rice cultivar was grown in the sand in a greenhouse and at electrical conductivity 1.9, 3.4, 4.5, 6.1, 7.9, and 11.5 dS m^{-1} the control and treatments amended with NaCl and $CaCl_2$ (2:1 molar concentration) were applied. After seeding, seedlings shoot dry weights were measured in the first month. At the lowest salinity treatment, 1.9 dS m^{-1} significantly decreased growth of seedling was observed. Grain weight per plant, grain weight per panicle, spikelet number per panicle, and tiller number per plant exhibited significantly linear response to salinity. Seedling survival and harvest index (HI) were reduced significantly at 3.40 dS m^{-1} and higher. Both fertility and pollen germination had a common lowest salt level. In M-202 reductions in seedling survival, tillers number per plant, and spikelet number per panicle were found to be the main sources of yield loss under salinity.

6.1.1.2 SEED VIGOR

Seed vigor is an index of seed quality desirable for the fast and uniform germination and the strong seedlings establishment in any environmental conditions. In rice, Xie et al. (2014) assumed identification and fine mapping of quantitative trait loci (QTLs) for seed vigor. Though, in traditional breeding programs seed vigor is not considered as a key breeding trait, in direct-sowing rice production systems under low-temperature conditions strong seed vigor is important. In this study, two rice cultivars, ZS97 and MH63 were crossed and their recombinant inbred population was utilized to find and map eight QTLs for seed vigor. From the conditional QTL analysis, for seedling establishment qSV-*1*, *qSV-5b*, *qSV-6a*, *qSV-6b*, and *qSV-11* and for germination *qSV-5a, qSV-5c*, and *qSV-8* QTL's were identified. Among these *qSV-1*, *qSV-5b*, *qSV-6a*, *qSV-6b*, and *qSV-8* QTLs were specific to low-temperature. Two most important and effective QTLs, *qSV-1*, and *qSV-5c* were constricted to 1.13-Mbp and 400-kbp genomic regions, respectively. For both normal and low-temperature germination environments to increase seed vigor, tightly linked DNA markers for the marker-assistant pyramiding of multiple positive alleles were specified in this study (Figure 6.1).

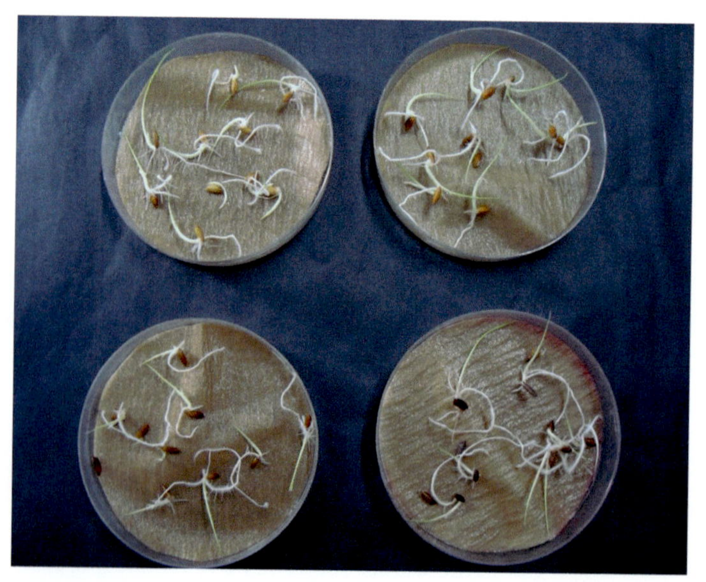

FIGURE 6.1 Germination of paddy seeds.

6.1.1.3 *SEED PRIMING*

Lee et al. (1998) examined the rice seeds priming effect on seedling establishment under excess soil moisture conditions. For priming seeds were soaked in –0.6 MPa polyethylene glycol (PEG) solution at 25°C for 4 days. Then the seeds were sown in soils with different soil moistures (60, 80, 100, 120, and 140% field capacity) at 17 and 25°C. At all the soil moisture levels and temperature tested, the primed seeds showed higher germination and emergence rates, plumule height, and radicle length than that of untreated seeds. The planting to 50% germination time was 0.9 ~3.7 days less for primed seeds. Among the soil moistures, the rate of germination, rate of emergence, plumule height, and radicle length were highest at the 80% soil moisture. Under the unfavorable soil moistures (60, 100, 120, and 140% field capacity), the rice seeds priming effects on rate of germination and emergence were more projecting than those under the optimum soil moisture condition (80% field capacity). But, compared with excessively lower or higher soil moistures priming effects on growth of seedling were higher at near optimum soil moisture. Thus, these results suggested that for better seedling establishment under the unfavorable soil condition priming of rice seeds maybe an effective method.

In direct seeding of rice (DSR), one of the major problems is poor seedling establishment. The effect of field-sown rice seed priming on germination, seedling establishment, allometry, and yield was investigated by Farooq et al. (2006). For this purpose, various seed priming strategies such as hydro priming for 48 hours, osmo hardening with KCl or $CaCl_2$ for 24 hours, ascorbate priming for 48 hours and seed hardening for 24 hours, pre-germination (traditional soaking for nursery raising) and untreated control were adopted. The primed seeds exhibited improved germination and emergence, allometry, kernel yield, and its quality, while poor and erratic emergence of seedling subsequent to poor plant performance was noticed in pre-germination. In primed kernels, the improved activity of α-amylase enhances the soluble sugars level leading to faster and uniform emergence. Osmo hardening with KCl provided greater kernel and straw yield and HI, followed by that of $CaCl_2$, hardening, and ascorbate priming. The mean emergence time and days to heading exhibited positive correlation, whereas kernel yield and HI showed negative correlation indicating the seed priming long-term effects on plant growth and development. The findings proposed that the osmo hardening produces physiological changes which in turn increases hydrolysis of starch and produces more sugars accessible for embryo growth; production of vigorous seedling and later, improved allometric, kernel yield and quality traits.

6.1.1.4 SAND PRIMING

In direct-sown rice sand priming, effects on germination and field performance were studied by Hu et al. (2005). In this study, sand was used as a priming solid matrix wherein, seeds were first mixed with sands containing 3.8% (v/w) water and sealed in plastic box then priming was performed at 18°C for 72 h. Four rice varieties were examined for the sand priming effects on germination of direct-sown rice in the laboratory. The results unveiled the improved germination energy, germination percentage, germination index (GI), and vigor index (VI) in all the four varieties. The primed seeds had significantly higher seedling height, root length, number of root and root dry weight than the non-primed controls. Further, significantly enhanced seed establishment and yield in sand primed seeds were revealed from field experiments. These findings proposed that in direct-sown rice the seedling establishment can be improved with the help of sand priming method and probable to be used in the field crop production.

6.1.1.5 SALINITY STRESS ON SEED GERMINATION

In rice, the germination and early seedling growth are highly susceptible to salinity. The result of salt stress at these stages of rice was studied by Hakim et al. (2010) where they exposed twelve rice varieties to six salinity levels (0, 4, 8, 12, 16 and 20 dS m^{-1}) at germination and early seedling stages. They applied SAS and with LSD, means were separated for final germination percentage (FGP), speed of germination (SG), germination energy percentage (GE%), plumule, and radical length and plumule and radical dry weight. By considering the reduction in dry matter yield, they categorized rice varieties as tolerant (T), moderately tolerant (MT), moderately susceptible (MS), or susceptible (S). At a salt concentration of 20 dS, m^{-1} germination was fully inhibited. In all the twelve varieties FGP, SG, and GE% were reduced due to salinity, leading to reduction in shoot and root length and dry weight. With increasing salinity, stress the extent of reduction increased. During germination at 12 dS m^{-1} salinity, the varieties MR211, IR20, BR40, and MR232 exhibited tolerance to salinity. However, based on reduction in dry matter yield MR211, MR232 and IR20 varieties were found better. The result put forward that, the crop salt tolerance at germination and early seedling stage may not be reliable with the tolerance at advanced growth stage. Therefore, additional study on effect of salinity on growth processes and physiological consequences at advanced growth stage is required, for which the varieties MR211, MR232, and IR20 might be used.

6.1.1.6 HIGH MOISTURE/FLOODING

Under lowland conditions, the seedling establishment mechanism in direct-seeded rice was investigated by Biswas and Yamauchi (1997). In the IRRI phytotron, (29/21°C) 36-hour-old seedlings of rice genotypes ASD1, IR41996-50-2-1-3, Mahsuri, and IR72 were raised under lowland and drained soil conditions for 10 days. After four days of seeding the genotypes ASD1 and IR41996-50-2-1-3, coleoptile tips penetrated the reduced soil layer to the oxygenated water to allow the establishment of O_2 transport system. Subsequently, the growth of the mesocotyl and first leaf take place, but the root growth was stopped up to 6 days after seeding. The identical sequence of growth was seen in genotypes ASD1 and IR41996-50-2-1-3, but these genotypes had clear variance in the tolerance of their tissue to oxygen deficit.

Flooding affects the direct-seeded rice by delaying seed germination and seedling establishment and imposes fatalities. In rice, the flooding tolerance mechanism was examined by Ismail et al. (2009). Seeds were sown in soil and watered normally and flooded with 10 cm of water. Then different parameters such as survival of seedling and shoot and root growth, activities of amylase and peroxidase, ethylene production and concentrations of soluble sugars and starch were assessed. Afterward, they determined the activities of enzymes related with anaerobic respiration and using seeds germinated under various amounts of dissolved oxygen in dilute agar, gene expression was analyzed separately. Under flooding, reduced seedling survival was observed, however, the reduction in seedling survival was less in tolerant genotypes. Flooding decreased the concentration of starch and increased the sugar concentration in germinating seeds. Amylase activity showed a positive association with shoot and root elongation and with plant survival. The amylase activity and *RAmy3D* gene expression was high in tolerant genotypes. Within 2 days after sowing, the ethylene was not found in seeds, but its concentration increased afterward, with an increment in tolerant genotypes from 3 days after sowing. Peroxidase activity showed a negative correlation with survival and its concentration was high in germinating seeds of sensitive genotypes. Finally, it was determined that the high amylase activity and anaerobic respiration in tolerant genotypes are responsible for their faster growth and more seedlings survival and helps to maintain stored starch reserves use capacity, further, the tolerant genotypes have high rates of ethylene production and lower peroxidase activity in germinating seeds and seedlings.

When the rice is sown under flooded or water-saturated condition it gets exposed to hypoxia or anaerobic situation. Thus, to elucidate the interrelationship between seed vigor and seedling growth in anaerobic soil and to find seedling features that forecast establishment in the field Yamauchi and Winn (1996) conducted an experiment. Using different cultivars, times of harvest and accelerated aging treatments various seed vigor levels were obtained. Then, germination rate in Petri dishes, coleoptile, and mesocotyl lengths in N_2 gas in flasks, and seedling establishment, VI, height, leaf development and shoot weight in flooded soil in trays and in water-saturated soil in the field were measured. In the field, the seedling establishment was not affected with harvest time. Except ASD1 cultivar, all the tested cultivars showed significantly reduced seedling establishment with aging. Seedling establishment was found to be strongly associated with VI in the field, establishment, and VI in the tray and coleoptile length in N_2 gas, signifying that the seedling establishment in anaerobic soil can be predicted with the help of these characteristics.

6.1.1.7 OTHER FACTORS

6.1.1.7.1 Rhizobium

Plant growth-promoting rhizobacteria (PGPR) stimulate plant growth by producing phytohormones. In recent times, it is found that the PGPR have the potential for growth stimulation of both legume and non-legumes. In this context, the rhizobia and plant growth-promoting bacteria inoculation effects on germination and seedling vigor of lowland rice were studied by Mia et al. (2012). Using UPMB10 (*Bacillus sphaericus*) PGPR strain and SB16, UPMR1006 and UPMR1102 *Rhizobium* strains, seeds of lowland rice variety MR219 were inoculated in Petri dishes. Then, the Petri dishes were kept in an incubator at 30 ± 2°C for 120 h. The results unveiled the enhanced seedling emergence, seedling vigor, root growth viz. root length, root surface area, and volume in inoculated seeds. The strain UPMB10 performed well in seedling growth and strain UPMR1006 produced profuse hair in the radical. Thus, their research findings have indicated that PGPR and Rhizobia strains promote the emergence of seed and seedling characteristics which helps in the early establishment of seedling and subsequently the crop growth and development.

6.1.1.7.2 Putrescine

The NaCl and putrescine induced changes on the germination and early seedling growth of rice were determined by Prakash and Prathapasenan (1988), where they exposed the seeds to salt stress (100, 150 and 200 mM NaCl) for 120 hours. All the concentrations of NaCl declined the seeds germination rates and growth of embryo axis. In seeds, water uptake and the mobilization rate of reserves were reduced significantly at 150 mM NaCl concentration. The putrescine (0.001, 0.01, 0.1, and 1.0 mM) was applied exogenously to seeds. The presence of both putrescine and NaCl increased the germination rate and seedling growth and the maximum response was observed at putrescine (0.01 mM) and NaCl solution (150 mM). The addition of putrescine (0.01 mM) to NaCl solution (150 mM) enhances the water uptake of seeds by reducing the sodium and chloride ions net accumulation in seeds. Thus, they reported that the addition of putrescine reduces the inhibition effects of NaCl during germination and early seedling growth of rice.

6.1.1.7.3 Brassinosteroids (BRs)

Anuradha and Seeta (2001) analyzed the 24-epibrassinolide and 28-homobrassinolide effect on the salinity stress-induced rice germination and seedling growth inhibition. They found that under salinity stress brassinosteroids (BRs) increase levels of nucleic acids and soluble proteins associated with seedling growth activation, thereby reversing the inhibitory effect of salinity stress on germination and seedling growth of rice.

6.1.1.7.4 *Abscisic Acid (ABA)*

Seed dormancy and stress tolerance are positively regulated by abscisic acid (ABA) and in *Arabidopsis thaliana* ABA-dependent gene expression was found to be controlled by PYL/RCARs intracellular ABA receptors. But, in monocot species, the function of these receptors has not been characterized yet. Kim et al. (2012) reported that PYL/RCAR orthologues act as a positive regulator of the ABA signal transduction pathway in seed germination and early seedling growth of rice. The rice orthologue PYL/RCAR was expressed in rice plants. These rice transgenic plants exhibited hypersensitivity to ABA during seed germination and early seedling growth. Further, using interaction assays and a transient gene expression assay a rice ABA signaling unit comprising OsPYL/RCAR5, OsPP2C30, SAPK2, and OREB1 for ABA-dependent gene regulation was detected. Therefore, in rice seed germination and early seedling growth modulating ABA-responsive gene expression has been resolved. This study contributes considerably towards the understanding of ABA signal transduction pathway in rice.

6.1.1.7.5 *Heavy Metal*

A study has been undertaken by Mishra and Choudhuri (1999) on phytotoxicity of lead and mercury on germination and early seedling growth of rice. Two rice cultivars, Ratna, and IR36 were treated separately with 10^{-5} and 10^{-4} M $PbCl_2$ and $HgCl_2$. In comparison to control, both the cultivars exhibited reduced germination percentage, GI, shoot, and root length, tolerance index (TI), VI and dry mass of shoot and root and increased percentage difference from control (% DFC) of germination and percentage phytotoxicity. At equal concentrations, the phytotoxic effect of mercury was found to be greater than lead and the cultivar IR36 was more tolerant than Ratna to these metals. In the estimation of relative toxicity of lead and mercury the monitoring indices TI, VI, and % phytotoxicity appeared to function as good biological monitoring methods particularly for rice cultivars.

6.1.1.7.6 *Sucrose Transporter*

By means of expression analysis, a comprehensive study on the function of the sucrose transporter OsSUT1 during germination and initial growth of rice seedlings was carried out by Scofield et al. (2007), over a time period varying from 1 to 7 days post-imbibition. The wheat orthologue, TaSUT1, is supposed to be directly engaged in transfer of sugar through the scutellar epithelium, but OsSUT1 is not expressed in the scutellar epithelial cell layer of germinating rice and hence, not engaged in transportation of sugar through the symplastic discontinuity between the endosperm and the embryo. Moreover, in aleurone cells the expression of OsSUT1 was not present, indicating that it is not involved in sucrose transport in this cell layer during germination. By 3 days post-imbibition, OsSUT1 expression was identified in the companion cells and sieve elements of the scutellar vascular bundle, where it may take part in phloem loading of sucrose for transport to the emerging shoot and roots. Most probably, the sucrose is obtained from hexoses bring in from the endosperm. Additionally, starch granules present at a high density in the scutellar ground tissues adjoining the vasculature and at the base of the shoot may remobilize the sucrose. OsSUT1 was also found in the coleoptile, the first and second leaf blades and within the phloem of the primary roots, where it could get involve in sugars recovery from the apoplasm in these tissues.

6.1.2 *VEGETATIVE GROWTH, FLOWERING (FL), AND FRUITING OF RICE*

Rice crop grown under cool climates and flooding can be subjected to suboptimal water temperature (T_w) at any phase of the crop cycle. Shimono et al. (2002) conducted an experiment to understand the response of biomass and grain yield (GY) to T_w at three different stages: vegetative and reproductive growth periods, and early grain filling. At variable temperature with year and developmental stage, irrigated water was supplied to the plants. The temperature treatments varying from 4–7°C (low) to 20 to 25°C (high) were applied for 20–34 days. The low temperature during the reproductive stage affected the GY severely as a result of the low spikelet fertility (SF). During the vegetative period, the low T_w decreased the GY by 20%. The magnitude of crop growth rate (CGR) reduction due to low

T_w varied according to period, which was highest during the vegetative period, followed by the reproductive and early grain-filling period. Before heading the decreased CGR was largely correlated with the reduced canopy radiation interception and restricted leaf area, while radiation use efficiency (RUE) was comparatively unaffected by T_w. After heading, though leaf area was also reduced by low $T_{w,}$ decreased CGR was mostly correlated with reduced RUE. This study possibly will assist in quantification of rice growth and yield as influenced by low T_w (Figure 6.2).

FIGURE 6.2 Rice crop at vegetative stage.

The root exudations play important roles in the rhizosphere of plants. Aulakh et al. (2001) characterized root exudates of ten rice cultivars at different growth stages. To collect root exudates of soil-grown rice plants three media (nutrient solution, deionized water and $CaSO_4$ solution) were tested for three time periods (2, 4, and 6 hours). Tall rice cultivars (Dular, B40 and Intan), high-yielding dwarf cultivars (IR72, IR52, IR64, and PSBRc 20), new plant type (NPT) cultivars (IR65598 and IR65600) and a hybrid (Magat) were placed in 0.01 M $CaSO_4$ solution for 2 hours, then at seedling, panicle initiation, flowering (FL), and maturity root exudates were collected. Total organic carbon and organic acids of root exudates were analyzed (Figure 6.3). Lowest amount of root exudate was found at seedling stage, which increased during FL, then reduced at maturity. Among organic acids, malic acid was found in the maximum concentration followed by tartaric, succinic, citric, and lactic acids. With progressing plant growth, the substitution of sugar exudates with the organic acid exudates was observed. Finally, they observed that the root exudates offer substrates for methanogenesis in rice fields, significant differences in root exudation by cultivars and at various growth stages can effect CH_4 emissions. Thus, to alleviate CH_4 emission from rice fields, breeding new rice cultivars with low exudation rates could be a feasible option.

FIGURE 6.3 Panicle initiation stage

Tao et al. (2006) assessed rice growth and yield in the water-saving ground cover rice production system (GCRPS). They applied three treatments: (i) low land rice cultivated without standing water using GCRPS and irrigated at soil water tension below 15 kPa, (ii) the soil surface covered with 14 μm thick plastic film ($GCRPS_{Plastic}$) or mulched with straw ($GCRPS_{Straw}$) and (iii) the soil surface without any covering ($GCRPS_{Bare}$). The treatments were compared with traditionally cultivated lowland rice. Additionally, one aerobic rice variety was cultivated in bare soil. In comparison to paddy control, plants grown in GCRPS were smaller, with fewer panicles and smaller leaf area index (LAI). $GCRPS_{Bare}$ and $GCRPS_{Straw}$ had significantly lower yield than the Paddy control yield, whereas the yield in $GCRPS_{Plastic}$ was only 8% lesser than the Paddy control yield. Significant correlation was found between the GY and maximum LAI. Number of productive tillers exhibited the greatest positive effect on yield, while the grains number per panicle, thousand-grain weight, and HI remained almost not affected. Under uncovered condition, significantly higher HI, yield, and WUE were noted in aerobic rice variety than the lowland rice variety ($GCRPS_{Bare}$). These results specified that using GCRPS substantial amount of water can be saved with relatively negligible yield penalties (Figure 6.4).

FIGURE 6.4 Rice field at active tillering stage.

To work out the profits gained in terms of yield and other physiological parameters from using SRI practices, Thakur et al. (2010) undertook comparison of system of rice intensification (SRI) practices and recommended rice cultivation practices in India. Some commonly followed practices of SRI were: transplanting single, young (10-day-old) seedlings in a square pattern; no continuous flooding (CF); and use of a mechanical weeder. Same amount of fertilizer and a combination of organic and inorganic nutrients were applied to all the plots; however, the spacing in SRI was 20% less than commonly recommended. The plants in SRI plots exhibited significant measurable changes such as longer panicles, more grains panicle with higher percentage of grain-filling and out-yielded RMP by 42%. Thus, the decreased plant density with SRI_m management was compensated by the increased per-plant productivity. Compared to the RMP hills with multiple plants, the SRI_m hills with single plants had deeper and better-distributed root systems, higher xylem exudation rates, more open plant architecture with more erect and larger leaves, and more tillers. In spite of having fewer hills and fewer tillers per unit area, SRI_m had higher the LAI due to larger leaves. All these plant architecture alteration in SRI_m contributed to more light interception, the greater fluorescence efficiency (Fv/Fm and ΦPS II), more efficient utilization of light and a higher rate of photosynthesis in SRI_m plants. The higher photosynthesis rate together with lower transpiration in SRI_m plants unveiled that these plants were using water more proficiently than did RMP plants. Altogether, these changes may possibly result in increased grain filling and heavier grains in SRI_m plants compared to RMP plants (Figure 6.5).

FIGURE 6.5 Anthesis stage (flowering).

Vijayakumar et al. (2006) studied the growth characters, days to FL, growth analysis and labor productivity of rice Influenced by SRI Practices. Significantly, tall plants with high total dry matter production and LAI were seen in the treatment combination of 14 days old seedlings established at 25x25 cm^2 spacing + water-saving irrigation and SRI weeding. While, the treatment combination of 14 days old seedlings established at 15x10 cm^2 spacing + water-saving irrigation and conventional weeding exhibited significantly greater tiller density per m^2. In addition to this, significantly improved CGR, RGR, and NAR were noticed between panicle initiation (PI) to FL and FL to maturity stage in the treatment combination of 14 days old seedlings established at 25x25 cm^2 spacing, limited irrigation of 2 cm and SRI weeding with rotary weeder (Figure 6.6).

FIGURE 6.6 Flag leaf clipping of female line in hybrid rice seed production field.

In rice, the genetic analysis of transparency and chalkiness area at different filling stages was assumed by Shi et al. (2002). A developmental genetic model for quantitative traits of triploid endosperm in cereal crops were used for the data analysis. The unconditional analysis showed that the accumulated genetic effects of genes expressed from the initial time (at flowering and fertilization) to the filling time t were all significant for transparency and chalkiness area. These results indicated that the genetic effects of the triploid endosperm, cytoplasmic and diploid maternal plant were all of importance for both traits at various filling stages, especially for maternal additive and dominance effects on transparency. The relatively high endosperm and maternal additive effects on transparency and chalkiness area indicated that the two traits could be improved by selection in early generations. From the conditional analysis for the net genetic effects of genes expressed during time t−1 to time t, new expression of genes in endosperm, cytoplasm and maternal plant for transparency and chalkiness area was found at most of the filling stages, especially from 8 to 14 days after flowering for transparency and from 1 to 14 days after flowering for chalkiness area. Predicted genetic effects and conditional genetic effects at different filling stages showed that transparency and chalkiness area of offspring could be improved by using some parents, such as Zuo 5, because of their better endosperm additive and cytoplasmic effects (Figure 6.7).

FIGURE 6.7 Hybrid rice seed production: pollination by using stick method.

In rice, the high temperature at grain filling stage reduces the yield. Wang et al. (2005) constructed QTL map for heat tolerance (HT) in rice. From a backcross of Nipponbare/Kasalath/Nipponbare 98 backcross inbred lines (BILs) were obtained and subjected to high and optimal temperature during grain filling, respectively. Using 245 RFLP markers, a genetic linkage map was constructed and QTLs and their main effects, epistatic interactions and QTL × environment interactions (Q × E) were detected with a mixed linear-model approach. On chromosomes 1, 4, and 7 three QTLs influencing HT during grain filling with LOD scores of 8.16, 11.08, and 12.8 were identified. In the C1100-R1783 region of chromosome 4, a QTL with no QTL × environment interaction and epistatic effect was detected. This QTL allele was derived from Kasalath, under heat stress it reduced 3.31% of the grain weight loss and could be expressed in diverse environments and genetic backgrounds. Therefore, it would be effective in rice breeding for HT improvement. Two QTLs derived from Nipponbare were located between R1613-C970 on chromosome 1 and between C1226-R1440 on chromosome 7, with additive effect 2.38 and 2.92%, respectively. Significant Q × E were observed in both the QTLs and second QTL was involved in epistatic interaction also. The findings of this study would help in understanding the genetic basis of heat-tolerance, to develop novel rice varieties with HT during grain filling phase.

Using soft X-ray continuous photo-taking Zhu et al. (2005) examined the percentage of grains of hybrid rice. In Nanyou No. 3 rice hybrid, the associations among the sequence in FL, the number of days from anthesis to the hull filled with caryopsis (filled stage) and the formation of abortive kernel were detected. The results unveiled that the increase in FL sequence (x) delays the mean number of days of the filled stage (y). This relation may be expressed by the exponential equation $y=a\ (b)x^2$. In some late- FL weak spikelets, the filled stage was more than 25—30 days and the grain dry weight was less than 19 mg, thus developing abortive kernels. This might be the reason for the development of more abortive kernels in the hybrid rice "Nanyou No.3." They noted that the application of grain fertilizer shortens the filled stage of the later-FL weak spikelets and sunlight shading prolongs the filled stage. Thus, they speculated that applying grain fertilizer, improving sunlight, and prolonging the fruiting stage will be helpful in increasing the percentage of ripened grains, the grain weight and the yield of "Nanyou No. 3" (Figure 6.8).

FIGURE 6.8 Selfing at boot leaf to panicle emergence stage.

6.1.2.1 FACTORS AFFECTING GRAIN PRODUCTION

6.1.2.1.1 Salinity

Different salinity resistant varieties of rice were explored by Lutts et al. (1995) to study the changes in plant response to NaCl during development. Tall *indica* landraces (Nona Bokra, Buhra Rata, Panwell, and Pokkali) appeared to be resistant during the vegetative stage, whereas the varied resistance was observed in *japonica* varieties (I Kong Pao (IKP) and Tainung 67) and elite breeding lines (IR 4630, IR 2153, and IR 31785). The *indica* variety, Panwell exhibited salt resistance during booting, heading, and grain maturation. While, the *indica* varieties IR 4630, IR 31785, and IKP showed greatest variability in salt stress response during their vegetative and reproductive phases. The mean level of relative resistance in different genotypes was not correlated to their variability in salt resistance during the vegetative and reproductive phases. It is suggested that there was an ontogenic evolution of salt resistance and during vegetative growth; seedling stage was the most sensitive to NaCI. Furthermore, they found that the variations in plant responses to NaCI vary with criterion used to measure salinity resistance at specific stages of development. Thus, their studies have indicated that understanding of the salt stress effects upon phenology is essential for the development of rice genotypes resistant to NaCI in future breeding programs.

6.1.2.1.2 Water Stress

Water deficits at the rice anthesis stage cause a high percentage of spikelet sterility and reduce grain yield. Ekanayake et al. (1989) undertook a study to give an insight on direct effects of water stress on panicle exertion, spikelet opening, and spikelet desiccation causing spikelet sterility. The IRAT 13 (upland cultivar) and IR20 (lowland cultivar) were subjected to well-watered treatment and two water stress levels at their FL stage under greenhouse conditions. The cultivars exhibited significant differences in FL response to water stress. The cultivar IR20 was found to be highly sensitive to water stress. Owing to the low panicle water status, the time course of panicle exertion had an inhibitory effect on number of opened spikelets. In both IR20 and IRAT 13 cultivars the spikelet opening was totally restricted at panicle water potentials below −1·8 MPa and −2·3 MPa, respectively. The peak spikelet opening time in a day was not affected by the stress treatment and the spikelets in stressed panicles remained open for a longer period than in the well-watered panicles. Therefore, they determined that the water stress at FL stage has adverse effects on the spikelet opening and SF leading to reduction in GY.

6.1.2.1.3 Photoperiod

The photoperiod controls the FL time in rice. This photoperiodic control of FL was discussed by Izawa et al. (2000). They mentioned that *photoperiodic sensitivity 5* (*se5*) mutant of rice, which is very early FL lacks photoperiodic response. They cloned the *SE5* gene using candidate cloning which revealed that it encodes a putative hemeoxygenase. Absence of coleoptile elongation responses by light pulses and photoreversible phytochromes in rough *se5* extracts unveiled that *SE5* may play a role in biosynthesis of phytochrome chromophore. They found that the photoperiodic response in the *se5* mutant can be restored with the ectopic expression of *SE5* cDNA by the CaMV 35S promoter. Their findings revealed that phytochromes play a role in the photoperiodic control of FL in rice and the different roles of phytochromes in the photoperiodic control of FL were verified by comparing the *se5* with *hy1*, a complement mutant of *Arabidopsis*.

As phytochromes contributes to the photoperiodic control of FL and mediates the external light signal to repress *FT* orthologs in rice, a short-day plant. Izawa et al. (2018) studied the interaction between phytochrome signals and circadian clocks in photoperiodic-FL mutants of rice. By monitoring behaviors of circadian clocks, they noticed that under both short-day (SD) and long-day (LD) conditions, phase setting of circadian clocks is not affected in phytochrome-deficient mutant. Non-24-hr-light/dark-cycle experiments showed that an *Arabidopsis gene CONSTANS* (*CO*), termed as *PHOTOPERIOD SENSITIVITY 1* (*Heading date 1*) [*SE1* (*Hd1*)], which is a counterpart of rice gene acts as an output of circadian clocks. Further, they found that upon floral transition, phytochrome deficiency does not have an effect on the diurnal mRNA expression of *SE1*. The RT-PCR data revealed that phytochrome signals suppress *FT* orthologs mRNA expression, while *SE1* promote and represses *FT* orthologs mRNA expression under SD and LD, respectively. This SE1 transcriptional activity may be regulated post-transcriptionally and may be influenced by the Pfr phytochromes. They proposed a model to understand the mechanism of day length recognition in photoperiodically FL short-day plants (Figure 6.9).

FIGURE 6.9 Emasculation of spikelet: Selection of panicle.

Simultaneously, Hayama et al. (2007) investigated Pharbitis, a commonly used short day model species to study the expression of two putative orthologs of *FT* (Pn*FT1* and Pn*FT2*) which induce FL under short-day condition. They found that orthologs of *FT* (Pn*FT1* and Pn*FT2*) expressed only under short-day, and light flashes provided during the night decrease their expression and inhibit FL. Moreover, Pharbitis accessions differing in critical night-length responses express Pn*FT* at different times after dusk. This indicated that natural genetic variation affecting the clock which regulates Pn*FT* expression changes the FL response. Finally, they reported that circadian rhythm set by dusk regulates the FT homologs expression and the short-day photoperiodic FL response in Pharbitis and for measuring day length in *Arabidopsis* and rice different mechanisms were proposed.

Successively, Yano et al. (2001) studied the genetic control of FL time in rice. They mentioned that the FL time in plants is under the control of some endogenous genetic factors and environmental signals. One of the environmental signals is photoperiod. They stated that using Arabidopsis as a model plant, the genetic control mechanisms for the photoperiodic response of long-day plants (LDPs) FL were widely investigated, but its mechanisms in short-day plants (SDPs) is yet to be clarified. Current progress in genome analysis has set forth a new approach for assessing the genetic control of FL in rice. The resemblance of photoperiodic response of FL in rice with those of the *Arabidopsis* was confirmed in several studies (Figure 6.10).

FIGURE 6.10 Sequential steps in emasculation of spikelet.

Later, the main photoperiod sensitivity QTL in rice was described by Yano et al. (2000). Using map-based cloning strategy a lowering time Gene *CONSTANSA* and most important quantitative trait locus (QTL) regulating response to photoperiod, *Hd1 was recognized.* For *Hd1* a candidate genomic region of ~12 kb was detected through High-resolution mapping of 1,505 segregants. The additional analysis unveiled that the *Hd1* QTL resembles a homologous gene of *CONSTANSA* in Arabidopsis. The assessment of two *se1* mutants, HS66, and HS110 determined that the *photoperiod sensitivity 1 (Se1)* is allelic to the *Hd1* QTL. Then the function of the candidate gene was verified using genetic complementation analysis. The amount of *Hd1* mRNA exhibited variations with the length of the photoperiod and they proposed that under short-day conditions *Hd1 promotes heading, while it inhibits* heading under long-day conditions (Figure 6.11).

FIGURE 6.11 Rice plants at grain filling stage.

6.2 MINERAL NUTRITION

Muhammed et al. (1987) conducted two experiments to study the effect of Na/ Ca and Na/K ratios on the rice growth and mineral nutrition in saline culture solution. In previous experiments rice plants grown on coastal saline-sodic soil showed rolling and bleaching of young leaves, the causes of these symptoms were investigated in this experiment. The young leaves of rice cultivar KS282 showed similar symptoms in saline culture solution with Na/Ca ratios of 100 or greater. They noticed that Ca deficiency causes these symptoms and as Cu concentration was greater in saline shoots than in the control and it is not involved in rolling and bleaching of young leaves. In the shoot reduction in Na and Cl concentration was noted with the decrease in Na/K or Na/Ca ratio in the saline solution. Further, they found that Na/Ca and Na/K ratios of the growth medium significantly affects the shoot and root growth of rice.

In rice grain to find out the QTLs for mineral contents Garcia (2009) undertook a study. From a cross between an elite *indica* cultivar Teqing and the wild rice (*Oryza rufipogon*) 85 introgression lines (ILs) were obtained. Then Fe, Zn, Mn, Cu, Ca, Mg, P, and K contents of these lines were measured using inductively coupled argon plasma (ICAP) spectrometry. Significant variation was noticed for all characters and except for Fe with Cu, significant positive associations were noticed among the mineral elements. Through single point analysis, total 31 putative QTLs were identified for these eight mineral elements. For most of the QTLs (26 QTLs) favorable alleles were found to be contributed by wild rice (*O. rufipogon*). For these traits, 14 QTLs (45%) were located on chromosomes 1, 9, and 12. On chromosome 8, one major QTL effect for zinc content was observed near the simple sequence repeats (SSR) marker RM152. In the mapping population, these co-locations of QTLs for some mineral elements revealed the molecular level association among these traits. This information would assist in improving these traits of rice grain through marker aided selection (Figure 6.12).

FIGURE 6.12 Diversity in morphology of rice panicle.

In rice plants, the effects of cadmium and nickel accumulation on mineral nutrition and its possible interactions with abscisic and gibberellic acids (GA) were investigated by Rubio et al. (1994). Rice plants were grown in a medium containing Cd and Ni for 10 days and the plants accumulated high quantities of these heavy metals. A reduction in shoot and root length was seen with the Cd and Ni uptake, however, the buildup of dry matter was not affected. These metals caused reduction in K, Ca, and Mg contents of plants, especially in the shoots. This specifies that Cd and Ni affect both nutrient uptake and nutrient distribution in plants. They found that, though the reducing effects of the metals on nutrient uptake were not overwhelmed with the addition of ABA or gibberellic acid (GA_3) to the solution, these hormones led to a reduction of Ca content and heavy metal incorporation into the plants. Furthermore, the hormones application influenced the transport of Cd and Ni to the shoots, leading to a high metals accumulation in the roots (Figure 6.13).

FIGURE 6.13 Rice plants at grain maturity stage.

6.3 MINERAL DEFICIENCY AND FORTIFICATION

A study has been undertaken by Vasconcelos et al. (2003) on the ferritin gene role in increasing iron and zinc accumulation in transgenic rice. It was found that the soybean *ferritin* gene expression under the control of endosperm-specific glutelin promoter increases levels of iron and zinc in transgenic *indica* rice grains. They mentioned that brown rice is rarely consumed, and the removal of outer layers causes substantial loss of micronutrients from polished rice. In this experiment, they noticed that the expression of soybean *ferritin* gene regulated by the glutelin promoter in rice was effective in improving grain nutritional levels, in both brown and polished grains. In transgenic rice grains, along with iron levels, zinc concentration was also analyzed and the higher concentration of zinc was found in transgenic seeds than in the control. Further, Prussian blue staining reaction unveiled the existence of iron in the endosperm cells of transgenic rice grains.

The significance of biofortified crops in alleviating micronutrient malnutrition was discussed by Mayer et al. (2008). They mentioned to reduce micronutrient malnutrition with positive outcomes concerted international and national fortification and supplementation efforts have been assumed. For dietary complements reaching rural populations, the transmission of micronutrients via micronutrient-dense crops i.e., Biofortification is a cost-effective and sustainable method. If the adequate genetic variation exists for a given trait by means of breeding or transgenic approaches, the concentrations of bioavailable micronutrients in the edible parts of staple crops can be increased (Figure 6.14).

FIGURE 6.14 Rice crop at pre-harvest stage.

Mineral deficiency is one of the most widespread dietary problems in the World. Lee et al. (2009) reported that nicotianamine synthase (*NAS*) gene can enhance mineral contents in rice grains. To verify this they selected activation-tagged mutant lines of rice and by introducing 35S enhancer elements, the expression of *NAS* gene, *OsNAS3*, was increased in these mutant lines. The higher accumulation of Fe and Zn was observed in shoots and roots of the *OsNAS3* activation-tagged plants (*OsNAS3-D1*). Seeds of these *OsNAS3-D1* plants had higher quantities of Fe (2.9-fold),

Zn (2.2-fold), and Cu (1.7-fold). Size exclusion chromatography analysis together with inductively coupled plasma mass spectroscopy unveiled the equal amounts of Fe bound to IP6 in WT and *OsNAS3-D1* seeds, while *OsNAS3-D1* had 7-fold more Fe bound to a low molecular mass, NA. Additionally, this activation enhanced the tolerance to Fe and Zn deficiencies and to excess metal (Zn, Cu, and Ni) toxicities in plants. The bioavailability of Fe was confirmed by feeding the engineered or WT seeds to anemic mice for 4 weeks. Then the concentrations of hemoglobin and hematocrit in mice were measured. It was found that mice fed with engineered seeds regained normal hemoglobin and hematocrit levels within 2 weeks, while those fed with WT seeds stayed anemic. Thus, the research findings suggested that the fortification of rice seeds by activating *NAS* gene could enhance the bioavailable mineral content in grains of rice.

The main causes of poor health and disability in the World are vitamin and mineral deficiency. So as to combat with these deficiencies and their associated harmful health effects there is a need to fortify crops by developing advanced technologies (Muthayya et al., 2012). Rice being a staple food has the potential to fill the gap in recent fortification programs. There are prospects to fortify an important share of rice that comes from large mills contributing centralized markets and national welfare programs in major rice-growing countries. The strongest proof for attaining wide coverage and impact has been reported in countries that mandate rice fortification. A worldwide network of public and private partners, rice fortification resource group (RiFoRG), proposes technical and advocacy support for rice fortification has a dream of promoting rice fortification globally. It has targeted the main countries where the prospects of early adoption of rice fortification are greatest. The challenges are real, the necessity to address them is strong, and the prospects to deliver the potential of rice fortification are clear.

6.4 MINERAL TOXICITY

6.4.1 IRON TOXICITY

In flooded soils, large concentrations of reduced iron (Fe^{2+}) causes iron toxicity and mostly affect the productivity of lowland rice. The iron toxicity symptoms in rice and its management concepts were discussed by

Becker and Asch (2005). The excessive uptake of Fe^{2+} by the rice roots and its acropetal translocation into the leaves causes iron toxicity symptoms in rice. The excess concentration of Fe^{2+} produces high levels of toxic oxygen radicals which can damage cell structural components and affect physiological processes. The "bronzing" of the rice leaves related with significant yield shortfalls are the typical visual symptoms of iron toxicity.

The constraints of iron toxicity in rice are well recognized and the effects of Fe^{2+} on geochemistry, soil microbial processes, and the physiological processes within the plant or cell have been recorded in literature. Iron toxicity and Zn deficiency are the most commonly detected micronutrient disorders in wetland rice.

To decrease the Fe^{2+} concentration in the soil or to increase the rice plants' capacity to adjust with surplus iron in either soil or the plant several agronomic management strategies have been proposed. Many accessions and cultivars in the existing rice germplasm are reported as tolerant to iron toxicity.

The iron toxicity interrupts the rice plant physiology in several ways. In lowland rice production, interspecific hybridization between African and Asian rice is deliberated as a probable solution to iron toxicity. To analyze the effects of iron toxicity on the growth and mineral composition of interspecific rice, *Oryza sativa* L. x *Oryza glaberrima,* Dorlodot et al. (2005) conducted hydroponic experiments. They applied different concentrations of ferrous iron (0, 125, 250, and 500 mg L^{-1} Fe^{2+}) at two different plant ages of interspecific rice line. In a hydroculture, system by adjusting the pH, oxygen content, associated iron redox state, and iron availability of nutrient solution symptom development was achieved. After 4 weeks of iron-toxicity stress, the symptoms of iron toxicity viz. bronzing, and reduction in plant growth and survival rate were appeared on hybrid line at and above 250 mg L^{-1} of Fe^{2+}. The hybrid line did not show iron toxicity symptoms at 125 mg L^{-1} Fe^{2+}, in spite of having an iron concentration of 3,356 mg kg^{-1} in its leaves which is above the normal critical toxicity concentration in rice (700 mg kg^{-1}). Under the different concentrations of applied ferrous iron, the mineral elements in the tissue of hybrid line were categorized into three groups based on their concentration. They observed that except iron, the concentrations of all mineral elements were retained between their critical deficiency and toxicity limits. This property might be accountable for the tolerance of hybrid line to leaf iron concentrations which are generally toxic to rice. The findings of this analysis can be effectually utilized while screening the rice cultivars for ferrous iron toxicity resistance in further breeding programs.

6.4.2 BORON TOXICITY

The boron toxicity effects on crop yields are higher than the B deficient soils in different regions of the World, but specific studies are not available on boron toxicity in rice. Nable et al. (1997) made a review on boron toxicity in rice. The major sources of soil B are marine evaporites and marine argillaceous sediment. Various anthropogenic sources with excess B, further increases soil B concentration to toxic levels for plants. The information on genetic variation in the response of plants to high concentrations of B has been provided in recent physiological and genetic studies. This genetic variation was observed in many plant species. Furthermore, these studies have assisted the breeding of tolerant genotypes for cultivation on high B soils. They found that several major additive genes controls tolerance mechanism and the specific chromosomal locations for these genes were identified in some species. Concerted research activities have been directed and achieved success in developing B toxicity tolerant varieties.

6.4.3 CADMIUM TOXICITY

Wup et al. (2006) investigated the genotypic variation in the seed germination, growth, and physiological characters responses of rice seedlings to Cd toxicity. The findings unveiled that low Cd concentration (0.01–1.5 mM Cd) stimulates germination to some extent, but higher Cd concentration (2.0 mM) reduces germination severely. The increased plant height, root volume biomass, and chlorophyll concentration were noted in seedlings exposed to 0.01 mM Cd, while these parameters were reduced significantly at 0.5 mM Cd concentration in the medium. Further, superoxide dismutase (SOD) and peroxidase (POD) activities, and MDA (malondialdehyde) were reduced at 1 mM Cd when compared with those under 0.5 mM Cd. Addition of cadmium decreased Fe, Cu, and Mn concentrations in roots and shoots. The response of these parameters to Cd stress exhibited significant genotypic difference. Under Cd stress, the minimum inhibition of growth and increase in MDA content, higher shoot Cd concentration and greatest increase in POD and SOD activities were noted in Xiushui 110. This indicates Xiushui 110 is tolerant to Cd toxicity. Whereas, the highest decrease of growth, and Zn, Cu, Fe, and Mn contents, maximum increase in MDA content, and minimum increase in antioxidative enzymes activities were observed in Bing 9914 signifying its sensitivity to Cd toxicity.

6.4.4 NICKEL TOXICITY

The essential micronutrient nickel is a heavy metal recovered at low concentrations in most natural soils. In some areas, nickel concentration may increase to toxic levels, affecting the biochemical and physiological processes in plants. The typical symptoms of nickel toxicity are wilting and drying of leaves. The effects of Ni^{2+} toxicity on membrane functionality and plant water content in rice were studied by Llamas et al. (2008). The primary barrier for the entry of heavy metals is plasma membrane (PM) of root cells. In this study, the association between instabilities of membrane functionality and the occurrence of the typical symptoms of Ni^{2+} toxicity were determined. They grew rice plants in a nutrient medium with 0.5 mM Ni^{2+}. As a result of the stress, significant reduction in plant water content was observed. Addition of Ni^{2+} to the solution bathing the roots caused a concentration-dependent PM depolarization but the presence of Ni^{2+} did not show any effect on the activity of the PM-H^+-ATPase.

In the short time (hours), the Ni^{2+} treatments did not show any influence on membrane permeability of root cells. However, in the long term (days), Ni^{2+} was found to bring about a severe loss of K^+ in roots and shoots. This loss of K^+ thought to be concerned with water content fluctuations, because conductance of stomata and the transpiration rate were not influenced by Ni^{2+} treatment. They mentioned that the effects caused by Ni^{2+} were not permanent and could be reversed, by shifting the plants to a medium deprived of Ni.

6.4.5 ARSENIC TOXICITY

The arsenic (As) from tailings at abandoned lead-zinc mine polluted over 500 hectares of paddy soils in China. Xie and Huang (2008) described control of arsenic toxicity in rice plants grown on an arsenic-polluted paddy soil. Many field trials were performed to find control measures for as toxicity in rice plants. The results acquired were like this; Fresh Chinese milk vetch (*Astragalus sinicus* L.) should not be applied as green manure in arsenic polluted paddy soils. Though liming (1,500 kg CaO hectare^{-1}) reduces water-soluble As (H_2O-As) in the soil, it adversely affects the rice plant growth. The $FeCl_3$ (25 mg Fe kg^{-1} soil) and MnO_2 (25 mg Mn kg^{-1} soil) treatments could significantly decrease the percentage of H_2O-As and arsenite [As (III)] in the soil and leads to better rice growth. Merely wetting and drying without addition of any materials to the soil, could increase redox potential of soil and decrease the H_2O-As and As(III) percentage in the paddy soil.

6.4.6 SILICON

To study the responses of rice seedlings to exogenous silicon addition and its effects on chromium uptake, growth, mineral elements, oxidative stress, antioxidant capacity, and leaf and root structures in rice Kumar et al. (2012) conducted hydroponic experiments. Rice seedlings were subjected to hexavalent chromium by growing under 100 μM Cr (VI) stresses without or with 10 μM Si. Reduced growth, photosynthetic pigments and protein together with a significant rise in Cr accumulation and lipid peroxidation (as MDA) were found in chromium treatment. Though, Si addition reduced the toxicity of Cr and improved growth of rice by declining Cr accumulation, root-to-shoot Cr transport, and MDA level. The Cr treatment decreased the antioxidant capacity, total phenolic contents macro (Mg, Ca, and K) and micronutrients (Zn and Fe) contents except Mn, whereas these reductions were prevented by the addition of Si. Further, Cr treatment decreased the leaf epidermal cells length and stomatal frequency, and unfavorably influenced chloroplasts containing mesophyll cells and xylem and phloem integrity. These abnormalities were reduced by Si addition. Thus, this study indicated that the silicon addition alleviates the harmful effects of chromium and improves Cr(VI) tolerance in rice seedlings.

6.5 MANAGEMENT OF MINERALS

6.5.1 SILICON

Rice is a typical silicon accumulating plant. For increase and sustainable yields in rice, Si nutrition and its management should be considered in research programs. In this context, Savant et al. (1996) discussed silicon management and sustainable rice production. Rice plant absorbs silicon as monosilicic acid [H_4SiO_4, or Si $(OH)^{\cdot}_4$] and then deposited as amorphous SiO. In the tops (leaves and hulls) of rice plant silicon is primarily related with cellulose and hemicellulose where it forms a cuticle-Si double layer. Soil Si has low solubility and its dissolution kinetics influences the Si in soil solution (intensity factor). Further, the soil factors like Al, Fe oxides, organic matter, redox potential, and moisture influences the Si in soil. In this study, to meet a critical need for Si and to improve rice yields on a sustained basis few future research strategies were suggested.

6.5.2 NITROGEN

The influences of various rice straw management practices and winter flooding on yield, N uptake, and N use efficiency of rice were studied by Alison et al. (2000) in California. The experiment was planned by allocating winter flooding and no winter flooding treatments to main plants and four straw management practices viz. straw burned, incorporated, rolled, and baled/removed as subplot treatments. In each plot, zero N fertilizer microplots were established yearly. Alternative straw management or winter flooding did not influence the grain yield (GY) at presently recommended N fertilization. In straw incorporated and rolled plots the GY was greater at zero N fertilization. At zero and recommended rate of N fertilizer application, the N uptake by rice was enhanced by an average of 19 kg N ha^{-1} and 12 kg N ha^{-1}, respectively when straw was retained. The winter flooding in straw retained plots further increased the crop N uptake. Thus, the straw increases N uptake by crops without increasing GY, which reduces N use efficiency and demands the re-evaluation of N fertilizer application rates.

Kundu and Ladha (1995) discussed the efficiency of soil and biologically fixed N_2 in intensively-cultivated rice fields. Since the early 1980s in some intensively-cultivated experimental farms of Asia, reduction in wetland rice productivity was observed. To uphold the original yield levels, high doses of N fertilizer were applied in both experimental and farmers' fields. The primary N source for rice is native soil N and biological N_2 fixation (BNF), which largely replenishes the N concentration in soil. But much importance is not given to judicious management of soil N. Here, different effects of long-term flooding and puddling related with intensive cultivation of wetland rice on soil N availability and BNF were reviewed. Further, some approaches for the efficient management of these two sources of N, so as to uphold high productivity of the wetland rice were proposed.

6.5.3 ORGANIC ACID

Root system plays a key role in uptake of water and nutrients from soil. Yang et al. (2004) investigated the organic manure effect on root growth and nutrient uptake of rice in continuously and alternately flooded paddy soils. They applied three nutrient systems: (i) combined application of chemical fertilizers with farmyard manure (CM), (ii) combined use of chemical fertilizers and wheat straw (CS), (iii) only chemical fertilizers (CK). Improved root morphological features were observed with the addition of organic manures into paddy soil. Further, enhanced root length density (RLD), and root weight density (RWD) by 30 and 40%, respectively were noted for organic fertilization treatments. In continuous waterlogging (CWL) water regime, combined application of organic and inorganic fertilizers showed harmful effects on root active absorption area (AAA), root oxidation ability (ROA) of alpha-naphthylamine (α-NA) (ROA), and root surface phosphatase (RSP) of rice plants. Whereas, in alternate wetting and drying (AWD) water regime, the uptake of N, P, and K by rice plants was enhanced significantly with the organic manure application. This assisted in the distribution and transfer of nutrient elements, particularly P to rice ears and grains leading to increased number of filled grains per panicle, 1000-grain weight, and GY. Thus, the study demonstrated that the CWL water regime significantly decreases the beneficial effects of organic and mineral fertilizers combined use on rice GY.

6.6 RESEARCH ADVANCES IN INCREASING RICE PRODUCTIVITY

In Madagascar, several issues were raised on SRI. Stoop et al. (2002) reviewed agricultural research problems and prospects for improving farming systems. The plant physiological and bio-ecological factors related with agronomic practices were studied. This could explain how the thoughtful management and interactions of the major crop production factors: time, space, water, plant nutrients and labor results in extraordinary yields. The findings emphasize the significance of integrated and interdisciplinary research, relating strategic and adaptive (on-farm participatory) methods that investigate and link bio-physical and socio-economic factors in crop production. Using such approaches, presently unexploited production potentials of rice and other major cereal grain crops can be unraveled without added costs to farmers or to the environment.

The green revolution (GR) role in increasing crop productivity was discussed by Gurdev (1999). He stated that in the 1960s there were large-scale concerns about the World's ability to feed itself. But major increases in food-grain production were realized with the extensive implementation of "GR" technology. Between 1966 and 1990, the population density of low-income countries was increased by 80% and food production was improved to more than double. In the last 30 years, significant achievements were made in World food production with the advancements of technologies. One of such achievement is the development of high-yielding wheat and rice varieties. These varieties are fertilizer responsive, lodging resistant, and their yield potential is 2–3 times that of varieties available prior to the GR. Further, these varieties exhibit multiple resistances to diseases and insects. Apart from this, irrigation facilities development, the accessibility of inorganic fertilizers, and good government policies have aided for the adoption of green-revolution technology. But in the 1990s, the population growth rate was much higher than the rate of growth in food-grain production. This will cause serious food shortages in the next century. To feed 5.0 billion rice consumers in 2030, the new possible strategies have to be formulated and together with the development of high yielding varieties, strategies for integrated nutrient management, integrated pest management (IPM), and efficient utilization of water and soil resources should be developed.

In tropical lowland rice, Kronzucker et al. (2000) made comparative kinetic analysis of ammonium and nitrate acquirement and its inferences for rice cultivation and yield. In cereals, the key factor affecting yield potential is nitrogen limitation. In rice, nitrogen is utilized only in ammonium form and the nitrate form is highly ignored. In this analysis, in a major variety of *Indica* rice (*Oryza sativa*) direct comparisons of root transmembrane fluxes and cytoplasmic pool sizes for nitrate and ammonium N were made using short-lived radiotracer ^{13}N. The findings revealed that nitrate acquisition has high capacity and more efficient than that of ammonium. Their research findings would assist in designing the water and fertilization management regimes, further new approaches to exploit the unexplored N-utilization potential were proposed.

In China, the accomplishments and prospects of super hybrid rice breeding were discussed by Cheng et al. (2007a). They stated that hybrid rice has contributed significantly to the self-sufficiency of food supply in China. In China, a national program on super rice breeding was started to meet the forthcoming rice production demand. NPT models adjustable to different rice growing regions were developed. After recognizing the significance of using parents with intermediary subspecies differentiation in improving F_1 yield, using DNA markers for subspecies differentiation intermediate type parental lines were chosen from populations obtained from inter-subspecies crosses. Thirty-four super hybrid rice varieties have been released for commercial cultivation, producing 6.7 million kg more rice in 1998–2005. Though outstanding advancement has been attained in super hybrid rice breeding in China, the addition of new biotechnological tools propose a great challenge.

Hybrid rice breeding strains are successfully produced in China and now the country is making efforts to develop new hybrids with high-yielding potential, superior grain quality, and tolerance to biotic and abiotic stresses. Cheng et al. (2007b) made a review on significant advancements in hybrid rice breeding in China, and conducted a study on fine-mapping QTLs for yield traits. In China, new hybrids with super high yielding potential are produced by applying inter-subspecies crosses of *indica*-inclined and *japonica*-inclined parental lines, which broadened the genetic diversity of hybrid rice. Later, to develop restorer lines having disease resistance genes marker-assisted selection (MAS) was conducted. Even though in highly heterotic hybrids the genetic basis of heterosis has been successfully produced, data from these studies are not adequate to come to certain deductions. In a fine-mapping QTLs study with stepwise residual heterozygous lines, two linked intervals harboring QTLs for yield traits were determined, one of which was positioned to a 125-kb region. In conclusion, they proposed molecular MAS as a competent tool for future breeding programs, and for extending this technique from major genes to QTLs much work remains to be done.

Zeiglerm and Barclay (2008) discussed the importance of rice. Research on rice is needed for the advancement of rice productivity-improving technologies and to benefit the farmers depending on rice for their livelihood. The profits of such improved productivity will flow through rice-growing countries' landless rural and urban poor, all of whom (1) are key consumers of rice, and (2) expend a large part of their income on rice. The investment in high-quality research for the intensive irrigated rice-based systems and the rainfed rice-based systems is needed, because of the current abrupt increases in rice price.

In Asia, rice is commonly grown by transplanting seedlings into puddled soil (wet tillage). This method of rice production is laborious-, water-, and energy-intensive and is less profitable for the dearth of these resources. Additionally, it decreases the soil physical properties, unfavorably affects the performance of following upland crops, and increases methane emission. Therefore, these considerations necessitate a major shift from puddled transplanting to DSR in irrigated rice ecosystems. Kumar and Ladha (2011) reported rice direct seeding as current development and forthcoming research needs which is extensively implemented in some countries and is extending to other Asian countries. However, to deal with increasing water and labor scarcity, and to improve system sustainability combining dry seeding (Dry-DSR) with zero/reduced tillage (e.g., conservation agriculture (CA)) is gaining importance. In previous studies, the benefits of direct seeding were compared with puddled transplanting: (1) comparable yields; (2) savings in irrigation water, labor, and production costs; (3) higher net economic returns; and (4) a decrease in methane emissions. In spite of, these benefits, dry seeding has variable yields in some regions, particularly when combined with reduced/zero tillage because of (1) irregular and poor crop stand, (2) poor weed control, (3) higher spikelet sterility, (4) crop lodging, and (5) poor knowledge of water and nutrient management. Moreover, rice varieties chiefly selected and bred for puddled transplanted rice are at present used for DSR. Threats related with a change from puddled transplanting to DSR are (1) a shift toward hard-to-control weed flora, (2) herbicide resistance development in weeds, (3) weedy rice evolution,

(4) enhances in soil-borne pathogens such as nematodes, (5) higher emissions of nitrous oxide—a potent greenhouse gas, and (6) nutrient disorders. Therefore, this study proposed an integrated package of technologies for Dry-DSR, together with the recognition of rice traits related with the achievement of optimal GY with Dry-DSR.

Recent studies on super hybrid rice were summarized by Zou et al. (2010). Under diverse ecological condition of South China, the China super hybrid rice can attain 12 t ha^{-1} of yield in one year. The super hybrid rice has growth duration of 130–150 d, spikelet number 5.5–6.0 ×104/m^2, LAI of 9–10 and plant height 100–110 cm. Further, it is characterized with stronger tillering ability, showing the probable yield of sufficient panicle number and large panicles, coordination between sink and source, fertilizer tolerant and lodging resistance. From the systematic studies on growth and development of super hybrid rice, its dry matter production, tillering ability, root activity, most favorable locations and seasons, high yielding path, the physiological basis of high yielding formation, cultivation environments and agronomy practices under the cultivation environments of single seedling and sparse planting and increase in nitrogen fertilizer is identified. Additional studies on the coordination between adequate panicle number, large panicles and their grain filling percentage from the enzyme, hormone, and molecular level prospective, along with the SRI intensive cultivation techniques optimization are required for the exploitation of high yielding potential of super hybrid rice.

Zhang et al. (2009) considered the effects of different application rates of nitrogen on hybrid rice by comparing Liangyoupeijiu, with Shanyou 63 under field conditions. The findings unveiled the increment in the N-uptake amount, biomass accumulation and the N-requirement per 100 kg grain with the N-application rate. Initially, with N-application, rate the yield and N-recovery increased, reached to highest level at moderate N level, and then declined considerably at high N level. At the same level of N, the variety Liangyoupeijiu had higher yield, N-uptake amount, biomass accumulation, and N-recovery than Shanyou 63. The N-requirement per 100 kg grain is around 2.0–2.2 kg and the N-recovery is about 40% at the yield level from 9,000 to 10,500 kg·ha^{-1}. With the N-application rate, the nutrition quality, milling quality, steaming, and taste quality become better, whereas the apparent features become poorer. Among varieties yield and quality, differences were significant.

In comparison to the variety Shanyou 63, the variety Liangyoupeijiu was superior in terms of yield and quality.

With the combined use of ideotype method and inter subspecific heterosis China's "super" hybrid rice breeding projects has developed many novel varieties. Whether these "super" hybrid varieties have improved the productivity of irrigated rice or not is still debatable. In Liuyang (moderate-yielding site) and Guidong (high-yielding site) countries, China, a field experiment was undertaken by Zhang et al. (2009) to compare GY and yield characteristics among "super" hybrid, ordinary hybrid, and inbred varieties. From each varietal group, two varieties were grown by applying moderate and high N rates. For each variety GY, yield components, aboveground total dry weight, HI, total N uptake, and crop RUE were evaluated. The results showed the significant yield difference between the varieties and varietal groups. The yield potential of "Super" hybrid varieties was 12% more than the ordinary hybrid and inbred varieties, which is mainly due the improvement of both source and sink. Further, the long growth duration and high accumulation of incident radiation contributed to the higher biomass of the "super" hybrid varieties. Crop RUE did not clarify the yield advantage of "super" hybrid rice. The researcher opined that more N fertilizer is not required for the production of "super" hybrid rice varieties with increased GY.

Wu (2009) discussed the potentials of developing hybrid rice with super high yield. Attempts made for the development of hybrid rice with super high yield involve improvement of plant type, intersubspecific heterosis exploitation, the pyramiding of heterosis genes in diverse rice ecotypes, and the utilization of promising genes from distant relatives. To improve the hybrid rice yield, three key plant types have been suggested in the literature: (i) heavy panicle type, (ii) super high yielding plant type, and (iii) super high yielding ideotype. In recent times, inter subspecific heterosis has been partly utilized using three approaches: (i) development of *indica-japonica* intermediate type parental lines using promising genes from both parents, (ii) introgressing inter subspecific gene using wide compatibility and thermosensitive genic male sterility (TGMS) genes, and (iii) breeding pro-*indica japonica* lines. By means of molecular marker-assisted reciprocal recurrent selection heterosis genes from different rice ecotypes can be pyramided into a single cultivar. Utilizing molecular biotechnology detection of yield improving genes from wild rice, the construction of autoregulated senescence delaying gene, and the cloning of main enzymes associated to C_4 pathway have been achieved. Thus, the heterosis of the hybrid rice can be significantly enhanced through the introduction of these genes into the parents of the hybrid rice.

Sundaram et al. (2008) undertook a study to find informative SSR markers which can differentiate hybrid rice parental lines and can be applied in seed purity assessment. Ten CMS lines and ten restorer (R) lines along with 10 popular Indian rice varieties were characterized using a set of 48 hyper polymorphic SSRs. All the SSR markers were polymorphic and bring about a total of 163 alleles, with an average of 3.36 ± 1.3 allelic variants per locus. Using multiplex PCR, twenty-seven SSR markers that were specific and unique to a particular parental line were identified. With the help of these SSR markers, all the public bred Indian rice hybrids with their parental lines could be clearly differentiated. Further, two dimensional bulked DNA sampling approach containing a 20×20 grow-out matrix was designed to use these markers for detection of impurities in parental lines effectively. In addition to this, using 2–3 markers for the assessments of hybrid seed purity a multiplex PCR strategy with single tube analysis was designed and its advantage over single marker analysis in revealing of impurities in hybrids was demonstrated. The consequences of parental and hybrid specific SSR markers and approaches to use these informative SSR markers for recognition of contaminants in a cost-effective manner were discussed.

6.7 CONCLUSION

The productivity of a crop is an outcome of various physiological and biochemical functions, mineral nutrition, biotic, abiotic factors, apart from genetic and agronomic practices employed. This chapter explains the various physiological functions contributing to rice productivity. Different factors affect different stages of crop growth starting from germination, seedling establishment, vegetative growth, FL, and yield. Besides, it discusses various research advances made in increasing rice productivity.

KEYWORDS

- **aboveground biomass**
- **abscisic acid**
- **crop growth rate**
- **germination**
- **mineral nutrition**
- **rice physiology**
- **root length density**

REFERENCES

Alison, J. E., Jeffrey, A., William, R. H., Bruce, A. L., Sylvie, M. B., James, E. H., & Chris, V. K., (2000). Rice yield and nitrogen utilization efficiency under alternative straw management practices. *Agronomy Journal, 92*(6), 1096–1103.

Anuradha, S., & Seeta, S. R. R., (2001). Effect of brassinosteroids on salinity stress induced inhibition of seed germination and seedling growth of rice (*Oryza sativa* L.). *Plant Growth Regulation, 33*, 151–153.

Aulakh, M. S., Wassmann, R., Bueno, C., Kreuzwieser, J., & Rennenberg, H., (2001). Characterization of root exudates at different growth stages of ten rice (*Oryza sativa* L.) cultivars. *Plant Biology, 3*, 139–148.

Becker, M., & Asch, F., (2005). Iron toxicity in rice—conditions and management concepts. *Journal of Plant Nutrition, 168*, 558–573.

Biswas, J.K., & Yamauchi, M., (1997). Mechanism of seedling establishment of direct-seeded rice (*Oryza sativa* L.) under lowland conditions. *Botanical Bulletin- Academia Sinica, 38*, 29–32.

Cheng, S. H., Cao, L. Y., & Zhuang, J. Z., (2007a). Super hybrid rice breeding in China: Achievements and prospects. *Journal of Integrative Plant Biology, 49*, 805–810.

Cheng, S. H., Zhuang, J. Y., Fan, Y. Y., Du, J. H., & Cao, L.Y., (2007b). Progress in research and development on hybrid rice: A super-domesticate in China. *Annals of Botany, 100*, 959–966.

Counce, P. A., Keisling, T. C., & Mitchell, A. J., (2000). A uniform, objective, and adaptive system for expressing rice development. *Crop Science, 40*, 436–443.

Dorlodot, S., Lutts, S., & Bertin, P., (2005). Effects of ferrous iron toxicity on the growth and mineral composition of an interspecific rice. *Journal of Plant Nutrition, 28*, 1–20.

Ekanayake, I. J., Datta, S. K., & Steponkus, P. L., (1989). Spikelet sterility and flowering response of rice to water stress at anthesis. *Annals of Botany, 63*, 257–264.

Farooq, M., Barsa, S. M. A., & Wahid, A., (2006). Priming of field-sown rice seed enhances germination, seedling establishment, algometry, and yield. *Plant Growth Regulation, 49*, 285–294.

Garcia, O. A. L., (2009). Genetic identification of quantitative trait loci for contents of mineral nutrients in rice grain. *Journal of Integrative Biology*, *51*, 84–92.
Gnyp, M. L., Miao, Y., Yuan, F., Ustin, S. L., Yu, K., Yao, Y., Shanyu, H., & Bareth, G., (2014). Hyperspectral canopy sensing of paddy rice aboveground biomass at different growth stages. *Field Crops Research*, *115*, 42–55.
Gurdev, S. K., (1999). Green revolution: preparing for the 21st century. *Genome*, *42*, 646–655.
Hakim, M. A., Juraimi, A. S., Begum, M., Hanafi, M. M., Ismail, M. R., & Selamat, A., (2010). Effect of salt stress on germination and early seedling growth of rice (*Oryza sativa* L.). *African Journal of Biotechnology*, *9*, 1911–1918.
Hayama, R., Agashe, B., Luley, E., King, R., & Coupland, G., (2007). A circadian rhythm set by dusk determines the expression of homolog's and the short-day photoperiodic flowering response in pharbitis. *The Plant Cell*, *19*, 2988–3000.
Hu, J., Zhu, Z. Y., Song, W. J., Wang, J. C., & Hu, W. M., (2005). Effects of sand priming on germination and field performance in direct-sown rice (*Oryza sativa* L.). *Seed Science and Technology*, *33*, 243–248.
Ismail, A. M., Ella, E. S., Vergara, G. V., & Mackill, D. J., (2009). Mechanisms associated with tolerance to flooding during germination and early seedling growth in rice (*Oryza sativa*). *Annals of Botany*, *103*, 197–209.
Izawa, T., Oikawa, T., Sugiyama, N., Tanisaka, T., Yano, M., & Shimamoto, K., (2018). Phytochrome mediates the external light signal to repress *FT* orthologs in photoperiodic flowering of rice. Genes repress *FT* orthologs in photoperiodic flowering of rice. *Gene and Developments*, *32*, 17–18.
Izawa, T., Oikawa, T., Tokutomi, S., Okuno, K., & Shimamoto, K., (2000). Phytochromes confer the photoperiodic control of flowering in rice (a short-day plant). *The Plant Journal*, *22*, 391–399.
Khush, G. S., (2005). What it will take to feed 5.0 billion rice consumers in (2030). *Plant Molecular Biology*, *59*, 1–6.
Kim, H., Hwang, H., Hong, J. W., Lee, Y. N., Ahn, P., Yoon, I. S., Yoo, S. D., Lee, S., Lee, S. C., & Kim, B. G., (2012). A rice orthologue of the ABA receptor, OsPYL/RCAR5, is a positive regulator of the ABA signal transduction pathway in seed germination and early seedling growth. *Journal of Experimental Botany*, *63*, 1013–1024.
Kronzucker, H. J., Glass, A. D. M., Siddiqi, M. Y., & Kirk, G. J. D., (2000). Comparative kinetic analysis of ammonium and nitrate acquisition by tropical lowland rice: Implications for rice cultivation and yield potential. *New Phytologist*, *145*, 471–476.
Kumar, D. T., Pratap, V. S., Kumar, D., & Kumar, D. C., (2012). Impact of exogenous silicon addition on chromium uptake, growth, mineral elements, oxidative stress, antioxidant capacity, and leaf and root structures in rice seedlings exposed to hexavalent chromium. *Acta Physiologiae Plantarum*, *34*, 279–289.
Kumar, V., & Ladha, J. K., (2011). Chapter six-direct seeding of rice: Recent developments and future research needs. *Advances in Agronomy*, *111*, 297–413.
Kundu, D. K., & Ladha, J. K., (1995). Efficient management of soil and biologically fixed N_2 in intensively-cultivated rice fields. *Soil Biology and Biochemistry*, *27*, 431–439.
Lancashire, P. D., Bleiholder, H., Boom, T., Langeluddeke, P., Stauss, R., Weber, E., & Witzenberger A., (1991). A uniform decimal code for growth stages of crops and weeds. *Annals of Applied Biology*, *119*, 561–601.

Lee, S. S., Kim, J. H., Hong, S. B., Yun, S. H., & Park, E. H., (1998). Priming effect of rice seeds on seedling establishment under adverse soil conditions. *Korean Journal of Crop Science*, *43*, 194–198.

Lee, S., Jeon, U. S., Lee, S. J., Kim, Y. K., Persson, D. P., Husted, S., et al., (2009). Iron fortification of rice seeds through activation of the nicotianamine synthase gene. *PNAS*, *106*, 22014–22019.

Llamas, A., Ullrich, C. I., & Sanz, A., (2008). Ni^{2+} toxicity in rice: Effect on membrane functionality and plant water content. *Plant Physiology and Biochemistry*, *46*, 905–910.

Lutts, S., Kinet, J. M., & Bouharmont, J., (1995). Changes in plant response to NaCl during development of rice (*Oryza sativa* L.) varieties differing in salinity resistance. *Journal of Experimental Botany*, *46*, 1843–1852.

Mayer, J. E., Wolfgang, H. P. F., & Beyer, P., (2008). Biofortified crops to alleviate micronutrient malnutrition. *Current Opinion in Plant Biology*, *11*, 166–170.

Mia, M. A. B., Shamsuddin, Z. H., & Mahmood, M., (2012). Effects of rhizobia and plant growth promoting bacteria inoculation on germination and seedling vigor of lowland rice. *African Journal of Biotechnology*, *11*, 3758–3765.

Mishra, A., & Choudhuri, M. A., (1999). Monitoring of phytotoxicity of lead and mercury from germination and early seedling growth indices in two rice cultivars. *Water, Air, and Soil Pollution, 114*, 339–346.

Muhammed, S., Akbar, M., & Neue, H. U., (1987). Effect of Na/Ca and Na/K ratios in saline culture solution on the growth and mineral nutrition of rice (*Oryza sativa* L.). *Plant and Soil, 104*, 57–62.

Muthayya, S., Hall, J., & Bagriansky, J., (2012). Rice fortification: An emerging opportunity to contribute to the elimination of vitamin and mineral deficiency worldwide. *Food and Nutrition Bulletin*, 33, 296–307.

Nable, R. O., Bañuelos, G. S., & Paull, J. G., (1997). Boron toxicity. *Plant and Soil*, *193*, 181–198.

Prakash, L., & Prathapasenan, G., (1988). Putrescine reduces NaCl-induced inhibition of germination and early seedling growth of rice (*Oryza sativa* L.). *Australian Journal of Plant Physiology*, *15*(6), 761–767.

Rubio, M. I., Escrig, I., Martínez, C. C., López, F. J. B., & Sanz, A., (1994). Cadmium and nickel accumulation in rice plants. Effects on mineral nutrition and possible interactions of abscisic and gibberellic acids. *Plant Growth Regulation*, *14*, 151–157.

Savant, N. K., Snyder, G. H., & Datnoff, L. E., (1996). Silicon management and sustainable rice production. *Advances in Agronomy*, *58*, 151–199.

Scofield, G. N., Aoki, N., Hirose, T., Takano, M., Jenkins, C. L. D., & Furbank, R. T., (2007). The role of the sucrose transporter, OsSUT1, in germination and early seedling growth and development of rice plants. *Journal of Experimental Botany*, *58*, 483–495.

Settle, W. H., Ariawan, H., Astuti, E. T., Cahyana, W., Hakim, A. L., Hindayana, D., Lestari, A. S., & Pajarningsih., (1996). Managing tropical rice pest through conservation of generalist natural enemies and alternative pray. *Ecology*, *77*, 1975–1988.

Shi, C. H., Wu, J. G., Lou, X. B., Zhu, J., & Wu, P. (2002). Genetic analysis of transparency and chalkiness area at different filling stages of rice (*Oryza sativa* L.). *Field Crops Research*, *76(1)*, 1–9.

Shimono, H., Hasegawa, T., & Iwama, K., (2002). Response of growth and grain yield in paddy rice to cool water at different growth stages. *Field Crops Research, 73*, 67–79.

Stoop, W. A., Uphoff, N., & Kassam, A., (2002). A review of agricultural research issues raised by the system of rice intensification (SRI) from Madagascar: Opportunities for improving farming systems for resource-poor farmers. *Agricultural Systems*, *71*, 249–274.

Sundaram, R. M., Naveenkumar, B., & Biradar, S. K., (2008). Identification of informative SSR markers capable of distinguishing hybrid rice parental lines and their utilization in seed purity assessment. *Euphytica*, *163*, 215–224.

Tao, H., Brueck, H., Dittert, K., Kreye, C., Lin, S., & Sattelmacher, B., (2006). Growth and yield formation of rice (*Oryza sativa* L.) in the water-saving ground cover rice production system (GCRPS). *Field Crops Research*, *95*, 1–12.

Thakur, A. K., Uphoff, N., & Antonyan, E., (2010). Assessment of physiological effects of system of rice intensification (SRI) practices compared with recommended rice cultivation practices in India. *Experimental Agriculture*, *46*, 77–98.

Vasconcelos, M., Datta, K., Oliva, N., Khalekuzzaman, M., Torrizo, L., Krishnan, S., Oliveira, M., Goto, F., & Datta, K. S., (2003). Enhanced iron and zinc accumulation in transgenic rice with the ferritin gene. *Plant Science*, *164*, 371–378.

Vijayakumar, M., Ramesh, S., Prabhakaran, N. K., Subbian, P., & Chandrasekaran, B., (2006). Influence of system of rice intensification (SRI) practices on growth characters, days to flowering, growth analysis and labor productivity of rice. *Asian Journal of Plant Sciences*, *5*, 984–989.

Wang, C., Jiang, L., Zhai, H., Wan, Z. C., Xiao, Y., & Jian, M., (2005). Mapping QTL for heat-tolerance at grain filling stage in rice. *Rice Science*, *12*, 33–38.

Wu, X., (2009). Prospects of developing hybrid rice with super high yield. *Agronomy Journal*, *101*, 688–695.

Wup, F., Dong, J., Jia, G., Zheng, S., & Zhang, G., (2006). Genotypic difference in the responses of seedling growth and Cd toxicity in rice (*Oryza sativa* L.). *Agricultural Sciences in China*, *5*, 68–76.

Xie, L., Tan, Z., & Zhou, Y., (2014). Identification and fine mapping of quantitative trait loci for seed vigor in germination and seedling establishment in rice. *Journal of Integrative Plant Biology*, *56*, 749–759.

Xie, Z. M., & Huang, C. Y., (2008). Control of arsenic toxicity in rice plants grown on an arsenic-polluted paddy soil. *Communications in Soil Science and Plant Analysis*, *29*, 2471–2477.

Yamauchi, M., & Winn, T., (1996). Rice seed vigor and seedling establishment in anaerobic soil. *Crop Science*, *36*, 680–686.

Yang, C., Yang, L., Yang, Y., & Zhu, O., (2004). Rice root growth and nutrient uptake as influenced by organic manure in continuously and alternately flooded paddy soils. *Agricultural Water Management*, *70*, 67–81.

Yano, M., Katayose, Y., Ashikari, M., Yamanouchi, U., Monna, L., Fuse, T., et al., (2000). Hd, a major photoperiod sensitivity quantitative trait locus in rice, is closely related to the *Arabidopsis* flowering time gene constants. *The Plant Cell*, *12*, 2473–2484.

Yano, M., Kojima, S., Takahashi, Y., Lin, H., & Sasaki, T., (2001). Genetic control of flowering time in rice, a short-day plant. *Plant Physiology*, *127*, 1425–1429.

Zeiglerm, R. S., & Barclay, A., (2008). The relevance of rice. *Rice*, *1*, 3–10.

Zeng, L., & Shannon, M. C., (2000). Salinity effects on seedling growth and yield components of rice. *Crop Science*, *40*, 996–1003.

Zhang, H. C., Wang, X. Q., Dai, Q. G., Huo, Z. Y., & Xu, K., (2009). Effects of N-application rate on yield, quality, and characters of nitrogen uptake of hybrid rice variety Liangyoupeijiu. Agricultural College of Yangzhou University, Yangzhou, China, www.cnk.com.cn (accessed on 29 January 2020).

Zhang, Y., Tang, Q., Zou, Y., Li, D., Qin, J., Yang, S., Chen, L., Xia, B., & Peng S., (2009). Yield potential and radiation use efficiency of "super" hybrid rice grown under subtropical conditions. *Field Crops Research, 114*, 91–98.

Zhu, Q. S., Cao, X. Z., & Guo, Z. F., (2005). Studies on the percentage of grains of hybrid rice. *Rice Science, 12*, 33–38.

Zou, Y. B., Zhou, S. Y., & Tang, Q. Y., (2010). Status and outlook of high yielding cultivation researches on china super hybrid rice. *Review of China Agricultural Science and Technology, 29,* 78–84.

CHAPTER 7

Research Advances in Abiotic Stress Management

ABSTRACT

Many abiotic stresses such as drought, high, and low temperature, salinity, flooding, etc., exert their effect on the growth and productivity of rice. Several concerted research studies were directed in studying their effects, developed techniques, and enabled in understanding physiological, biochemical, and molecular mechanisms to some of these stress factors. The research advances that occurred are briefly reviewed in the following sections.

7.1 DROUGHT EFFECT

Drought reduces the yield of rice grown under rainfed conditions in Asia, Africa, and Latin America. In the context of current and predicted water scarcity, increasing irrigation is generally not a viable option for alleviating drought problems in rainfed rice-growing systems. Extensive genetic variation for drought resistance exists in rice germplasm. However, the current challenge is to decipher the complexities of drought resistance in rice and exploit all available genetic resources to produce rice varieties combining drought adaptation with high yield potential, quality, and resistance to biotic stresses. The research advances made in this area has been presented here.

7.1.1 TECHNIQUES AND GENOTYPIC VARIABILITY

A variety of techniques are used to monitor rice for drought resistance and screen tolerant varieties. A few techniques are mentioned. In most of the breeding programs that are involved in the development of rice tolerant varieties, generally adopt the technique of seedling leaf death score for assessment of the genotypes, as a selection index of drought resistance. Mitchell et al. (1998) used this technique for investigating the factors that are affecting the drought score of rice cultivars. They have also established an association between the yield and drought score in rice cultivars which were exposed to drought stress at the vegetative stage. In two different field experiments that were conducted, they found that the genotypes exhibited large variations in plant size. Further, it was observed that those genotypes which exhibited a good plant size also exhibited a better drought score of the highest value. There existed a noteworthy positive association between drought score and light interception. However, covariance analysis has shown that there exist genotypic differences in drought score on plant size. Similarly, a positive and significant genotypic variation was seen in leaf water potential and this has shown a negative association with drought score. Similarly, it was found that there was no decline in the dry matter either before or after the stress period in those cultivars which possessed large leaf areas and tended to lose more number of green leaves at the time of the stress period. The experimental findings have shown that the amount of green leaf area that was retained after a drought stress period is more influential on the dry matter accumulation and growth rather than the dead leaf or the drought score. No correlation occurred between the grain yield (GY) and the drought score of the seedlings. Therefore, they suggested that estimates of plant size are to be considered for usage in a vegetative screening program for drought resistance and in the assessments of the drought score (Figure 7.1).

Water deficits cause significant yield declines. More than one-half of the World's rice land area is to a large extent under rainfed conditions. These areas are prone to severe water shortages that take place during the reproductive stage.

Garrity and O′Toole (1994) undertook the selection of rice for drought resistance at the reproductive phase, in a three-year field study and developed a method for screening the rice varieties for drought resistance when drought stress takes place at the reproductive-phase. It is mentioned that an effectual method to attain synchronization of flowering (FL) during the stress treatment in the rice entries is the adoption of the method of staggered planting date. Their results revealed a negative association with the number of days beyond the commencement of stress period when a cultivar of rice began FL and the GY, relative GY, and spikelet fertility

(SF). They have proposed a mass screening method. They have developed a drought index, based on the variation of entry performance from the regression of the plant character on FL date (stated as days from the use of the drought stress treatment) to regulate cultivar performance. Then rankings were given alike for the index values of the GY, relative GY, and SF. There was a positive significant association between percent SF and GY, when there was water stress during the reproductive-stage. They proposed that a practical and important character in rice for scoring its performance is the SF. The suggested mass screening method includes staggered planting, interruption of irrigation during the FL period, visual scoring for SF, and regression analysis for adjusting the variations in dates of FL.

FIGURE 7.1 Variation in response to drought stress among rice genotypes (lines), drought susceptible line showing complete dried seedlings under drought treatment.

7.1.2 MECHANISMS OF RESISTANCE

7.1.2.1 PHYSIOLOGICAL

The build-up of polyamines such as putrescine, spermine was found to take place in plants during drought stress conferring resistance in those against drought stress. The function of polyamines in plant defense to water stress differs with the forms of polyamine and also the stages during which there is an occurrence of water stress. Yang et al. (2007) stated the contribution of a few polyamines in conferring drought resistance of rice cultivars which were subjected to water stress during their reproductive stage. The outcomes unveiled that there was an increment in the activities of arginine decarboxylase, *S*-adenosyl-L-methionine decarboxylase, and spermidine (Spd) synthase in the leaves with water stress. These have resulted in an increase in putrescine (Put), spermidine (Spd), and spermine (Spm) contents, which exhibited positive associations in yield maintenance ratio of the cultivars. The yield maintenance ratio is expressed as the ratio of GY under water-stressed conditions to GY under well-watered conditions. A positive relation was found with reference to the accumulation of free putrescine at early stages of water stress, and a negative association at later stage with the yield maintenance ratio. There were no noteworthy variances in the contents of soluble-conjugated polyamines and insoluble-conjugated spermidine and spermine amongst the cultivars. Free polyamines revealed a noteworthy build up in drought-resistant cultivars when their leaf water potentials have reached −0.51 MPa to −0.62 MPa and the same was noticed in drought susceptible cultivars when the leaf water potentials reached −0.70 MPa to −0.84 MPa. Their results revealed that rice has an immense capability to develop an increase in polyamine biosynthesis in leaves in retort to water stress. In adapting to drought, it would be excellent for rice to sustain the physiological character of elevated levels of free Spd/free Spm and insoluble-conjugated Put, and early build-up of free PAs, under water stress.

7.1.2.2 MOLECULAR

Price and Courtois (1999) discussed mapping QTLs linked with drought resistance in rice and its: advancement, drawbacks, and potentials. They specified that the application of molecular markers in the mapping of characters of agronomic significance offers immense assurance for enhancing the expansion of improved plant varieties and mounting our understanding of the physiological or molecular mechanisms at the back of biological phenomena. The technique that is being used in many crops for drought resistance screening has to be used particularly the molecular genetic analysis of drought resistance in rice also, as it serves as one of the model species of the monocots. Further, within its germplasm a variety of drought resistance mechanisms are found. They reviewed the improvements in locating quantitative trait loci (QTLs) for individual character mechanisms of drought resistance in regulated controlled environment conditions. They compared the reported QTLs related to root morphological characters. They analyzed the search for QTLs linked with field performance under drought stress, the problems that are related with understanding the genetic control of multifaceted physical and physiological events under situations of large environmental disparities. They emphasized that usage of near isogenic lines may be better in the solving the problems that are likely to arise in future.

Zhang et al. (2001) located genomic regions linked with mechanism of drought resistance in rice. They constructed virtual comparative mapping within and across species. They mentioned that the research on drought resistance with the use of molecular-marker technology in crop plants particularly has led to a shift from physiological basis of the observable facts to genetic basis of the phenomenon and the mechanisms that are involved in conferring drought resistance. Here, they made a wide-ranging study of mapping the drought resistance components (osmotic adjustment and root traits) in doubled-haploid rice population of 154 lines. A genetic linkage map that consisted of 315 DNA markers was made. They identified 41 of the QTLs for osmotic adjustment and root traits. These indicated the occurrence of the phenotypic variance of 8–38% in these lines. They observed that a region on chromosome 4 enclosed most important QTLs for a number of root traits. They detected many reliable QTLs for drought responses throughout the genetic backgrounds of these lines. They are of much help in marker-aided selection (MAS) for the addition of a characteristic trait of concern into an elite line. Three preserved genomic regions which were related with a variety of physiological responses to drought in numerous grass species were found by their comparative mapping. Further, it is suggested that these regions which report drought adaptation and have been conserved across grass species during genome evolution, which can be accurately applied across species for the improvement of drought resistance in cereal crops (Figure 7.2).

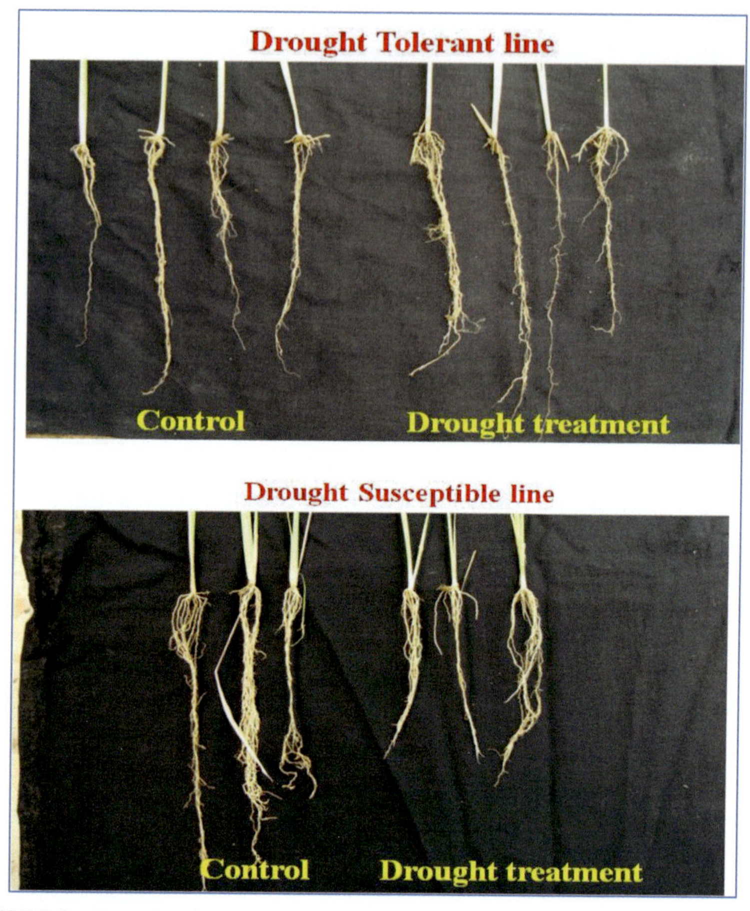

FIGURE 7.2 Drought tolerant line showing increased root length under drought treatment compared to drought susceptible line.

Chezhian et al. (2003) in their genetic analysis of drought resistance in rice with molecular markers specified that identification of genomic regions contributing to drought resistance is very much essential as it enables in the development of suitable rice cultivars for rainfed regions using molecular marker assisted breeding. They mapped QTLs related to plant water stress indicators, phenology, and production characters of a doubled-haploid (DH) population of 154 rice lines obtained from the cross CT9993-5-10-1-M/IR62266-42-6-2. Water stress was imposed before anthesis in these lines. There was a large variability in these lines for the above characters. They identified 47 QTLs for different plant water stress indicators, phenology, and production characteristics. These individually exhibited 5 to 59% of the phenotype variation. They detected a region on chromosome 4. It contained key QTLs for plant height, GY, and grains number per panicle under drought stress. They compared the coincidence of QTLs with specific traits and genetically dissected the nature of association of root characteristics and capacity of these for osmotic adjustment and with rice production under drought stress. These root traits exhibited positive associations with yield and yield traits under drought stress. They suggested that the region RG939-RG476-RG214 on chromosome 4 was recognized for root-related drought resistance constituent QTLs also had shown pleiotropic effects on yield components under stress.

Yue et al. (2005) investigated the genetic basis of drought resistance at reproductive stage. They analyzed the QTLs for drought response index (DRI, standardized by potential yield and FL time), relative yield, relative SF, etc., in a cross between an *indica* rice and upland rice. For these traits, 39 QTLs were identified. Individual QTL elucidated 5.1–32.1% of phenotypic variation. Only two QTLs for plant water status were found to be common and these were located in two environments. This suggests that in these two types of soil conditions different mechanisms might have existed. As the DRI did not exhibit an association with potential yield and FL time under control treatment, they suggested that under field conditions it could be effectively utilized as a better drought resistance index. There was co-location of QTLs for canopy temperature and delay in FL time. This has proposed that these two characters are useful as indexes in screening for drought resistance. Correlation and QTL similarity between root traits and putative drought tolerance traits suggest that drought avoidance (via thick and deep root traits) was the key genetic basis of drought resistance in sandy soil condition. In paddy soil condition, drought tolerance might play a key role in the genetic basis of drought resistance. Therefore, in improvements of drought resistance, especially at the reproductive stage in rice, drought mechanisms and soil textures must be considered equally.

Hu et al. (2006) report that overexpression of a NAM, ATAF, and CUC (NAC) transcription factor increases drought resistance and salt tolerance in rice. They demonstrated that in transgenic rice which showed 22–34% higher seed setting had an overexpression of stress responsive gene *SNAC1* (*stress-responsive NAC 1*). The expression of this gene was found to improve drought resistance in transgenic rice in the field under extreme drought stress conditions at the reproductive stage. There were no phenotypic variations or yield penalty with its expression in transgenic rice. It was further observed that this transgenic rice had exhibited better drought resistance and salt tolerance at the vegetative stage. The transgenic rice was found to be more susceptible to abscisic acid (ABA). It had slow water loss because of more number of stomatal pores closure. However, any noteworthy variation was not observed in photosynthesis rate. *SNAC1* is induced mostly in guard cells by drought. It encodes a NAM, ATAF, and CUC (NAC) transcription factor. This factor exhibits transactivation activity. The DNA chip analysis has confirmed that a large magnitude of stress-related genes were up-regulated in the *SNAC1*-overexpressing rice plants. Their data has suggested that *SNAC1* possesses potentially valuable in recuperating drought and salinity tolerance in rice.

7.1.2.3 GENETIC BASIS

The capability of rice plants to endure drought stresses is related with root system traits. Though a number of studies were focused on understanding the genetics of roots traits, it's not well understood. Ekanayake (1985) investigated the inheritance of six root characters of rice, by utilizing the parents, F_1, F_2, and F_3 populations of the cross IR20×MGL-2. They observed a polygenic system of root characters. The root systems of F_1 plants had thick, deep roots. These also had a superior lateral and vertical distribution than the low parent (IR20), with a positive heterosis also. The expression of all root characters was equally contributed by additive and dominant genetic effects. Both progeny parent regression of F_3 and F_2, and the narrow sense heritability assessments in F_3 were high for root thickness (0.61 and 0.80), root dry weight (0.56 and 0.92) and root length density (RLD) (0.44 0.77). They specified that if selection is based on these traits of root types in the individual plants of early segregating generations the performance of the plants would be more successful.

The genetic enhancement of adaptation to drought is focused mostly by the use of traditional method. In this approach, preference is given to selection for yield and its stability over varied areas and years. Under stress conditions yield has a low heritability, programs oriented towards yield selection attain a slow progress and are much expensive as there is prevalence of less inherent variation in the field. Nguyen et al. (1997) reviewed the progress of the information in physiology and molecular genetics, which have a key effect on the breeding programs involved in breeding resistant varieties of rice for drought stress. Earlier drought resistance selection studies have concentrated more on selections of the abilities of the root systems in providing the evapotranspirational requirement from subterranean soil moisture and capability for osmotic adjustment. These were emphasized and selected as key drought resistance features in rice. Huge investments in field nurseries or greenhouse facilities are required when one goes for the selection of these traits, further it also brings in the problems of repeatability due to environmental inconsistency. Thus, they asserted that modern improvements that take place in the development of molecular linkage maps of rice and other progresses in molecular biology open novel vistas for drought resistance breeding. Molecular markers linked to root characters and osmotic adjustments are being recognized. These help out in MAS. Transgenic rice plants with tolerance to water shortage and osmotic stresses have been identified. An excellent advancement has occurred on genetic engineering of osmoprotectants, such as proline and glycine betaine, into the rice plant for drought tolerance enhancement. For attaining required goals, close cooperation between molecular geneticists, plant physiologists, and breeders has to be necessitated for vital assessment of the role of specific genes and the appliance of molecular genetics to drought resistance breeding in rice.

Chang et al. (2008) undertook genetic analyses on the drought components in rice with the help of diverse techniques in three rice germplasm sources namely, upland, bulu, and aus cultivars. They investigated the genetic control of component traits contributing to drought escape and avoidance. They also studied the genetic association among these three rice germplasm sources. A small number of anisomeric genes were found to control the earliness trait in Aus rice. This was responsible for the contribution to drought escape in these. Though lowland varieties had high tillering, the dominant trait in all three variety groups was observed as low to moderate tillering ability. There was occurrence of modifying genes and plant height was under multigenic control. Root number was also under multigenic control. In many upland rice varieties, long roots seem to be under the control of dominant alleles, however, in two semi-dwarfs these dominant alleles also conferred the existence of small roots. In upland parents, the presence of thick roots was to a little degree under control of dominant allele, while in semi-dwarfs, these dominant alleles controlled thin roots and low root-shoot ratio. A strong character relationship occurred between plant stature, tiller number, root length, and root thickness. Based on the studies on pollen fertility, meiotic behavior and electrophoresis analysis, it was observed that there was a strong correlation between the three variety groups. The upland rice varieties exhibited added superior characters in seed protein bands and in plant morphology.

Serraj et al. (2009) discussed about the development of drought resistance in rice. It is known that drought is the major restraint to rice production in Asia and sub-Saharan African rainfed areas. It is thus crucial that the strategies of genetic management adopted for drought improvement must concentrate on greatest extraction of accessible soil moisture and its resourceful utilization in crop establishment and growth. This may bring about the maximization of biomass and yield. The rice germplasm encompasses huge genetic dissimilarities for drought resistance. Thus, currently it poses a big challenge in understanding the drought resistance complexities in rice and use of all existing genetic resources to bring into being rice varieties which combine drought adaptation related with towering yield potential, quality, and biotic stresses resistance. They proposed that strategies adopted must be directed towards development of a channel for generation of best breeding lines and hybrids. These developed ones can later be incorporated with competent management methods. They can later be given to farmers for better yield productions. Thus, it includes the requisite of the improvement of high-throughput, high-precision phenotyping systems. These systems must enable in permitting the genes for yield traits under stress. These component genes are to be well mapped and their special effects are to be assessed on a variety of drought-associated characters. Once this has been achieved, these promising genes are to be incorporated into extensively grown rice mega-varieties. Simultaneously there should also be scaling up of gene recognitions and their release for application in marker-aided breeding.

7.2 TEMPERATURE STRESS

7.2.1 HEAT STRESS

7.2.1.1 EFFECTS

Rice germplasms exhibit varietal differences towards temperature sensitivities and high temperature injuries. An increase in temperature of above

35°C often causes heat injuries in rice plants. Rice plants also exhibit variations to injuries of high temperature at different growth stages. Satake and Mackill (1981) in their discussions of high-temperature stress in rice mentioned that the ears of rice are most susceptible to high temperature injuries prevailing at the time of flowering (FL). Similarly, the stage of 9 days before FL also appears to be second sensitive stage. It is observed that there is a higher percentage of spikelet sterility by prevalence of high temperatures particularly at the stage of anthesis. This may be owing to a disturbance caused in pollen shedding and also because of impaired pollen germination. The effect of inactivated pistil in bringing about spikelet sterility may have a meager role compared to the above two factors. In general, anthesis occurring during the early morning hours evades high temperature stress. It is found that few crosses of *Oryza glaberrima* flowers open 3 hrs prior than IR 36 and *O. sativa*, however, there are also some *O. sativa* and *O. glaberrima* crosses which commence their anthesis in advance than IR 36. Their studies which were undertaken on genetics of high temperature-induced sterility has thus unveiled that the trait of heat tolerance (HT) shows a quite high heritability, with additive type of genetic variation in many cases. Rice variety stocks reveal the existence of apparent varietal differences in high temperature-induced sterility at anthesis, with the occurrence of both the early maturing and also heat-tolerant varieties. They suggested that phytotron is more effective for studies on the screening of rice varieties or lines for HT rather than field experiments.

Jagadish et al. (2008) in their study on phenotyping of rice parents and mapping of rice populations for HT during the stage of anthesis has stated that even the occurrence of short periods of high temperature at time of anthesis drastically affect the seed setting in rice. Keeping in view the future climate changes and occurrence of high temperatures, the need of the hour is to breed heat tolerant rice varieties suitable to the future changing climatic conditions. In their experiments conducted in few prominent mapping populations of rice at varied temperatures (30°C daytime, 35°C and 38°C air temperature), they observed that there was a decline in SF when these were exposed to high temperatures at anthesis for more than 6 hours, with a drastic decline in SF at 38°C. Rice cultivar CG14 (*O. glaberrima*) attained peak anthesis prior in the morning (1.5 hours (h) after dawn) than *O. sativa* genotypes (≥3 h after dawn) at control and high temperatures. They observed the response of SF to be reproducible and consistent. The check cultivar N22 exhibited SF of 64–86% even at 38°C and was in particular selected as the most heat tolerant genotype. On the other hand, Azucena, and Moroberekan cultivars which exhibited SF of <8% were selected as most liable ones to high temperature stress.

Likewise, Prasad et al. (2006) investigated in rice the responses to high temperature stress in species, ecotype, and cultivars to trace out the variations in SF and harvest index (HI) in these. A key constituent of yield is SF. It is very much susceptible to high temperatures. To trace out the variabilities in responses and gain an understanding of HTs, they conducted their study in fourteen rice cultivars of diverse species (*Oryza sativa* and *Oryza glaberrima*), ecotypes (*indica* and *japonica*) and origin (temperate and tropical), in Florida. The findings unveiled that though high temperature have resulted in a decline in SF in all cultivars, the variations in SF declines were not significant. By considering the fall in SF at high temperature, N-22 cultivar was mainly selected as tolerant. The cultivars L-204, M-202, Labelle, Italica Livorna, WAB-12, CG-14, and CG-17 were exceedingly vulnerable and cultivars M-103, S-102, Koshihikari, IR-8, and IR-72 were reasonably susceptible to high temperature. They could not trace out clear cut differences between species and ecotypes; however, they noticed that a few cultivars in each species or within ecotypes of tropical and temperature derivation were found evenly susceptible to high temperature (M-202 temperate *japonica*, Labelle tropical *japonica*, CG-14 *O. glaberrima*, and WAB-12 interspecific). The decline in SF and cultivar variations at elevated temperatures were ascribed chiefly to condensed pollen production and pollen reception (pollen numbers on stigma). Inferior SF at high temperature has bring about in production of less number of filled grains, poorer grain weight per panicle, and a decrease in HI. There subsist a prospective for genetic enhancement for HT; therefore, much emphasis has to be laid in identifying and screening heat tolerant rice cultivars keeping in view the screening at the sensitive stage i.e., the reproductive stage.

Similarly, Shah et al. (2011) in their study on the effect of high-temperature stress on rice plant mentioned that the expected 2–4°C rise in temperatures by the last part of the 21st Century may pose a menace to rice production. They also mentioned about the distressing blow of high temperatures at night rather than the daily mean or day temperatures. Most sensitive stages to high temperature are booting and flowering (FL) stages. The prevalence of high temperatures during these stages may also result in complete sterility. It is suggested that along with the occurrence of high temperatures if there is the presence of humidity, there is a possibility of increase in the rate of spikelet sterility. Rice germplasms have considerable variations of responses to temperature stress. Several phenotypic markers of high temperature tolerance include the production of FL at cooler periods of the day, production of large sized anthers, with high pollen viability and longer duration of basal dehiscence of pollen, and also the existence of lengthy basal pores. Apart from these, phenotypic markers, a few of the biochemical processes viz., protection of structural proteins, the synthesis of enzymes and membrane polarities, the expression of large heat shock proteins (HSPs) are also responsible for imparting thermo-tolerance. They suggested that each and every trait of this HT should be vigorously subjugated in the upcoming breeding programs for development and improvement of heat-resistant cultivars. They specified a number of adaptive measures in rice for overcoming the effects of temperature stress in the coming years. A few of these include the inclusion of heat tolerant cultivars in cultivation with due replacement of the heat-sensitive cultivars, altering the sowing times, choosing of the tolerant varieties which have the capacity to put forth such a type of growth so that there is an avoidance of sensitive stages to heat stress. Apart from these, application of some plant growth regulators exogenously on the plants may also be able to escape the influence of heat stress and yield reductions.

7.2.1.2 SILICON IN HEAT STRESS

Agarie et al. (1998) evaluated the helpful effects of silicon on the stress tolerance of rice plants. Measurements of the electrolyte leakage (El) from leaf tissues were considered. Leaf tissues were desiccated with polyethylene glycol (PEG) and by high temperature. The cell membranes integrity was estimated on the basis of the El measurements. The findings unveiled that the increased contents of silicon in the leaf tissues reduced the declines in the Els caused by desiccation of PEG of 30% and 40% solutions. They also observed an increment in the levels of polysaccharides in the cell walls of leaves of those plants which were grown in 100 ppm SiO_2. The build-up of polysaccharides within cell walls, appears as one among the factor linked with tolerance to desiccation. Thus, in the cell walls of leaves of this silicon grown plants there was an increase in polysaccharide levels by 1.6 folds high. The ultrastructural interpretations of leaves revealed only the polymerized Si accumulation in the cell walls of epidermal cells. This was not to be seen in the mesophyll cells, even though they were thought to be possibly the chief sites of El. Their research findings have put forward that the existence of silicon in rice leaves is most chiefly concerned in the maintenance of water relations of cells viz., mechanical properties, and water permeability. It has also an important part to play in the inhibition of the El through the synthesis and functions of cell walls. Even at high temperatures of 42.5°C, the Silicon grown plants had less El, thus, their results has suggested that silicon has its involvement in the maintenance of thermal stability of lipids in cell membranes and prevent the structural and functional worsening of cell membranes of rice plants when subjected to environmental stresses.

7.2.2 MECHANISM

7.2.2.1 PHYSIOLOGICAL

Roy and Ghosh (1996) reported the production and accumulation of free and conjugated polyamines in rice callus under heat stress. Further, there was an increment in the enzyme activities of those concerned with the biosynthesis of these compounds. There were high levels of activities of the enzymes arginine decarboxylase and polyamine oxidase in callus of tolerant cultivars with a concomitant high build-up of free and conjugated polyamines in callus of N 22 a heat tolerant cultivar. Apart from these, some rare polyamines norspermidine and norspermine were also accumulated in cv. N22. Though the concentration levels of these compounds were higher during the stress, their levels could not be noticed in callus of cv. IR8. There was also an elevated expression of transglutaminase activity, involved in the build-up of conjugated polyamine.

Most of the higher plants that grow under natural environments are often exposed to the occurrences of a range of abiotic stresses. Under most of these conditions, there is a production of H_2O_2 and nitric oxide (NO) free radicals in plants causing oxidative damages. Uchida et al. (2002) considered the effect of hydrogen peroxide and nitric oxide in their study on heat and salt tolerance in rice. They noted that a pre-treatment of rice seedlings with and <10 of H_2O_2 or NO resulted in the survival of large green leaf tissue, and elevated quantum yield for photosystem II, under salt and heat stresses. It also resulted in the production of active oxygen scavenging enzymes activities and elevated expression of transcripts for stress-related genes which were involved in encoding of sucrose-phosphate synthase, Δ′-pyrroline-5-carboxylate synthase, and also the production of small heat shock protein 26. These findings suggested that H_2O_2 and NO act as signal molecules for the response of heat and salt stress and can increase rice seedling tolerance to both salt and heat stresses.

Cao et al. (2008) studied the underlying physiological mechanisms heat stress during meiosis on GY of two indica rice cultivars viz., Shuanggui 1 (heat-sensitive) and Huanghuazhan (heat-tolerant), varying in the HTs. There was a considerable decline in the dehiscence of anthers under heat stress in susceptible cultivar Shuanggui 1. It had a low pollen fertility rate at 38°C of heat stress. Though in both the cultivars there was a considerable decline in the spikelet number per panicle, rate of seed setting, test weight and GY, this reduction in GY was higher in susceptible cultivar Shuanggui 1. This cultivar also had a considerable narrowness of grain width and large enlargements of grain length (GL)/width ratio. This effect was meager in heat tolerant cultivar Huanghuazhan. Some physiological parameters such as oxidation activity in roots, RNA contents in young developing panicles were affected by heat stress. There was a substantial decrease in these. Further, it was observed that there was a considerable enhancement in the contents of malondialdehyde (MDA) in leaves and ethylene evolution pace in young panicles. These physiological features were more susceptible to heat stress in Shuanggui 1 than in heat tolerant cultivar, Huanghuazhan. These outcomes revealed that the heat tolerant in rice is shown due to the occurrence of high activity of roots, strong antioxidative defense system, elevated RNA content, modest ethylene synthesis, and little MDA content in plants during meiosis.

Several simulation modeling studies and other investigators have specified that in upcoming years, rice crops will be recurrently subjected to water shortfall and heat stress; at the most susceptible FL stage, leading to spikelet sterility and yield losses. Krishna et al. (2011) investigated on the spikelet proteomic response to collective stresses of water and heat in rice. It was assessed that water deficit only and in association with heat stress considerably declines the panicles elongation, trapping 32% and 55% of spikelets within the leaf sheath, respectively. These trapped spikelets had poorer SF of 66% than normal exerted ones which had a SF of more than

93%. In the exerted spikelets, the average weighted fertility with heatlessness was lowest 35%, which was more i.e., 44% under combined stress. Their research outcomes put forward that rice acquired thermo-tolerance when it is led by water-deficit stress. Apart from above, it was also found that there was a greater upregulation of those proteins that favored pollen germination such as pollen allergens and beta expansion, with water deficits, while these were found to be presently maintained at normal levels under combined stress. Similarly, under combined stress treatments there was a considerable up-regulation in the transcripts of the chaperonic HSPs than in either heat or water deficit situations alone.

In a similar study, to understand the physiological and proteomics of HT during anthesis in rice, Jagadish et al. (2010) studied the reproductive organs morphology and pollen number, and variations in anther protein expression, in response to high temperature at anthesis in three rice genotypes, exposed to 6 hours of high (38°C) and control (29°C) temperature at anthesis. They observed that Moroberekan had only 18% SF at 38°C and recorded as most heat-sensitive genotype. Similarly, in IR64 and N22, the SF was 48% and 71% respectively at 38°C and these were grouped as moderate and highly heat tolerant, cultivars. Noteworthy differences were also found amongst these genotypes with respect to the anther length and width, apical and basal pore lengths, apical pore area, and stigma and pistil length. These characters were more influenced by temperature. High temperature enhanced the anther pore size and decreased the stigma length. However, this disparity in the number of pollen on the stigma was not associated to precise morphological characters. Difference in SF was exceedingly related with the percentage of spikelets with ≥20 germinated pollen grains on the stigma. A 2D-gel electrophoresis, there were 46 protein spots which were found to be varying in abundance, of which 13 differentially expressed protein spots were studied by MS/MALDI-TOF. A cold and a heat shock protein were found extensively up-regulated in N22. These proteins might have contributed to the larger HT of N22.

Rice (*Oryza sativa* L.) plants control their proteomes in reaction to high temperature stress. Chen et al. (2011) stated that phosphoproteins are synchronized in regulation by heat stress in rice leaves. Furthermore, phosphorylation is the most common form of protein post-translational modification (PTM). But, the differential expression of phosphoproteins which were caused by heat in rice was not explored. In their study, they recognized differentially expressed phosphoproteins in leaves under heat stress by MALDI-TOF-TOF-MS/MS and established by Western blotting. They recognized 10 heat-phosphoproteins from twelve protein spots. This includes ribulose bisphosphate carboxylase large chain, 2-Cys peroxiredoxin BAS1, putative mRNA binding protein, Os01g0791600 protein, OSJNBa0076N16.12 protein, putative H(+)-transporting ATP synthase, ATP synthase subunit beta and three putative uncharacterized proteins. The identification of ATP synthase subunit beta was additionally validated by Western-blotting. They concluded that heat stress-induced the dephosphorylation of RuBisCo and the phosphorylation of ATP-β, but it reduced the activities of RuBisCo and ATP synthase.

7.2.2.2 MOLECULAR

Lee et al. (2000) isolated a rice (*Oryza sativa* L. cv. Nakdong) cDNA clone, *Oshsp26*, encoding the chloroplast. Screening of a cDNA library has given away that the *Oshsp26* gene is set in its encoding by a single gene in the rice genome. The transcripts of this gene were more in expression in leaves of rice in those plants which were exposed to heat stress of 42°C. These transcripts were detectable even after 20 min of heat stress and had reached the highest level. This attained a maximum level later than 2 hours. Its expression was also seen after exposure to oxidative stress in the dearth of heat stress in rice. There was a considerable build-up of the transcript and protein on the treatment of rice plants with methyl viologen (MV) in the light and with hydrogen peroxide (H_2O_2), either in the light or in the dark. MV treatment resulted in the generation of H_2O_2 inside the chloroplast, it is thought that H_2O_2 function itself to provoke the expression of the *Oshsp26* gene. Based on these results, they proposed that the chloroplast smHSP has an essential function in the safeguarding of the chloroplast against injury affected by oxidative and heat stress.

Yamanouchi et al. (2002) reported that rice spotted leaf gene, *Spl7*, encodes a heat stress transcription factor (HSF) protein. They identified this by use of a high-resolution mapping. By using several cleaved amplified polymorphic sequence markers, they identified a genomic region of 3 kb as a candidate for *Spl7*. An ORF presenting more resemblance with a HSF was detected. The role of gene was tested by transgenic analysis. The analysis has shown that the leaf spot development was undeveloped in *spl7* mutants that had a wild-type *Spl7* transgene. Therefore, their results concluded that this gene transcript encodes the HSF protein. The transcript of *spl7* was observed in mutant plants. Under heat stress, it was found that there were higher levels of mRNAs (*Spl7* in wild type and *spl7* in mutant). The sequence analysis revealed a substitution in only one base in the HSF DNA-binding domain of the mutant allele. This base substitution resulted in a change from the amino acid tryptophan to cysteine.

Ye et al. (2012) using SNP markers mapped the QTL for HT at FL stage in rice progenies of BC_1F_1 and F_2 populations resulting from an IR64 × N22 crosses which were subjected to 38/24°C for 14 days at the FL stage. Later the SF was evaluated. By usage of selective genotyping and single-marker analysis, they observed that there were four single nucleotide polymorphisms (SNP) that were found to be related to HT in the BC_1F_1 population, while in F_2 populations four putative QTL were found to be linked with HT. The two main QTLs were observed to be placed on chromosome 1 (qHTSF1.1) and chromosome 4 (qHTSF4.1). These two chief QTLs explained 12.6% (qHTSF1.1) and 17.6% (qHTSF4.1) of the disparities in SF under high-temperature stress. A tolerant allele of qHTSF1.1 has been obtained from the susceptible parent IR64, and similarly, the allele qHTSF4.1 was obtained from tolerant parent N22. The influence of qHTSF4.1 on chromosome 4 had been established in certain BC_2F_2 progeny from the same IR64 × N22 cross, and the plants with qHTSF4.1 confirmed considerably the presence of greater SF than other genotypes.

7.2.2.3 GENETIC

High temperature tolerance is imperative in breeding heat tolerant rice cultivars for high temperature stress. Mackill et al. (1982) studied the pollen shedding and its combining ability for high temperature tolerance in six rice lines in a diallele cross of rice in a phytotron to establish the projections for breeding heat tolerant cultivars. These lines were exposed to 38/27°C for 10 days during anthesis while control plants were at a 29/21°C temperature regime. A HT index was computed by using the formula:

HT Index: Percentage of filled grains of treated plants/percentage of filled grains of control plants.

They observed considerably high general and specific combining ability effects for this index. The tolerant lines, N22, IR2006 and IET4658 showed general combining ability effects of 6.80, 4.08, and 3.02, respectively. The susceptible lines, IR28, IR1561 and IR52, showed general combining ability effects of –3.40, –4.92 and –5.58, respectively. Tolerant cultivars were found to shed additional pollen grains on the stigma in both the temperature regimes, whereas in susceptible genotypes there was a considerable decrease in the pollen shed amount. In cultivar IET4658, there was a decline in pollen germination in the temperature regime of 38/27°C. However, the percent fertility had shown a positive correlation with the amount of pollen on the stigma at 38/27°C. Even though there was reduced pollen growth under high temperatures in the tolerant genotypes, this was found to be compensated with the presence of more amount of pollen on stigma.

Cheng et al. (2012) studied the genetic basis of cold tolerance (CT) at seedling stage and HT at anthesis in a set of 240 introgression lines (ILs) resultant from the advanced backcross population of a cross between a *japonica* cultivar, Xiushui 09, and an *indica* breeding line, IR2061. They recognized QTLs affecting CT at the seedling stage and HT at anthesis. There were considerable variations between two parents under cold and heat stress in the survival rate of seedlings (SRS) and SF, the index traits of CT and HT. They detected four QTLs (qSRS1, qSRS7, qSRS11a, and qSRS11b) for CT on chromosomes 1, 7, 11, and the Xiushui 09 alleles enhanced SRS at all loci excluding qSRS7. They also recognized four QTLs for SF on chromosomes 4, 5, 6, and 11. These QTLs were categorized into two main types on the basis of their behaviors under normal and stress conditions: 1) the first QTL expressed only under normal condition; and the second QTL was apparently stress-induced and only expressed under stress. Amongst these, two QTLs (qSF4 and qSF6) which decreased the trait variation between heat stress and normal conditions were found to contribute to HT, mostly owing to their clear contribution to trait stability. IR2061 allele at the qSF6 and the Xiushui 09 allele at the qSF4 also elevated HT. A comparably similar QTL between CT at seedling stage and HT at anthesis was not recognized. As a result, they suggested that a new variety with CT and HT can possibly be bred by pyramiding the favorable CT- and HT-improved alleles at above loci from Xiushui 09 and IR2061, respectively, using marker-assisted selection (MAS).

7.3 COLD TOLERANCE (CT)

Ranjan et al. (2009) reported the variation in the pattern of CT at the early growth stage in genetic bases of wild and cultivated Asian rice in 57 strains which included cultivated rice (*Oryza sativa* ssp. *indica* and ssp. *japonica*) and its wild progenitor (*Oryza rufipogon*). They investigated genetic variation at the germination, plumule, and seedling stages among these strains. It was found that among these taxonomically varied groups there was also the existence of considerable demarcation in CT. Both *indica* and *japonica* subspecies at germination stage were observed to be more tolerant than *O. rufipogon.* However, at plumule and seedling stages, ssp. *japonica* appeared to be more tolerant than ssp. *indica* and *O. rufipogon.* Additionally, at plumule stage, in CT, the clinical dissimilarity was noted within *O. rufipogon* and ssp. *Japonica* across the latitude of origins. This suggests that the existing form of dissimilarity appears to have been framed by both their phylogenetic histories and on-going adaptation to the local environments. QTL analysis carried between *O. sativa* ssp. *japonica* (tolerant) and *O. rufipogon* (susceptible) unveiled five putative QTLs for CT at plumule and seedling stages except germination stage.

7.3.1 MECHANISMS

7.3.1.1 MOLECULAR MECHANISMS

Significant research inputs have directed on molecular basis of CT of rice. Qian et al. (2000) undertook QTL analysis of the rice seedling CT in a double haploid population obtained from anther culture of a hybrid between *indica* and *japonica* rice. By vigorously analyzing every day's continued existence parents and DH lines percentages under 7 days cold plus 9 days normal temperature condition, they mapped the QTLs for seedling CT. From this population a molecular linkage map was built. The findings have shown that two parents considerably varied in seedling cold tolerance (SCT). Further, they observed that segregation of SCT in DH lines was chiefly an uninterrupted distribution. A very high severe injury was observed on the 6th day of the cold treatment. Four QTLs for SCT were located on chromosomes 1, 2, 3, and 4 correspondingly. The additive effects of qSCT-1, qSCT-2 and qSCT-3 have been donated by the *japonica* cv JX17, although that of qSCT-4 was found to be contributed by the *indica* cv ZYQ8. They observed that SCT mechanism was complicated, as the above 4 QTLs were noticed at different stages during the cold treatment. Furthermore, the study unveiled that these lines have stronger

SCT in excess of JXI7, have 3–4 loci for SCT. Thus, integration of these QTLs into a suitable variety may bring about a booming in rice breeding program for CT.

Similarly, Andaya and Mackill (2003) constructed mapping of QTLs related with CT using 181 microsatellite marker loci during the vegetative stage in 191 recombinant inbred lines (RILs) which is obtained from a cross of a cold-tolerant temperate *japonica* cultivar (M202) with a cold-sensitive *indica* cultivar (IR50) of rice. These 191 RILs were maintained in growth chambers at varied temperature regimes which simulated cold stress injuries at the seedling and late vegetative stages. A chief QTL was recognized on chromosome 12. It was named as qCTS12a. This QTL was found to be intimately linked with cold-induced necrosis and wilting tolerance. The presence or absence of this QTL was responsible for the explanation of the occurrence of 41% of the phenotypic variation. They also identified a number of QTLs with slighter influences on eight rice chromosomes.

Andaya and Tai (2006) developed a fine mapping of the qCTS12 locus. This was a key QTL for seedling CT in rice. They mentioned that the temperate *japonica* rice cultivar M202 is the principal variety grown in California. This cultivar is more tolerant to low-temperature stress, with better grain quality and yield. Former investigation of a recombinant inbred line mapping population resulting from a cross between M202 and IR50, an *indica* cultivar which is extremely susceptible to cold stress, led to the recognition of many QTLs which conferred tolerance to cold-induced wilting and necrosis. It is suggested that 40% of the phenotypic variance that was produced was due to a chief QTL, *qCTS12*, mainly present on the short arm of chromosome 12. For the detection of the gene(s) primary to*qCTS12*, they undertook fine mapping of this locus. The saturation of microsatellite markers on the short arm of chromosome 12 indicates that*qCTS12* is very close to RM7003. In addition, supplementary microsatellite markers were known from widely available genomic sequence and these were used to fine map *qCTS12* to a region of about 87 kb located on the BAC clone OSJNBb0071I17. This region possessed ten open reading frames (ORFs). It consisted of five hypothetical and expressed proteins of unidentified role, a transposon protein, a putative NBS-LRR disease resistance protein, two zeta class glutathione S-transferases (GSTs) (*OsGSTZ1* and *OsGSTZ2*), and a DAHP synthetase. Additionally, fine mapping with markers developed from the ORFs delimited the QTL to a region of about 55 kb. The most probable candidates for the gene(s) fundamental to*qCTS12* are *OsGSTZ1* and *OsGSTZ2*.

Hurab et al. (2004) reported that stress-inducible OsP5CS2 gene is needed for salt and CT in rice. They secluded a rice T-DNA tagging line, in which T-DNA was introduced into the sixth intron of *OsP5CS2*. This gene deciphered for a protein which is very much homologous to Δ^1-pyrroline-5-carboxylate synthetase (P5CS), a proline biosynthesis enzyme. The T-DNA had the promoterless *gus* gene. It promoted in the creation of a gene fusion between *OsP5CS2* and *gus*. So, they were able to monitor the expression pattern of *OsP5CS2* easily by *in situ* GUS assay. At the seedling stage, they detected low level of the transcript. Nevertheless, they observed that this gene expression was specially induced by salt, cold, or ABA treatments in the dividing zone of the roots. In mature spikelets, this gene expression was in stamens, with its transcript also found to be present in stamens. These transcript levels were found to increase in roots of seedlings on treatment with 250 mM NaCl, 4°C cold stress, or 0.5 μM ABA. *OsP5CS2* knockout (KO) plants were further found to be susceptible to salt and cold stresses with reduced growth of root and shoots by 250 mM NaCl. Similarly, there was an inhibition in seedlings growth by cold treatment for more than 12 hours.

Komatsu et al. (2007) reported that over-expression of calcium-dependent protein kinase 13 (CDPK13) and calreticulin which interact with protein 1 confers CT on rice plants. They mentioned that calcium, which is an omnipresent signaling molecule involved in bringing out different plant responses to several stimuli. The changes in its cytosolic concentrations are responsible for these varied responses in plants. The rice CDPK13 and calreticulin interacting protein 1 (CRTintP1) are found to be engaged in cold stress response in rice. The authors produced rice lines transformed with sense CDPK13 or CRTintP1 constructs and used these to examine the role of these proteins. They observed that incubation of plants at 5°C for 3 days resulted in wilting and curling of leaf blades of both the sense transgenic and vector control rice plants. On transferring back of these plants to non-stress conditions after cold treatment, though there was death of leaf blades, in sense transgenic rice plants the leaf sheaths still remained green. They found that the build-up of these proteins was much higher in cold-tolerant rice variety than that of intermediate tolerance. For verifying the result of over-expression of CDPK13 and CRTintP1 on the proteins, they analyzed sense transgenic rice plants using proteomics. The

2D-PAGE profiles of proteins from the vector control and sense transgenic rice plants were compared. Two of the proteins showing difference between these lines were calreticulin. It is recommended that CDPK13, calreticulin, and CRTintP1 could function as key signaling components for cold stress response in rice.

Lou et al. (2007) utilized 146 microsatellite markers and identified a key QTL related closely to CT at seedling stage in rice. Their mapping population comprised of 193 doubled haploid (DH) lines, obtained from a cross between a cold-tolerant *japonica* variety (AAV002863) and a cold-sensitive *indica* cultivar (Zhenshan97B). They recognized five chief effect QTLs using a composite interval mapping approach with LOD > 4.0 on chromosomes 1, 2, 8. The collected influence of the five QTLs was 62.28%, and a main QTL (LOD = 15.09) was recognized on chromosome 2 flanked by RM561 and RM341, which elucidated 27.42% of the total phenotypic variation. Further, four considerably significant epistatic interactions with a total contribution of 20.14% were identified.

Plants illustrate response to unfavorable environment by inducting a series of signaling processes which includes activation of transcription factors (TFs). These TFs can control the expression of an array of genes for stress response and adaptation. NAC (NAM, ATAF, and CUC) is a plant specific transcription factor family which has with different functions in development and stress regulation. Hu et al. (2008) reported the characterization of transcription factor gene *SNAC2* from upland rice *IRA109* that confers cold and salt tolerance in rice. SNAC2 was established to have transactivation and DNA-binding activities in yeast and the SNAC2-GFP fusion protein was localized in the rice nuclei. Northern blot and *SNAC2* promoter activity analyses have suggested that this gene was induced by drought, salinity, cold, wounding, and ABA treatment. It was overexpressed in *japonica* rice Zhonghua 11 to test its influence in enhancement of stress tolerance. Transgenic plants of more than 50% remained vigorous, even after severe cold stress (4–8°C for 5 days). The transgenic plants had greater cell membrane stability than wild type during the cold stress. These plants also had more germination percentage and growth rate even under saline conditions. Its Over-expression also improved the tolerance to PEG treatment. However, the *SNAC2*-overexpressing plants were found to be considerably more susceptible to ABA. DNA chip profiling analysis of these transgenic plants has shown the presence of a number of up-regulated genes, associated to stress response and adaptation. Some of these were peroxidase, ornithine aminotransferase, heavy metal-associated protein, sodium/hydrogen exchanger, heat shock protein, GDSL-like lipase, and phenylalanine ammonia lyase. They specified that not a bit of the up-regulated genes in the *SNAC2*-overexpressing plants corresponded to the genes up-regulated in the transgenic plants over-expressing other stress responsive NAC genes described earlier. It is recommended that SNAC2 is a new stress responsive NAC transcription factor that possesses prospective benefit in getting better stress tolerance in rice.

Rice seedlings are highly susceptible to chilling in temperate and subtropical zones and in high-elevation locations. Enhancement of chilling tolerance in rice may boost appreciably rice production. MYBS3 is a single DNA-binding repeat MYB transcription factor that mediates sugar signaling in rice. Su et al. (2010) noted that MYBS3 also has a central role in CT in rice. Analysis of gain- and loss-of-function revealed that MYBS3 was improving CT in rice. Transgenic rice that constitutively over expressed MYBS3 was able to tolerate a cold temperature of 4°C for at least 1 week. It did not exhibit any yield penalty in regular field conditions. Transcription profiling of transgenic rice over expressing or under expressing MYBS3 contributed to the detection of many genes in the MYBS3-mediated cold signaling pathway. They found that MYBS3 repressed the well-known DREB1/CBF-dependent cold signaling pathway in rice. The repression appeared to exhibit its function at the transcriptional level. *DREB1* exhibited quick response while *MYBS3* responded gradually to cold stress. This proposes that distinct pathways function sequentially and complementarily for adaptation of short- and long-term cold stress in rice. Their findings, therefore, disclose an up till now undiscovered new pathway that regulates cold adaptation in rice.

Koseki et al. (2010) undertook detection and fine mapping of a major quantitative trait locus (QTL) derived from wild rice, which controls CT at the seedling stage. They developed 331 SNP markers. These were utilized alongside with phenotypic evaluation to recognize QTLs linked with CTSS from a mapping population of 84 F_2 plants obtained from cold-tolerant wild rice, W1943 (*Oryza rufipogon*), and a sensitive *indica* cultivar, Guang-Lu-Ai 4 (GLA4). Three QTLs on chromosomes 3, 10 and 11 were recognized. A main locus, *qCtss11* (*Q*TL for CT at *s*eedling *s*tage, was located on the long arm of chromosome 11 which explains about 40% of the phenotypic variation. Introduction of the W1943 allele of *qCtss11* to the GLA4 genetic background amplified CTSS. Based on the phenotypic and genotypic estimation of advanced backcross progenies, *qCtss11* was dissected a single Mendelian factor. Using 23 markers over the *qCtss11* locus a high-resolution genetic map was made. Then they constructed fine mapping of *qCtss11* to a 60-kb candidate region defined by marker AK24 and GP0030 on chromosome 11, in which six genes were located. Expression and resequence analyses of the six candidates favored the hypothesis that Os11g0615600 and/or Os11g0615900 are fundamental gene(s) of the CTSS.

Tao et al. (2011) reported that *OsWRKY45* alleles play diverse roles in ABA signaling and salt stress tolerance, but alike roles in drought and CT in rice. They stated that though allelic diversity of genes are reported to give way to many phenotypic variations related with diverse physiological processes in plants, information on allelic diversity of abiotic stress-responsive genes is limited. In this study, it is revealed that the alleles *OsWRKY45-1* and *OsWRKY45-2* play diverse roles in ABA signaling and salt stress tolerance in rice. The two alleles contained different transcriptional responses to ABA and salt stresses. *OsWRKY45-1*-overexpressing lines exhibited declined ABA sensitivity, while *OsWRKY45-1*-KO lines exhibited elevated ABA sensitivity. *OsWRKY45-1* transgenic plants exhibited no clear difference from negative controls in response to salt stress. On the other hand, *OsWRKY45-2*-overexpressing lines presented increased ABA sensitivity and decreased salt stress tolerance, and *OsWRKY45-2*-suppressing lines presented decreased ABA sensitivity and increased salt stress tolerance. *OsWRKY45-1* and *OsWRKY45-2* transgenic plants exhibited differential expression of a set of ABA- and abiotic stress-responsive genes, but they exhibited alike responses to cold and drought stresses. These findings implied that *OsWRKY45-1* negatively and *OsWRKY45-2* positively regulates ABA signaling and, further, *OsWRKY45-2* negatively regulates rice response to salt stress. The diverse roles of the two alleles in ABA signaling and salt stress may be because of their transcriptional mediation of diverse signaling pathways.

It is stated that plant responses to abiotic stresses are harmonized by chain of growth and developmental processes. Indole-3-acetic acid (IAA) and ABA play fundamental roles in developmental programs and environmental responses, through intricate networks of signaling and metabolism networks. But, crosstalk between the two phytohormones in the stress responses is not well known. It is reported in Du et al. (2012) study that a GH3 family gene, *OsGH3-2*, which encodes an enzyme and catalyzes

IAA conjugation to amino acids, is involved in the modulation of ABA level and stress tolerance. They observed that drought-induced the expression of *OsGH3-2* while cold suppressed it. Overexpression of *OsGH3-2* in rice caused considerable morphological anomalies associated to IAA deficiency, such as dwarfism, smaller leaves, and fewer crown roots and root hairs. There was a considerable decline in carotene, ABA, and free IAA levels in the overexpressing line. Further, the lines had wider stomata aperture, and quicker water loss, and thus they became hypersensitive to drought stress. However, the overexpressing line also exhibited better CT. This was attributed to the combined effects of reduced free IAA content, lessened oxidative damage, and condensed membrane penetrability. Additionally, expression levels of some ABA synthesis- and stress-related genes had altered considerably the overexpression line. It was concluded that OsGH3-2 modulates both endogenous free IAA and ABA homeostasis and is responsible for the occurrence of differential effects on drought and CT in rice.

Yang et al. (2013) report the overexpression of microRNA319, its effect on morphogenesis of leaf and on CT in rice. They mention that microRNA319 (miR319) family is one of the conserved microRNA (miRNA) families amongst varied plant species. It is stated that miR319 controls plant development in dicotyledons, but currently much information is not available on its role in monocotyledons. In rice, the *MIR319* gene family is composed of two members, *OsaMIR319a* and *OsaMIR319b*. Here, the authors report an expression pattern analysis and a functional characterization of the two *OsaMIR319* genes in rice. They found that overexpressing *OsaMIR319a* and *OsaMIR319b* in rice both produced wider leaf blades. The transgenic plants with overexpression of OsamiR319 had an increased number of longitudinal small veins in the leaves. This might have possibly been responsible for increased leaf blade width. Further, they noticed that overexpressing OsamiR319 in transgenic rice seedlings also contributed to an increase in CT (4°C) after chilling acclimation (12 °C). Remarkably, under both 4 and 12 °C low temperatures, *OsaMIR319a* and *OsaMIR319b* were down-regulated while the expression of miR319 targeted genes were induced. Besides, genetically downregulating the expression of either of the two miR319 targeted genes, *OsPCF5*, and *OsPCF8*, in RNA interference (RNAi) plants also resulted in enhanced CT after chilling acclimation. Their research outcomes have specified that miR319 plays a main role in the morphogenesis of leaf and CT in rice.

Pereira et al. (2013) discussed on the methods to avoid damage and attaining CT in rice plants. Low temperatures affect rice plants during germination, vegetative growth, and reproductive stages, and rice production. They made a review on the efforts made to attain CT in rice through breeding, the major tools employed for evaluation of CT in rice plants, the detection of QTLs and genes associated to this tolerance, and the outcomes achieved on genetic transformation of rice plants with potential CT genes. They emphasized the need of combined efforts from breeders and plant biologists to enhance the production of rice plants to cold stress.

7.4 FLOODING/HIGH MOISTURE TOLERANCE

Direct seeding of rice (DSR) is progressively being taken up in both rainfed and irrigated areas, due to the prevalence of labor shortages for taking up timely transplanting and also a number of opportunities are nowadays available for intensification of crop cultivation. Inspite of its progressive adoption in large areas, its being faced with the obstacle of reduced crop establishment, which is responsible for its limitation to be taken up on large scare in many areas which are susceptible to flooding. Abdolhamid et al. (2010) studied the QTLs that are strongly correlated with the flooding tolerance in rice. More than 8,000 gene bank accessions and breeding lines were screened and a few flooding tolerant lines were identified. They selected KhaoHlan On, for mapping QTLs linked with tolerance by using a backcross population with IR64 as a recurrent parent. There was 0–68% variation in the survival rate of these BC_2F_2 lines towards flooding. However, the average survival rate was only 28%. A linkage map of 1475.7 cM was constructed, that has a mean interval of 11.9 cM. This was made with the use of 135 polymorphic SSRs and one indel marker. Five putative QTLs, on chromosomes 1 (*qAG-1-2*), 3 (*qAG-3-1*), 7 (*qAG-7-2*), and 9 (*qAG-9-1* and *qAG-9-2*) were detected, it explained for 17.9 to 33.5% of the phenotypic variation. The LOD scores were of 5.69–20.34. It was observed that for all the QTLs the alleles of KhaoHlan On exhibited increased tolerance of flooding during germination. The identified QTLs were detected by graphical genotyping of the lines with highest and lowest survival for flooding tolerance.

Singh et al. (2011) stated that for longer-term partial stagnant flooding tolerance in rice SUB1 locus has a crucial role. They mentioned that longer-term partial stagnant flooding, after a transient total submergence frequently causes serious damage in current rice varieties. Some developments were made and a few tolerant varieties for complete submergence were developed with the introduction of *Submergence-1* (*SUB1*) gene into a few popular accepted varieties. Nevertheless, *SUB1* may not be efficient under partial stagnant flooding. This may be because of the suppression of elongation which might be mediated by SUB1 under stagnant flooding. They studied the effect of stagnant flooding in a set of rice genotypes, a pair of near-isogenic lines (NILs), Swarna, and Swarna-Sub1. There was higher survival rate and yield in Swarna-Sub1 after stagnant flooding followed by 12 days of submergence. There was a drastic decline in all the lines when the stagnant flooding was followed with 15–30 cm of complete submergence. This decline was many folds high in sensitive lines. They mentioned that it is much important to combine *SUB1* with tolerance of Stagnant flooding particularly for those areas where there is possibility of occurrence of both these stresses in the season. They found that for long term partial stagnant flooding Swarna and Swarna-Sub1 are additionally sensitive rather than IR49830 and IR42 owing to their short stature. There was to some extent a large decline in tillering in Swarna-Sub1 than Swarna when exposed to deeper stagnant flooding only. This decline was attributed to the promoted inhibition of elongation by *SUB1*, which might have been induced in submerged tissue. Thus, they suggested that stagnant tolerance is more reliant on the genetic background of the recipient genotypes that have an improved performance rather than the introgression of SUB1. There are some recipient genotypes which were taller genetically, as IR42 and IR49830. The *SUB1* donor landrace FR13A and its derivative breeding line IR49830 has shown an improved good survival rate with comparatively low decline in the GY under stagnant flooding next to complete submergence and under longer-term partial stagnant flooding. Their research study has indicated that apart from the presence of SUB1 these genotypes might have some other genes involved for tolerance for stagnant flooding. The research findings have shown that introgression of SUB1 into genotypes having better tillering abilities with the production of taller shoots would be more beneficial for adaptation of these to long term partial stagnant flooding.

Alternate wetting and drying (AWD) irrigation is an efficient water-saving technology for irrigated rice system. Yao et al. (2012) undertook research study to understand the varietal differences that existed between a "super" hybrid rice varieties Yangliangyou 6 and a water-saving and drought-resistance rice (WDR) variety Hanyou 3 (HY3) in AWD conditions in China. It was noticed that there was nearly 24–38% saving of the irrigation water. The super hybrid rice variety had 21.5% higher yield with consistent high water productivity and nitrogen use efficiencies under AWD. There was an increase in total dry weight and HI and this had led to recognition of higher GY of YLY6. Apart from the above, it also had a larger sink size with increased number of spikelets per panicle. The research study has specified that for AWD the high yielding varieties that are to be developed must be considered the trait of presence of more number of spikelets per panicle to get improved better yields.

7.5 SALINITY TOLERANCE IN RICE

Heenan et al. (1998) studied salinity tolerance of some Australian and overseas rice varieties to find out the consistency of selection at germination, early vegetative growth, and reproductive development in a temperature-controlled glasshouse. Varietal dissimilarities were detected at all the stages in the degree of tolerance; however, this degree of tolerance varied noticeably between stages. The Australian long-grain variety Pelde found tolerant at germination, while was most fanatical all through the early vegetative growth and reproductive development. Similarly, though the Japanese variety Somewake was salinity susceptible at germination and vegetative growth, it appeared much tolerant at reproductive development. Linear regressions have shown a converse association between sodium concentrations in the shoots during early growth and shoot dry weight for most varieties. However, for Pelde, it was observed that there was an intimate relationship between potassium concentration of the shoots and dry matter production.

Gregorio et al. (2002) discussed the developments that took place in breeding for salinity tolerance and other related abiotic stresses in rice. They mentioned that the advancements in breeding for rice for saline environments involves the improvement of quick and consistent techniques of selection for elongation ability and tolerance for salinity, submergence, iron toxicity, aluminum toxicity and phosphorus efficiency. Many donor germplasms were recognized and enhanced, and its inheritance for tolerance for most soil-related stresses has been studied. Many of the underlying physiological mechanisms of a number of stresses are now reasonably well known. Many of the rice lines that were developed and improved with tolerance for multiple stresses have been productively used as donor parents, and at large released as cultivars for salt-affected areas by national agricultural research and extension systems (NARES). They developed and genotyped F_8 recombinant inbred line (RIL) mapping populations for different abiotic traits. They also mapped key genes and QTLs for extending capacity tolerance for salinity, submergence, P deficiency, and Al and Fe toxicities. They mentioned that it is now possible to use the refined mapping with NIL for the development of polymerase chain reaction (PCR)-based MAS, and these methods are being developed for elongation ability and tolerance of salinity, submergence, Al toxicity, P deficiency, and Zn deficiency.

7.5.1 PHYSIOLOGICAL MECHANISM

Lutts et al. (1996) mentioned that putrescine is involved in salinity tolerance in rice. They measured the evolution of ethylene from leaves of 5 rice cultivars that varied in salinity tolerance after 15 and 30 days of stress exposure in plants sprayed every 2 days with deionized water or putrescine 100 μM. It was observed there was an increase in ethylene synthesis in salt-tolerant cultivar due to NaCl. Similarly, it was observed that there was also an induced enhancement in the content of 1-aminocyclopropane-1-carboxylic acid (ACC) by salt stress in the salt-tolerant cultivars. A decline in ACC conversion to ethylene has been suggested due to a decrease in ACC oxidase activity. Putrescine improved the growth and the leaf tissue viability of salt-treated plants in all cultivars. This positive effect was related with a raise in ethylene biosynthesis with an enhancement in ACC content and an inhibition of NaCl-induced inhibition of ACC conversion to ethylene and suggested that putrescine is involved in rice salinity tolerance (Figure 7.3).

FIGURE 7.3 Increasing the level of salinity decreased seedling emergence (%). Entry 620 is showing seedling emergence in 0.2 M NaCl but Entry 624 Failed to emergence.

Plett et al. (2010) reported the cell type specific expression of AtHKT1 which improved salinity tolerance in rice. It was reported earlier, that cell type-specific expression of *AtHKT1;1*, a sodium transporter, improved sodium (Na^+) exclusion and salinity tolerance in *Arabidopsis*. In their research study*AtHKT1;1*, was expressed explicitly in the root cortical and epidermal cells of an *Arabidopsis* GAL4-GFP enhancer trap line. These transgenic plants exhibited considerable improvement in Na^+ exclusion under conditions of salinity stress. They explored the feasibility of an alike biotechnological advance in crop plants through the use of a GAL4-GFP enhancer trap rice line to drive expression of *AtHKT1;1* specifically in the root cortex. Compared with the background GAL4-GFP line, the rice plants which expressed, *AtHKT1;1 formed* a high fresh weight under salinity stress, which was found to be associated to a lesser concentration of Na^+ in the shoots. The root-to-shoot transport of $^{22}Na^+$ was also declined and this was concurrent with the up-regulation of *OsHKT1;5*, the native transporter responsible for Na^+ retrieval from the transpiration stream. Fascinatingly, in the transgenic *Arabidopsis* plants overexpressing *AtHKT1;1* in the cortex and epidermis, the native *AtHKT1;1* gene accountable for Na^+ retrieval from the transpiration stream, was also upregulated. Additional Na^+ released from the xylem was stored in the outer root cells and was related with a noteworthy raise in expression of the vacuolar pyrophosphatases (in Arabidopsis and rice). The activity of these is essential to move the extra stored Na^+ into the vacuoles of these cells.

7.5.2 MOLECULAR

Zhang et al. (1995) report RFLP tagging of a salt tolerance gene in rice. They obtained a salt tolerant rice mutant (M-20) via *in vitro* selection. Its tolerance was steadily inherited over eight generations and most traits between M-20 and its sensitive original 77–170 (*Oryza sativa*) were very alike. An F_2 population of M-20 × 77–170 was split up and was studied for the inheritance of salt tolerance. There was no noticeable segregation among F_2 individuals, even though there was apparent segregation in the traits under saline conditions. In their study, the ratio of salt sensitive: moderately-tolerant:tolerant plants was 25:42:18, in agreement with a 1:2:1 ratio. It is proposed that the advance of salt tolerance in their rice materials was induced by the mutation of a chief tolerant gene which exhibited incomplete dominance. Using 130 RFLP probes dispersed all through the rice genome, the gene was tagged by a single copy DNA probe, RG4, which was located on chromosome 7. The genetic distance between the salt tolerant gene and RG4 was 7.0 ± 2.9 cM. On the basis of this, they proposed the split method for better evaluation of effects of salt stress.

Gregorio (1997) undertook tagging of salinity tolerance genes in rice by means of amplified fragment length polymorphism (AFLP). They determined the genetic basis of salinity tolerance of the indica cross IR66946 (IR29-salt susceptible improved variety x Pokkali-tolerant traditional tall variety) through PCR-based AFLP. They used the F8 RILs of IR66946 to map the genes for salinity tolerance in Pokkali. They undertook phenotyping at seedling stage by using salinized (EC=12 dS m^{-1}) culture solution at IRRI phytotron. They constructed a linkage map for rice where in every linkage group was placed to analogous chromosome. The AFLP map consisted of 206 AFLP markers and was generated from 32 primer combinations of Pst 1 and Mse 1 using the 80 RILs. These RILs were chosen from the extreme tails of the population in response to salinity tolerance. The overall polymorphism was 19.1% while it was 0–39.3% for individual primer combination. The map length was 181.4 cM with a mean interval size of 10.53 cM. For assigning the linkage groups to chromosome, 42 published-mapped AFLP markers were utilized as anchor markers which are dispersed over the 12 chromosomes. The left over AFLP markers were then allocated to specific chromosomes based on their linkage to anchor markers. The AFLP map was correspondent to the reference RFLP/AFLP, having the anchors in the same order with the reference maps. Additional, tests with an STS marker exhibited that it mapped in the estimated position in the AFLP map.

Kawasaki et al. (2001) reported gene expression profiles during the early phase of salt stress in rice. They investigated transcript regulation in response to high salinity for salt-tolerant rice (Var Pokkali) with microarrays comprising 1728 cDNAs from libraries of salt-stressed roots. There was a decline to the extent of one tenth of prestress value of photosynthesis by NaCl at 150 mM, in a few minutes. An up regulation of transcripts was observed in pokkali. Further, they detected that within an hour of salt stress more or less 10% of the transcripts in Pokkali were considerably upregulated or downregulated. Though the initial disparities prevailed for some hours between the stressed and non-stressed plants, these became less evident as the plants adapted over time. The upregulated functions that were experiential with Pokkali at altered time points during stress adaptation have been altered over time. There was an enhanced protein synthesis and protein turnover at early time points. This was later followed by the induction of known stress-responsive transcripts within hours, and the induction of transcripts for defense-related functions later. After 1 week, the nature of upregulated transcripts (e.g., aquaporins) designated recovery (Figure 7.4).

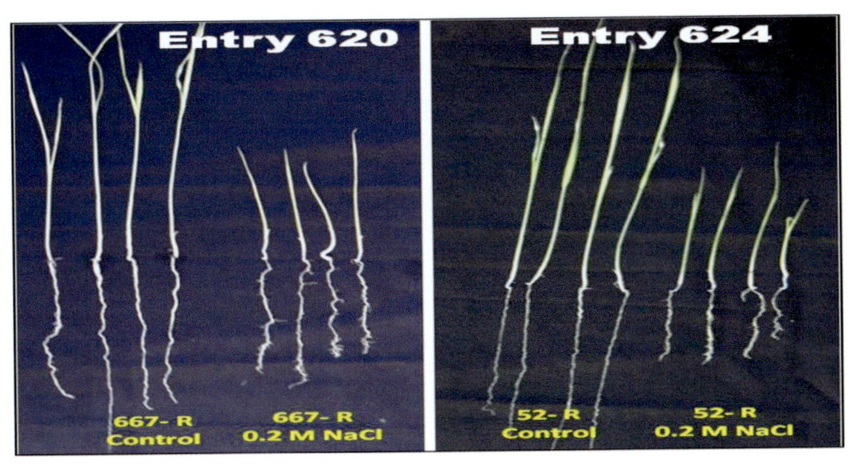

FIGURE 7.4 Twenty-five days aged seedlings: High reduction over control of seedling growth was observed in variety entry 624 under 0.15 M NaCl compared with variety entry 620.

Dubouz et al. (2003) stated that in rice, *OsDREB* genes encode transcription activators which play a role in drought-, high-salt-, and cold-responsive gene expression. The TFs DREBs/CBFs explicitly interact with the dehydration-responsive element/C-repeat (DRE/CRT) *cis*-acting element (core motif: G/ACCGAC) and regulate the expression of many stress-inducible genes in *Arabidopsis*. In rice, five cDNAs for *DREB* homologs: *OsDREB1A*, *OsDREB1B*, *OsDREB1C*, *OsDREB1D*, and *OsDREB2A*were isolated. Expression of *OsDREB1A* and *OsDREB1B* was induced by cold, while expression of *OsDREB2A* was induced by dehydration and high-salt stresses. The OsDREB1A and OsDREB2A proteins precisely bound to DRE and triggered the transcription of the *GUS* reporter gene driven by DRE in rice protoplasts. Over-expression of OsDREB1A in transgenic *Arabidopsis* induced over-expression of target stress-inducible genes of *Arabidopsis* DREB1A giving plants with higher tolerance to drought, high-salt, and freezing stresses. This unveiled that OsDREB1A exhibits functional resemblance to DREB1A. But, in microarray and RNA blot analyses, several stress-inducible target genes of the DREB1A proteins that have only ACCGAC as DRE were not over-expressed in the OsDREB1A transgenic *Arabidopsis*. The OsDREB1A protein bound to GCCGAC more favorably than to ACCGAC whereas the DREB1A proteins bound to both GCCGAC and ACCGAC proficiently. The structures of DREB1-type ERF/AP2 domains in monocots are directly associated to each other as compared with that in the dicots. *OsDREB1A* is possibly beneficial for producing transgenic monocots that are tolerant to drought, high-salt, and/or cold stresses.

Lin et al. (2004) identified QTLs for Na^+ and K^+ uptake of the shoots and roots regulating rice salt tolerance. They produced an F_2 and an equivalent F_3 population obtained from a cross between a high salt-tolerance *indica* variety, Nona Bokra, and a susceptible elite *japonica* variety, Koshihikari. They conducted QTL mapping for physiological characters associated to rice salt-tolerance. They detected three QTLs for survival days of seedlings (SDSs) under salt stress on chromosomes 1, 6, and 7, correspondingly, which elucidated 13.9% to 18.0% of the total phenotypic variance. Based on the associations amid SDSs and other physiological characters, they measured that injury of leaves was due to build-up of Na^+ in the shoot by transport of Na^+ from the root to the shoot in external high concentration. They set up eight QTLs comprising three QTLs for three traits of the shoots, and five for four traits of the roots at five chromosomal regions, controlled compound physiological characters associated to rice salt-tolerance under salt stress. Among these QTLs, the two chief QTLs with incredibly huge effect, qSNC-7 for shoot Na^+ concentration and qSKC-1 for shoot K^+ concentration, explained 48.5% and 40.1% of the total phenotypic variance, respectively. The QTLs noticed between the shoots and the roots practically did not occur on the same map locations. This suggests that the different genes control the transport of Na^+ and K^+ between the shoots and the roots.

Niones (2004) report mapping of the salinity tolerance gene on chromosome 1 of rice (*Oryza sativa* L.) using NIL. They mentioned that existing methods for QTL analyses gives fairly accurate locations of QTLs on chromosome. The region possesses three common salinity QTLs (Na+, K+ and Na+/K+ absorption) in chromosome 1. They evaluated the merits of a near-isogenic lines (NIL) for five mapping of the major QTLs gene for salt tolerance in chromosome 1 of rice using microsatelite markers, to saturate the chromosome 1 segment where the major gene controlling the Na^+:K^+ ratio responsible for salt tolerance, to phenotype the NIL (BC3F4) with salinity tolerance of Pokalli in IR29 background, and to perform a linkage analysis of NIL to the salinity tolerance gene. The NILs were used to five map quantitative trait loci for salinity tolerance in target region of chromosome 1 segment. Then they utilized phenotyping and genotyping in the NILs population. Eight microsatellite markers were used to generate a high-resolution genetic linkage map with an average interval size of 1.2 cM, and the total length of 8.234 cM. CP6224 marker in the single marker analysis was found with noteworthy (P.001) relationship in QTL. Two QTL peaks were detected in the interval analysis at the RM140 marker locus and between the flanking CP03970 and CP06224 markers with LOD value of 5.02 and 3.34, respectively. Microsatellite markers flanking with the QTL peak also reveal significant correlation to the trait (EC18).

Lee et al. (2007) undertook mapping of QTLs linked to salinity tolerance of rice at the young seedling stage. They used a population of RILs of the 164 genotypes resultant from a cross between 'Milyang 23' (*indica*) and 'Gihobyeo' (*japonica*) in rice (*Oryza sativa* L.), for salt tolerance at a young seedling stage in concentrations of 0.5% and 0.7% NaCl. They carried out mapping QTLs linked to salt tolerance by interval mapping with Qgene 3.0. Two QTLs (*qST1* and *qST3*) which conferred salt tolerance at young seedling stage were mapped on chromosome 1 and 3, respectively, and elucidated 35.5–36.9% of the total phenotypic variation in 0.5% and

0.7% NaCl. The favorable allele of *qST1* was donated by 'Gihobyeo,' and that of *qST3* by 'Milyang 23.' The results got in 0.5% and 0.7% NaCl for 2 years were alike in flanked markers and phenotypic variation.

Koh et al. (2007) reported the T-DNA tagged KO mutation of rice*OsGSK1*, which is an orthologue of *Arabidopsis BIN2*, with increased tolerance to different abiotic stresses. They screened T-DNA-tagged rice plants under cold- or salt-stress conditions to find out the genes concerned in the molecular mechanism for their abiotic-stress response. Line 0-165-65 was recognized as a salt-responsive line. The gene accountable for this GUS-positive phenotype was exhibited by inverse PCR as *OsGSK1* (*Oryza sativa glycogen synthase kinase 3-like gene 1*), a member of the plant GSK3/SHAGGY-like protein kinase genes and an orthologue of the *Arabidopsis brassino steroid in sensitive 2 (BIN2), AtSK21*. Northern blot analysis revealed that *OsGSK1* as mainly exceedingly noticed in the developing panicles. This proposes that its expression is a developmental stage-specific. KO mutants of*OsGSK1* exhibited better tolerance to cold, heat, salt, and drought produced a stunted growth phenotype comparable to the one observed with the gain-of-function *BIN/AtSK21* mutant. This proposes that*OsGSK1* could be a functional rice orthologue serving as a negative regulator of brassinosteroid (BR)-signaling. Thus, they suggest that stress-responsive *OsGSK1* may play physiological roles in stress signal-transduction pathways and floral developmental processes.

Wang et al. (2008) report overexpression of a rice *OsDREB1F* gene increases salt, drought, and low-temperature tolerance in both *Arabidopsis* and rice. They mention that DREB TFs play important roles in plant stress signaling transduction pathway. Further, they can exclusively bind DRE/CRT element (G/ACCGAC) and activate the expression of several stress inducible genes. They cloned a new rice DREB transcription factor, *OsDREB1F* and characterized via subtractive suppression hybridization (SSH) from upland rice. Expression analysis exhibited that *OsDREB1F* gene was induced by salt, drought, cold stresses, and also by ABA application, but not by pathogen, wound, and H_2O_2. Subcellular localization of this gene specified that this OsDREB1F is localized in nucleus. Transgenic plants hosting *OsDREB1F* gene improved tolerance to salt, drought, and low temperature in both rice and *Arabidopsis*. The additional characterization of *OsDREB1F*-overexpressing *Arabidopsis* revealed that, besides activating the expression of COR genes which has DRE/CRT element in their upstream promoter regions, the expression of *rd29B* and *RAB18* genes were also activated, suggesting that OsDREB1F may also take part in ABA-dependent pathway.

Pandit et al. (2010) combined QTL mapping and transcriptome profiling of bulked RILs for recognition of functional polymorphism for salt tolerance genes in rice (*Oryza sativa* L.). They state that detection of genes for quantitative traits is hard by usage of any particular method, because of composite inheritance of the traits and partial capability of the individual techniques. In this study, they used a combination of genetic mapping and bulked transcriptome profiling to narrow down the number of differentially expressed salt-responsive genes in rice for identification of functional polymorphism of genes basic to the QTLs. They used a population of RILs resultant from cross between salt-tolerant variety CSR 27 and salt-sensitive variety MI 48 for mapping QTL for salt ion concentrations in various tissues and salt stress susceptibility index (SSI) for SF, grain weight, and GY. They mapped eight important QTL intervals on chromosomes 1, 8, and 12 for the salt ion concentrations and a QTL controlling SSI for SF was co-located in one of these intervals on chromosome 8. However, there was totality of 2,681 genes in these QTL intervals, which makes it hard to locate the genes accountable for the functional variations for the traits. Likewise, transcriptome profiling of the seedlings of tolerant and sensitive parents grown under control and salt-stress conditions exhibited 798 and 2,407 differentially expressed gene probes, respectively. By analysis of pools of RNA extracted from ten each of extremely tolerant and extremely sensitive RILs to regularize the background noise, the number of differentially expressed genes under salt stress was severely declined to 30 only. They observed that two of these genes, an integral transmembrane protein DUF6 and a cation chloride co-transporter, were not only co-located in the QTL intervals but also had predictable distortion of allele frequencies in the extreme tolerant and sensitive RILs, and thus are suitable for future validation studies and improvement of functional markers for salt tolerance in rice to assist marker-assisted breeding.

Xiong et al. (2014) report that overexpression of *OsMYB48-1*, a new MYB-related transcription factor, and increases drought and salinity tolerance in rice. They mention that MYB-type TFs have important roles to play in plant growth, development, and response to environmental stresses. But the role of MYB-related TFs of rice in drought stress tolerance is not well known. In this paper, the authors report the isolation and characterization of a novel MYB-related TF, *OsMYB48-1*, of rice. Expression of *OsMYB48-1* was highly induced by PEG, ABA, H_2O_2, and dehydration, while being slightly induced by high salinity and cold treatment. The OsMYB48-1 protein was located in the nucleus with transactivation activity at the C terminus. Overexpression of *OsMYB48-1* in rice significantly improved tolerance to simulated drought and salinity stresses produced by mannitol, PEG, and NaCl, respectively, and drought stress was produced by drying the soil. On the contrary, to wild type plants, the overexpression lines had decreased rate of water loss, lower MDA content and higher proline content under stress conditions. Furthermore, overexpression plants were hypersensitive to ABA at both germination and post-germination stages and collected more endogenous ABA under drought stress conditions. Additional analyses proved that overexpression of *OsMYB48-1* could control the expression of several ABA biosynthesis genes (*OsNCED4*, *OsNCED5*), early signaling genes (*OsPP2C68*, *OSRK1*), and late responsive genes (*RAB21*, *OsLEA3*, *RAB16C*, and *RAB16D*) under drought stress conditions. Altogether, these outcomes proposed that *OsMYB48-1* functions as a new MYB-related TF which plays a positive role in drought and salinity tolerance by regulating stress-induced ABA synthesis.

7.5.3 GENETIC

Thomson et al. (2010) characterized Pokkali-derived QTLs for seedling stage salinity tolerance for use in marker-assisted breeding. They made an analysis of 100 SSR markers on 140 IR29/Pokkali RILs which established the location of the *Saltol* QTL on chromosome 1 and also recognized extra QTLs linked with tolerance. Analysis of a series of backcross lines and NIL were developed to illustrate the effect of the *Saltol* locus. The results unveiled that *Saltol* primarily functioned to regulate shoot Na^+/K^+ homeostasis. Multiple QTLs were needed to find a high level of tolerance. It is unpredictably observed that multiple Pokkali alleles at *Saltol* were noticed within the RIL population and between backcross lines, and representative lines were then compared with seven Pokkali accessions for better characterization of this allelic variation. Thus, in the event of the *Saltol* locus presenting a complex scenario, it offers an opportunity for marker-assisted backcrossing to increase salt tolerance of popular varieties which is followed by pointing multiple loci through QTL pyramiding for areas with higher salt stress.

7.6 HEAVY METAL TOLERANCE

Wang et al. (2000) report silicon induced cadmium tolerance of rice seedlings. Cadmium (Cd^{2+}) toxicity and effects of silicon (Si) applications on the cellular and intracellular accumulations and distributions of Cd were studied by conventional electron microscopy and EDX analysis. The Si-deprived rice (*Oryza sativa* L.) plants (-Si) varied largely from Si-replete ones in cell walls and vacuoles distributions of Ca in their leaves and roots. Energy dispersive x-ray microanalysis indicated that huge quantities of Cd could be found in the cytoplasm, vacuole or cellular organelles in-Si rice plants, while very little to be found in +Si ones. From the nanochemical and nano biological viewpoints, cell wall templates facilitated the formation of colloidal silica with the high specific adsorption property to prevent the uptake of Cd into the cell.

Hall (2002) discussed cellular mechanisms for heavy metal detoxification and tolerance. They mention that heavy metals such as Cu and Zn are important for normal plant growth, though increased concentrations of both essential and non-essential metals can affect growth inhibition and toxicity symptoms. Plants have an array of possible cellular mechanisms that may be involved in the detoxification of heavy metals and thus impart tolerance to metal stress. These involve roles for mycorrhiza and for binding to cell wall and extracellular exudates; for decreased uptake or efflux pumping of metals at the plasma membrane (PM); for chelation of metals in the cytosol by peptides such as phytochelatins; for the repair of stress-damaged proteins; and for the compartmentalization of metals in the vacuole by tonoplast-located transporters. This study offers a wide synopsis of the proof for a participation of each mechanism in heavy metal detoxification and tolerance.

Xiong et al. (2009) reported that exogenous nitric oxide improves cadmium tolerance of rice by increasing pectin and hemicellulose contents in root cell wall. In this study, the processes of exogenous NO contribution to alleviate the cadmium (Cd) toxicity in rice (*Oryza sativa*), rice plantlets exposed to 0.2-mM $CdCl_2$ and treated with different concentrations of sodium nitroprusside (SNP, a NO donor). Cd toxicity was assessed by the reductions in plant length, biomass production, and chlorophyll content. The findings revealed that 0.1 mM SNP alleviated Cd toxicity more clearly. Atomic absorption spectrometry and fluorescence localization revealed that treatment with 0.1 mM SNP reduced Cd accumulation in both cell walls and soluble fraction of leaves; although treatment with 0.1 mM SNP increased Cd, accumulation in the cell wall of rice roots obviously. Treatment with 0.1 mM SNP in nutrient solution exhibited slight effect on the transpiration rate of rice leaves, but this treatment improved pectin and hemicellulose content and decreased cellulose content significantly in the cell walls of rice roots. Based on these results, they deduce that reduced distribution of Cd in the soluble fraction of leaves and roots and improved distribution of Cd in the cell walls of roots are accountable for the NO-induced increase of Cd tolerance in rice. It seems that exogenous NO increases Cd tolerance of rice by increasing pectin and hemicellulose content in the cell wall of roots, increasing Cd accumulation in root cell wall and declining Cd accumulation in soluble fraction of leaves.

Kumar et al. (2013) report expression of a rice Lambda class of glutathione S-transferase, *OsGSTL2*, in *Arabidopsis* confers tolerance to heavy metal and other abiotic stresses. They mention that global industrial growth has polluted the soil and water with several harmful compounds, including heavy metals. These heavy metals besides being toxic to plants cause severe human health when leach out into the food chain. One of the methods employed for the decontamination of the environment involves the identification and overexpression of genes involved in the detoxification mechanism of plants. GSTs are a superfamily of enzymes, mainly known for their role in detoxification reactions. Diverse classes of GSTs have utilized to cultivate plants with enhanced detoxification mechanisms, but much information is not available for Lambda-class of GSTs. In this paper, the authors studied the expression of *OsGSTLs* in various rice genotypes under arsenic stress. The study proposes a differential expression of these genes in arsenic sensitive and tolerant genotypes. Besides, the role of one member of Lambda-class *OsGSTL2* was studied by expressing in a heterologous system, *Arabidopsis*. Transgenic lines developed were examined for their response to different abiotic stresses including heavy metals. Analysis indicated that *OsGSTL2* offers tolerance for heavy metals and other abiotic stresses like cold, osmotic stress and salt. They determine that *OsGSTLs* can be used for developing plant varieties tolerant to various abiotic stresses comprising heavy metals.

7.7 CONCLUSION

Under field conditions, plants encounter various abiotic stresses; their effect on plant and research advances that have were undertaken for understanding the effects of these stress factors has been discussed. Abiotic stresses caused by temperature and water are being intensified as a consequence of climate change. To overwhelm this problem and meet the worldwide demand for food, tolerant crops must be developed by means of suitable technologies. Physiological aspects, mechanisms of tolerance and management strategies for better crop production must be widely studied to effectively use such technologies.

KEYWORDS

- **abiotic stress**
- **alternate wetting and drying**
- **amplified fragment length polymorphism**
- **brassinosteroid**
- **drought**
- **heat stress transcription factor**

REFERENCE

Abdolhamid, S. A., Septiningsih, E. M., Mackill, D. J., & Ismail A. M., (2010). QTLs associated with tolerance of flooding during germination in rice (*Oryza sativa* L.). *Euphytica*, *172*, 159–168.

Agarie, S., Hanaoka, N., Ueno, O., Miyazaki, A., Kubota, F., Agata, W., & Kaufman, P. B., (1998). Effects of silicon on tolerance to water deficit and heat stress in rice plants (*Oryza sativa* L.), monitored by electrolyte leakage. *Plant Production Science*, *1*, 96–103.

Andaya, V. C., & Mackill, D. J., (2003). Mapping of QTLs associated with cold tolerance during the vegetative stage in rice. *Journal of Experimental Botany*, *54*, 2579–2585.

Andaya, V. C., & Tai, T. H., (2006). Fine mapping of the qCTS12 locus, a major QTL for seedling cold tolerance in rice. *Theoretical and Applied Genetics*, *113*, 467–475.

Cao, Y. Y., Duan, H., Yang, L. N., Wang, Z. Q., Zhou, S. C., & Yang, J. C., (2008). Effect of heat stress during meiosis on grain yield of rice cultivars differing in heat tolerance and its physiological mechanism. *Acta Agronomica Sinica*, *34*, 2134–2142.

Chang, T. T., Armenta, S. J. L., Mao, C. X., Peiris, R., & Loresto, G. C., (2008). Genetic studies on the components of drought in rice (*Oryza sativa* L.). *Rice Genetics Collection, 1*, 387–398.

Chen, X., Zhang, W., Zhang, B., Zhou, J., Wang, Y., Yang, Q., Ke, Y., & He, H., (2011). Phosphoproteins regulated by heat stress in rice leaves. *Proteome Science, 9*, 305–312.

Cheng, L. R., Wang, J. M., Uzokwe, V., Meng, L. J., Wang, Y., Sun, Y., Zhu, L., Xu, J. L., & Li, Z. K., (2012). Genetic analysis of cold tolerance at seedling stage and heat tolerance at anthesis in rice (*Oryza sativa* L.). *Journal of Integrative Agriculture, 11*, 359–367.

Chezhian, P., Jeyaprakash, P., Ganesh, S. K., Palchamy, A., Sadasivam, S., Sarkarung, S., et al., (2003). Genetic analysis of drought resistance in rice by molecular markers. *Crop Science, 43*, 1457–1469.

Du, H., Wu, N., Fu, J., Wang, S., Li, X., Xiao, J., & Xiong, L., (2012). A GH3 family member, OsGH3-2, modulates auxin and abscisic acid levels and differentially affects drought and cold tolerance in rice. *Journal of Experimental Botany, 63*, 6467–6480.

Dubouze, J. G., Sakuma, Y., & Ito, Y., (2003). *OsDREB* genes in rice, *Oryza sativa* L. encode transcription activators that function in drought, high salt, and cold-responsive gene expression. *The Plant Journal, 33*, 751–763.

Ekanayake, I. J., O´Toole, J. C., Garrity, D. P., & Masajo, T. M., (1985). Inheritance of root characters and their relations to drought resistance in rice. *Crop Science, 25, 927–933.*

Garrity, D. P., & O´Toole, J. C., (1994). Screening rice for drought resistance at the reproductive phase. *Field Crops Research, 39*, 99–110.

Gregorio, G. B., (1997). *Tagging Salinity Tolerance Genes in Rice Using Amplified Fragment Length Polymorphism (AFLP)*. Philippines Univ. Los Banos, College, Laguna (Philippines).

Gregorio, G. B., Senadhira, D., Mendoza, R. D., Manigbas, N. L., Roxas, J. P., & Guerta, C. Q., (2002). Progress in breeding for salinity tolerance and associated abiotic stresses in rice. *Field Crops Research, 76*, 91–101.

Hall, J. L., (2002). Cellular mechanisms for heavy metal detoxification and tolerance. *Journal of Experimental Botany, 53*, 1–11.

Heenan, D. P., Lewin, L. G., & McCaffery, D. W., (1988). Salinity tolerance in rice varieties at different growth stages. *Australian Journal of Experimental Agriculture, 28*, 343–349.

Hu, H., Dai, M., Yao, J., Xiao, B., Li, X., Zhang, Q., & Xiong, L., (2006). Over expressing a NAM, ATAF, and CUC (NAC) transcription factor enhances drought resistance and salt tolerance in rice. *PNAS, 103*, 12987–12992.

Hu, H., You, J., Fang, Y., Zhu, X., Qi, Z., & Xiong, L., (2008). Characterization of transcription factor gene *SNAC2* conferring cold and salt tolerance in rice. *Plant Molecular Biology, 67*, 169–181.

Hurab, J., Junga, K. H., Lee, C. H., & An, G., (2004). Stress-inducible OsP5CS2 gene is essential for salt and cold tolerance in rice. *Plant Science, 167*, 417–426.

Jagadish, S. V. K., Craufurd, P. Q., & Wheeler, T. R., (2008). Phenotyping parents of mapping populations of rice for heat tolerance during a thesis. *Crop Science- Crop Physiology and Metabolism, 48*, 1140–1146.

Jagadish, S. V. K., Muthurajan, R., Oane, R., Wheeler, T. R., Heuer, S., Bennett, J., & Craufurd, P. Q., (2010). Physiological and proteomic approaches to address heat tolerance during a thesis in rice (*Oryza sativa* L.). *Journal of Experimental Botany, 61*, 143–156.

Kawasaki, S., Borchert, C., Deyholos, M., Wang, H., Brazille, S., Kawai, K., Galbraith, D., & Bohnert, H. J., (2001). Gene expression profiles during the initial phase of salt stress in rice. *The Plant Cell*, *13*, 889–905.

Koh, S., Lee, S. C., Kim, M. K., Koh, J. H., Lee, S., An, G., Choe, S., & Kim, S. R., (2007). T-DNA tagged knockout mutation of rice *OsGSK1*, an orthologue of *Arabidopsis BIN2*, with enhanced tolerance to various a biotic stresses. *Plant Molecular Biology*, *65*, 453–466.

Komatsu, S., Yang, G., Khan, M., Aruko, O., Seiichi, T., & Masayuki, Y., (2007). Over-expression of calcium-dependent protein kinase 13 and calreticulin interacting protein 1 confers cold tolerance on rice plants. *Molecular Genetics and Genomics*, *277*, 713–723.

Koseki, M., Kitazawa, N., Yonebayashi, S., Maehara, Y., Wang, Z. X., & Minobe, Y., (2010). Identification and fine mapping of a major quantitative trait locus originating from wild rice, controlling cold tolerance at the seedling stage. *Molecular Genetics and Genomics*, *284*, 45–54.

Krishna, J. S. V., Muthurajan, R., Rang, Z. W., Malo, R., Heuer, S., Bennett, J., & Craufurd, P. Q., (2011). Spikelet proteomic response to combined water deficit and heat stress in rice (*Oryza sativa* cv. N22). *Rice*, *4*, 1–11.

Kumar, S., Asif, M. H., Chakrabarty, D., Tripathi, R. D., Dubey, R. S., & Kumar, P. T., (2013). Expression of a rice Lambda class of glutathione S-transferase, *OsGSTL2*, in *Arabidopsis* provides tolerance to heavy metal and other a biotic stresses. *Journal of Hazardous Materials*, *248*, 228–237.

Lee, B. H., Won, S. H., Lee, H. S., Miyao, M., Chung, W. I., Kim, I. J., & Jo, J., (2000). Expression of the chloroplast-localized small heat shock protein by oxidative stress in rice. *Gene*, *245*, 283–290.

Lee, S. Y., Ahn, J. H., Cha, Y. S., Yun, D. W., Lee, M. C., Ko, J. C., Lee, K. S., & Eun, M. Y., (2007). Mapping QTLs related to salinity tolerance of rice at the young seedling stage. *Plant Breeding*, *126*, 43–46.

Lin, H. X., Zhu, M. Z., Yano, M., Gao, J. P., Liang, Z. W., Su, W. A., Hu, X. H., Ren, Z. H., & Chao, D. Y., (2004). QTLs for Na^+ and K^+ uptake of the shoots and roots controlling rice salt tolerance. *Theoretical and Applied Genetics*, *108*, 253–260.

Lou, Q., Chen, L., Sun, Z., Xing, Y., Li, J., Xu, X., Mei, H., & Luo, L., (2007). A major QTL associated with cold tolerance at seedling stage in rice (*Oryza sativa* L.). *Euphytica*, *158*, 87–94.

Lutts, S., Kinet, J. K., & Bouharmont, J., (1996). Ethylene production by leaves of rice (*Oryza sativa* L.) in relation to salinity tolerance and exogenous putrescine application. *Plant Science*, *116*, 15–25.

Mackill, D. J., Coffman, W. R., & Rutger, J. N., (1982). Pollen shedding and combining ability for high temperature tolerance in rice. *Crop Science*, *22*, 730–733.

Mitchell, J. H., Siamhan, D., Wamala, M. H., Risimeri, J. B., Chinyamakobvu, E., Henderson, S. A., Fukai S., (1998). The use of seedling leaf death score for evaluation of drought resistance of rice. *Field Crops Research*, *55*, 129–139.

Nguyen, H. T., Babu, R. C., & Blum, A., (1997). Breeding for drought resistance in rice: Physiology and molecular genetics considerations. *Crop Science*, *37*, 1426–1434.

Niones, J. M., (2004). Five mapping of the salinity tolerance gene on chromosome 1 of rice (*Oryza sativa* L.) using near-isogenic lines. *Agris*. http://agris.fao.org/agris-search/search.do?recordID=PH2005000108

Pandit, A., Rai, V., & Bal, S., (2010). Combining QTL mapping and transcriptome profiling of bulked RILs for identification of functional polymorphism for salt tolerance genes in rice (*Oryza sativa* L.). *Plant Science*, *284*, 121–136.

Pereira, R. C., Sperotto, R. A., & Cargnelutti, D., (2013). Avoiding damage and achieving cold tolerance in rice plants. *Food and Energy Security Food and Energy Security*, *2*, 96–119.

Plett, D., Safwat, G., Gilliham, M., Møller, I. S., Roy, S., Shirley, N., Jacobs, A., Johnson, A., & Tester, M., (2010). Improved salinity tolerance of rice through cell type-specific expression of AtHKT1;1. *PLoS ONE, 5*(9), e12571. https://doi.org/10.1371/journal.pone.0012571.

Prasad, P. V. V., Boote, K. J., Allen, L. H., Heehy, J. E., & Thomas, J. M. G., (2006). Species, ecotype, and cultivar differences in spikelet fertility and harvest index of rice in response to high temperature stress. *Field Crops Research*, *95*, 398–411.

Price, A., & Courtois B., (1999). Mapping QTLs associated with drought resistance in rice: Progress, problems, and prospects. *Plant Growth Regulation*, *29*, 123–133.

Qian, Q., Dali, Z., Ping, H., Xianwu, Z., Ying, C., & Lihuang, Z., (2000). QTL analysis of the rice seedling cold tolerance in a double haploid population derived from another culture of a hybrid between *indica* and *japonica* rice. *Chinese Science Bulletin*, *45*, 448–453.

Ranjan, A. B., Ishigo, N. O., Adachi, M., Oguma, Y., Tokizono, Y., Onishi, K., & Sano, Y., (2009). Cold tolerance at the early growth stage in wild and cultivated rice. *Euphytica*, *165*, 459–470.

Roy, M., & Ghosh, B., (1996). Polyamines, both common and uncommon, under heat stress in rice (*Oryza sativa*) callus. *Physiologia Plantarum*, *98*, 196–200.

Satake, Y. S., & Mackill, T. D. S., (1981). *High-Temperature Stress in Rice*. International rice research inst., college, Laguna, Philippines.

Serraj, R., Nally, K. L., Loedin, I. S., Kohli, A., Haefele, S. M., & Atlin, G., (2009). Drought resistance improvement in rice: An integrated genetic and resource management strategy. *Plant Production Science*, *14,* 1–14.

Shah, F., Huang, J., Cui, K., & Nie, L., (2011). Impact of high-temperature stress on rice plant and its traits related to tolerance. *Journal of Agricultural Science*, *149*, 545–556.

Singh, S., Mackill, D. J., & Ismail, A. M., (2011). Tolerance of longer-term partial stagnant flooding is independent of the SUB1 locus in rice. *Field Crops Research*, *121*, 311–323.

Su, C. F., Wang, Y. C., Hsieh, T. H., Lu, C. A., Tseng, T. H., & Yu, S. M., (2010). A novel MYBS3-dependent pathway confers cold tolerance in rice. *Plant Physiology, 153(1)*, 145–158.

Tao, Z., Kou, Y., Liu, H., Li, X., Xiao, J., & Wang, S., (2011). *OsWRKY45* alleles play different roles in abscisic acid signaling and salt stress tolerance but similar roles in drought and cold tolerance in rice. *Journal of Experimental Botany*, *62*, 4863–4874.

Thomson, M. J., Ocampo, M., Egdane, J., & Rahman, M. A., (2010). Characterizing the saltol quantitative trait locus for salinity tolerance in rice. *Rice*, *3*, 148–160.

Uchida, A., Jagendorf, A. T., Hibino, T., Takabe, T., & Takabe, T., (2002). Effects of hydrogen peroxide and nitric oxide on both salt and heat stress tolerance in rice. *Plant Science*, *163*, 515–523.

Wang, L., Wang, Y., Chen, Q., Cao, W., Li, M., & Zhang, F., (2000). Silicon induced cadmium tolerance of rice seedlings. *Journal of Plant Nutrition*, *23*, 1397–1406.

Wang, Q., Guan, Y., Wu, Y., Chen, H., Chen, F., & Chu, C., (2008). Over expression of a rice *OsDREB1F* gene increases salt, drought, and low temperature tolerance in both *Arabidopsis* and rice. *Plant Molecular Biology*, *67*, 589–602.

Xiong, H., Li, J., & Liu, P., (2014). Over expression of *OsMYB48–1*, a Novel MYB-related transcription factor enhances drought and salinity tolerance in rice. *PLoS One*. *9*(3), e92913. doi: 10.1371/journal.pone.0092913

Xiong, J., An, L., Lu, H., & Zhu, C., (2009). Exogenous nitric oxide enhances cadmium tolerance of rice by increasing pectin and hemicellulose contents in root cell wall. *Planta*, *230*, 755–765.

Yamanouchi, U., Yano, M., Lin, H., Ashikari, M., & Yamada, K., (2002). A rice spotted leaf gene, *Spl7*, encodes a heat stress transcription factor protein. *PNAS*, *99*, 7530–7535.

Yang, C., Li, D., & Mao, D., (2013). Over expression of microRNA319 impacts leaf morphogenesis and leads to enhanced cold tolerance in rice (*Oryza sativa* L). *Plant Cell and Environment*, *36*, 2207–2218.

Yang, J., Zhang, J., Liu, K., Wang Z., & Liu, L., (2007). Involvement of polyamines in the drought resistance of rice. *Journal of Experimental Botany*, *58*, 1545–1555.

Yao, F., Huang, J., Cui, K., Nie, L., Xiang, J., Liu, X., Wu, W., Chen, M., & Peng, S., (2012). Agronomic performance of high-yielding rice variety grown under alternate wetting and drying irrigation. *Field Crops Research*, *126*, 16–22.

Ye, C., Argayoso, M. A., & Redoña, E. D., (2012). Mapping QTL for heat tolerance at flowering stage in rice using SNP markers. *Plant Breeding*, *131*, 33–34.

Yue, B., Xiong, L., Xue, W., Xing, Y., Luo, L., & Xu, C., (2005). Genetic analysis for drought resistance of rice at reproductive stage in field with different types of soil. *Theoretical and Applied Genetics*, *111*, 1127–1136.

Zhang, G. Y., Guo, Y., Chen, S. L., & Chen, S. Y., (1995). RFLP tagging of a salt tolerance gene in rice. *Plant Science*, *110*, 227–234.

Zhang, J., Zheng, H, G., Aarti, A., & Pantuwan, G., (2001). Locating genomic regions associated with components of drought resistance in rice: Comparative mapping within and across species. *Theoretical and Applied Genetics*, *103*, 19–29.

CHAPTER 8

Biotic Stress Affecting Crop Productivity

ABSTRACT

The yield of rice is greatly affected by biotic stress because of insects and diseases. In this chapter, a brief account of major pests and diseases is provided. Significant advancement has been attained in understanding physiological, biochemical, and molecular mechanisms to these stress factors. The research advances that occurred are briefly presented in the following sections.

8.1 BIOTIC STRESSES

8.1.1 INSECT PESTS

8.1.1.1 PEST PROBLEM IN RICE

A book is published by Grist et al. (1969) makes a broad review of pests and the problems that arose due to these in rice grown all over the World. This includes an outline classification of the major insect pests, with consideration of the effects of climate and cultural practices on them, their life-cycles and their population dynamics.

8.1.1.2 BROWN PLANTHOPPER (BPH)

Preetinder et al. (2016) collected 1,989 wild accessions from 11 wild species and African cultivated rice and sown in seed boxes under greenhouse conditions. Then, 1,003 accessions with good germination were screened against brown planthopper (BPH). After one year of screening, 159 accessions were recognized as resistant. These selected resistant accessions were screened for one more year. From the two years screening data, seven accessions of *O. nivara* (AA), one accession of *O. officinalis* (CC), seven accessions of *O. australiensis* (EE), five accessions of *O. punctata* (BB and BBCC) and nine accessions of *O. latifolia* (CCDD) were proved to be resistant to BPH. In this study, BPH resistance genes from *O. nivara* and *O. punctate* were identified for the first time which would act as new sources of resistance (Figure 8.1).

FIGURE 8.1 Damage of BPH (*Nilaparvata lugens*): Farmer's field photo showing the severity of the pest attack.

8.1.2 STEM BORER AND GALL MIDGE

Breeding for insect-resistant varieties is a feasible and ecologically sound approach to control the pests. Status of advancement achieved in breeding and acceptance of resistant varieties against stem borers vs. gall midge displays two opposing situations. Owing to the dearth of resistance sources for yellow stem borer in cultivated rice gene pool and genetically complex inheritance resistance, the traditional resistance breeding for yellow stem borer has not got much stimulus. Therefore, other approaches such as wide hybridization involving integration of resistant genes from other *Oryza* species, transgenic approach to employ

Bt cry and other insecticidal genes and RNAi methods are being actively practiced. Contrary to this, for gall midge high level of resistance sources are accessible in the crossable gene pool, resistance genetics is well studied, molecular markers linked to resistant genes are developed and many resistant varieties have been released for commercial cultivation. So far, 7 gall midge biotypes (GMB) and 11 plant resistance genes have been detected. However, the diversity in gall midge populations and continuous selection of virulent biotypes demand supplementation of conventional breeding methods with molecular and transgenic approaches. Current progress in the molecular breeding methods and transgenic rice biotechnology offers a great possibility for improved varietal tolerance to biotic stresses (Gurpreet and Bentur, 2017).

8.1.2.1 RICE STEM BORER

The yellow stem borer is one of the harmful pests of rice (Figure 8.2). But, limited knowledge is available on the biology and processes of YSB growth and differentiation. Therefore, to locate the genes involved in YSB development and infestation, de novo transcriptome analysis was assumed by Renuka et al. (2017). They employed Illumina Hi-seq at the first, third, fifth, and seventh larval developmental stages. During the developmental stages, genes related with different metabolic processes, i.e., detoxification mechanism, RNA interference (RNAi) machinery, chemoreception, and regulators (transcription factors (TFs) and hormones) were differentially regulated. With the help of stage-specific transcripts, the essential processes of larval development were determined. Comparative transcriptome analysis showed that YSB has distinct RNAi machinery, but the detoxification mechanism is not much evolved in YSB. The monophagous nature of YSB was confirmed by the presence of strong specific visual recognition together with chemosensory mechanisms. In this study, the YSB transcriptome characterization and the identification of genes concerned with important processes were undertaken for the first time. This will assist researchers to formulate new pest control approaches and will help in further advancement of the breeding programs meant for developing YSB resistant rice varieties.

FIGURE 8.2 Rice stem borer and damage symptom.

Prasad et al. (2013) conducted a study, where they screened 202 semi deepwater rice genotypes and check varieties Jalpriya and Madhukar against yellow stem borer under natural field conditions. Their results revealed that the genotypes, Madak 13, WAB 878-4-2-2-3-P1-HP and NDGR 268 were highly resistant to YSB infestation with an average infestation of 0.67, 1.33 and 1.67%, respectively. These resistant genotypes may be utilized as donors for yellow stem borer resistance in future breeding programs.

8.1.2.2 GALL MIDGE

The most efficient and economic, measure of gall midge control is host plant resistance. The characterization of a new GMB 4M highlighted some distinctive features related with gall midge resistance in the breeding line CR57-MR1523. Further, a new dominant gene withholding resistance against GMB4 was identified in F_3 families obtained from the cross between TN1 and CR57MR1523. This gene has been termed as Gm11t. Even though, CR57-MR1523 has been widely employed in breeding gall midge resistant rice varieties like Suraksha, the genetic basis of resistance and chromosomal location of the resistance genes is still unknown. In this analysis paper, Himabindu et al. (2010) undertook tagging and mapping of gall midge resistance gene, Gm11t, on chromosome 12, with SSR markers. 466 F_{10} generation recurrent inbred lines (RILs) obtained from the cross between TN1 9 and CR57-MR1523 were employed for mapping the gene locus. Out of 471 SSR markers distributed throughout the rice genome, only six SSR markers viz. RM28706, RM235, RM17, RM28784, RM28574, and RM28564 on chromosome 12 were found to be primarily linked with resistance and susceptibility. In selected 158 RILs, the gene locus was mapped between two flanking SSR markers, RM28574 and RM28706, within 4.4 and 3.8 cM on chromosome 12 using linkage analysis. Additional, two NILs with 99% genetic resemblance and differing in gall midge resistance were identified from the RILs, which could be utilized in functional analysis of the identified candidate gene.

To estimate combining ability and gall midge resistance in hybrid rice are search was carried out by Begum et al. (2018), where they crossed six cytoplasmic male sterile (CMS) lines and seven testers in Line x Tester mating design to produce 42 hybrids. Then, the parental lines and hybrids were assessed for combining ability and resistance to gall midge. The analysis of variance unveiled the significant differences and predominance of non-additive gene action among all the yield components and resistance to gall midge. Based on GCA and SCA effects, some parental lines and hybrids with good yield potential and resistance to gall midge have been identified.

8.1.3 INVERTEBRATE PESTS

Plant protection against insect pests is crucial for maintaining good yields of staple food crop rice. Grigarick et al. (1984) discussed about the general problems with rice invertebrate pests attacking rice in USA and also about their control. They mentioned that most of the rice stands are declined in California, due to the attack of a small number of invertebrate pests viz., *Triops longicaudatus,* very familiar pest, though is easily controllable through insecticidal applications, has developed resistance to parathion in few areas; *Procambarus clarkia* is visible in sporadic occurrence in rice stands with a number of species of seed midges causing infestation of rice stands; *Lissorhoptrus oryzophilus*, usually found to occur widely in many of the rice growing locations and observed to cause a decrease in tillering, is controlled with carbofuran; and *Hydrellia griseola,* another significant pest, relegated as an occasional pest of low down water culture, in

California; *Pseudaletia unipuncta* [*Mythimna unipuncta*] and *Spodoptera praefica* involved in causing impaired panicle development on attack, can be brought down by Carbaryl; and *Macrosteles fascifrons,* whose populations can effectively brought down from their build up through before time control of broad-leafed weeds. Similarly, it is noted that the grain development may be affected or turn out to be unfit for sale in southern states, due to attacks by *Oebalus pugnax,* which is easily controllable through the sprays of parathion-methyl, carbaryl, malathion.

8.2 RICE INSECT MANAGEMENT

8.2.1 MANAGEMENT OF TROPICAL RICE PESTS

The cultivation of tropical Asian rice appears to have its origin 9,000 years past. It is a representative of an unsurpassed ecologically complex agricultural ecosystem. Settle et al. (1996) discussed about the possibility of managing the rice pests of tropical origin through conservation of their substituted preys and their natural usual enemies. In their study on community ecology of the irrigated tropical rice fields of Indonesia, Java, there is the copious and sound spread of the predator populations that are mostly seen during the beginning of the season of numerous tropical paddy fields. An analysis of samples taken from surface, subsurface water and plants indicated that these elevated populations of generalist predators are supported to a large extent by plankton feeding insects and by detritus feeding in the near the beginning season. These populations reach a constant peak and are later turned down in the 1st third of the season. They assumed that this wealth of substitute prey gives the predator populations a huge start on shortly developing pest populations and as such, these developments powerfully curb the pest populations. Thus, it gives stability to the existing rice ecosystems. The stability is realized mostly because the predator populations are decoupled from their stringent dependence on herbivorous populations. They tested their assumption of trophic linkages amongst organic matter, detritivores, and plankton feeders, and generalist predators and discovered that by rising organic matter in test plots they could build up populations of detritivores and plankton feeders. This has in turn extensively increased the loads of generalist predators. They also established the relation between early-season natural enemy populations and afterward season pest populations by experimentally dropping early-season predator populations with insecticide applications, so that the pest populations resurge afterward in the season.

Matteson (2000) made a comprehensive of the pest management practices in irrigated rice of Tropical Asia. He mentions that in these regions there is the occurrence of copious innate enemies, which can considerably bring down the insect pest populations and problems. He emphasizes that an integrated pest management (IPM) extension education in depth and quality has to be called for discouraging the needless insecticide use. The indiscriminate use of insecticides is affecting the natural balance in paddy ecosystems. Farmers are to be trained so that they turn out to be as expert managers and be able to preserve a healthy rice ecosystem. Farmers' skill and alliance are predominantly vital for sustainable utilization of the probably new, higher-yielding, and pest-resistant rice varieties. The training and visit extension programs though failed in the transfer of this IPM technology, mass media campaigns are successful to some extent in the encouragement of the farmer participatory research and enabled in bringing down the insecticide use to some extent. The "farmer first" method of participatory non-formal education in farmer field schools (FFS), followed by community IPM activities given the emphasis on farmer-training-farmer and research by farmers, can bring greater success in attaining IPM execution. Extension challenges need to be undertaken which are a crucial subject for rice IPM research, and new pest management technology should encourage, rather than endangering, ecological balance in rice.

Berg (2001) discussed pesticide application in rice and rice-fish farms and undertook an inspection on pest management practices among rice and rice-fish farmers and their awareness of problems associated to pests and pesticides. Non-IPM farmers used double the pesticides rather than those used by IPM farmers. Similarly, the application frequencies and quantities of active ingredient used per crop were also 2–3 times high. Calculations of the usage of pesticides by IPM and non-IPM farmers all through the very last three years has shown that there was a decline in amount of pesticide usage by 65% by IPM farmers, while it was increased by 40% by non-IPM farmers. Further, their survey results have indicated that the pesticide usage in rice fields was low in those farmers field who had taken up fish growing also along with rice in fields, as the pesticidal application effects the fish cultures. In view of above observation, they mentioned that a long-term outlook of integration of rice-fish farming with IPM practices offers a sustainable choice to demanding rice mono-cropping, both from an economic and an ecological standpoint.

Adesina et al. (2008) discussed the problems of rice pests in the Ivory Coast, West Africa from the farmers' perceptions and its managing strategies. They mention that knowledge of these can give way in the progress of efficient IPM strategies. They mention that the moderately thriving rice IPM technologies in Asia have been widely achieved. However, such studies on the West African rice farmers are infrequently less. The success of the IPM strategies in this region appears limited. The authors narrate problems of rice pests, pest perceptions, and management practices of smallholder rice farmers in the Ivory Coast. These small-hold farmers encountered the problems from key rice pests, weeds, vertebrates (birds and rodents) and insects and couldn't effectively diagnose the diseases of rice. They never considered them as the main constraints of rice production. Therefore, they suggested that there is a need of necessitation of an IPM strategy that includes farmers' perceptions and builds on their existing cultural control practices.

8.2.2 MECHANISM OF RESISTANCE TO PLANTHOPPERS

Panda and Heinrichs (1983) determined the mechanisms of resistance by evaluating the levels of tolerance and antibiosis in a few rice varieties that had moderate resistance to the BPH, *Nilaparvata lugens*. They deliberate the host plant-insect interactions in the greenhouse and insectary to find out the levels of antibiosis and tolerance in the moderately resistant varieties. They evaluated parameters, such as plant damage, plant weight loss, and yield decline due to insect feeding measured as the tolerance indicator. They used the insect biomass production, growth index, and population growth rate on the test varieties as antibiosis indicators. A regression of plant weight loss, due to *N. lugens* feeding, on *N. lugens* weight for five varieties unveiled that the variety 'UtriRajapan' has a high level of tolerance and no antibiosis. The resistance in 'IR46,' 'Kencana,' and 'Triveni' is accredited to a combination of antibiosis and tolerance. The virtual quantity of each of the two resistance mechanisms was evaluated by utilizing the tolerance-antibiosis index (Figure 8.3).

FIGURE 8.3 Healthy rice seedling prior to release of BPH nymphs.

Alam and Cohen (1998) undertook identification and analysis of QTLs for resistance to the brown planthopper in a doubled-haploid rice population. They utilized a mapping population of 131 doubled-haploid lines, obtained from a cross between a developed *indica* rice variety (IR64) and a traditional *japonica* variety (Azucena), for detection of quantitative trait loci (QTLs) for resistance to the brown planthopper (BPH), *Nilaparvata lugens*. They assessed the parents and mapping population in six trials that determine varying combinations of the three essential mechanisms of insect host plant resistance, such as antixenosis, antibiosis, and tolerance. In order to find out the effect of the major resistance gene *Bph1* from IR64, they undertook the testing of two BPH populations from Luzon Island, The Philippines, that are almost fully adapted to this gene. They identified a total of seven QTLs related with resistance, which were located on 6 of the 12 rice chromosomes. Individual QTLs explained for between 5.1 and 16.6% of the phenotypic variance. Two QTLs were principally contributed a single resistance mechanism: one with antixenosis and one with tolerance. Many QTLs were obtained from IR64, which has exhibited a comparatively durable level of moderate resistance under field conditions. The authors recommend that these results would be helpful in transferring this resistance to other rice varieties (Figure 8.4).

FIGURE 8.4 Screening for BPH tolerance in net house.

Powell et al. (1998) studied immune histochemical and developmental studies to ascertain the mechanism of snowdrop lectin action on the rice brown planthopper. In this study, rice brown planthoppers were fed on artificial diet having snowdrop lectin (*Galanthus nivalis* agglutinin; GNA), toxic towards this insect pest. It was observed that along with decreasing survival, the lectin affected development, reduced the growth rate of nymphs by about 50% when present at a concentration of 5.3 μM. Immunolocalization studies revealed that lectin binding was concentrated on the luminal surface of the midgut epithelial cells within the planthopper. This suggests that GNA is fixed to cell surface carbohydrate moieties in the gut. Immunolabeling at a lesser level was also detected in the fat bodies, the ovarioles, and all through the hemolymph. These observations put forward that GNA can surpass through the midgut epithelial barrier, and exceed into the insect's circulatory system, causing a systemic toxic effect. Electron microscope studies revealed morphological alterations in the midgut region of planthoppers that were fed on a toxic dose of GNA, with anomalies of the microvilli brush border region. The existence of glycoproteins which bind GNA in the gut of the brown planthopper was proved using digoxigen-labeled lectins to probe blots of extracted gut polypeptides (Figures 8.5–8.7).

FIGURE 8.5 Second instar nymphs of BPH insects were released into trays and covered with net bags.

FIGURE 8.6 Variation in paddy varieties in response to BPH screening.

- C = control (normal condition, without BPH).
- B = BPH infested (BPH insects released).
- Highly BPH susceptible CHECK = TN1.
- Resistant to BPH= 808 (BHAVYA).
- Moderately resistant to BPH = 806 (MTU 1153).

FIGURE 8.7 Clear variation in varieties response to BPH infestation.

8.3 INTEGRATED PEST MANAGEMENT (IPM)

Teng (1994) has reported that IPM in rice is chiefly employed to rice ecosystems of irrigated, lowland paddy. These ecosystems utilize more inputs and simultaneously produce increased yields per hectare compared to rainfed ones. IPM was implemented on a large scale in many Asian countries in 1970s with the occurrences of pest resurgences due to unsystematic insecticide uses and in five countries, the legislations were enacted by 1992. By this period, roughly about 0.05% of Asian farmers had received training in IPM. Two countries adopted IPM in their national policies and in those countries, the imports of insecticides were declined much and led to enough savings and enhancements in rice productivity outputs. The inter-country programs on rice IPM of the Food and Agriculture Organization (FAO), the United Nations, the International Rice Research Institute (IRRI) and national programs have together donated to a greater extent in the promotion of the IPM in rice in several regions of the World. As host plant resistance is mostly influenced by the cultural practices followed which appear as location-specific, the basis for the implementation of low-cost sustainable IPM relies on the conservation of the natural enemies of insects pests in that region. Farmer's practices in the tropics are less influential by the monitoring of thresholds or other forecasting systems, but their group learning appears to be an efficient way for adoption of IPM.

Roling and Fliert (1994) had discussed in their model about the IPM in rice areas of Indonesia and the significance of transformation of the extension technologies for its better practice and implementation. Agricultural technology developed is transferred to the users, i.e., through some experts of agricultural extension. The agriculture extension model negates local knowledge and creativity. It ignores the self-confidence of the farmers' and also the importance of social energy. Both these are the main sources of change. Information from and about farmers is very crucial for anticipation of the utilization of the extension technology, however, the model in its largest part linear expression, does not pay much attention to them. Practically, it is observed that farmers generally rely on the knowledge that they gained and developed through practical experiences and very often reinvent ideas that were brought from outside. They then try to integrate them actively in the multifarious farming decisions. Thus, the effectiveness of the extension is mostly relied on the checks and balances that are likely to match with the farmer's intervention powers as well as the countervailing power and effectively be able to mobilize the creativity of the farmers and increase their adoption and participation in technology development and exchange of information. They suggested alternate models for informing extension investment, design, and practice stress adult learning and its easing. The farm development is mostly relied on the farmers energy, communication as he himself is the expert of farming. It proposes that a move to information intensive sustainable practices needs a learning process based on participation and empowerment.

Mangan and Mangan (1998) studied in China the effectiveness of two IPM training programs for farmers. Their study has shown that one of it is more effective in bringing down the pesticide applications and was also effective in imparting knowledge of the rice ecosystems to the farmers in a better manner. They used a triangulated method of measuring concept attainment among farmer trainees in China. Their results indicated that the FFS model of training, which was based upon a new ecology-based IPM model, is additionally efficient than the 3 Pests 3 Diseases (3P3D) model which was centered upon an ancient economic threshold IPM model.

Settle et al. (1996) explored the significance of natural enemies and alternate preys in management of the insect pest of the Tropical ecosystems. They have undertaken an investigation of the community ecology of irrigated tropical rice fields on Java, Indonesia under Indonesian National IPM Programme, whose chief concern was to impart training to the farmers so that they turn out into better agronomists and adopt at a quick pace the principles of IPM in their farming.

The review of Savary et al. (2012) reports four principles of basis for sustainable pest management in rice. The four principles are biodiversity, host plant resistance, landscape ecology, and hierarchies. The analysis indicates that the measured framework—human-pest environment crop applies, with all of its summits holding a diverse bearing depending on the pest considered. In addition to this, the review emphasizes the need for basic research across an array of disciplines, with new approaches and methods, as well as the need for connecting hierarchy levels, from farmers, to consumers, to societies, to environment, and to policies.

Pretty and Bharucha (2015) discussed about the significance of IPM for bringing sustainable agriculture in Asia and Africa. IPM is therefore most accompaniments and an option to the usage of synthetic pesticides. It strengthens the sustainable intensified agriculture, particularly, the farmers holding small farm holdings in the tropical regions. The statistical data indicated that in the last 20 years the global pesticide use has been more than 3.5 billion kg year^{-1}, with $45 billion worth of the global pesticide market. The IPM methods suggest that due to the consumption and usage of lower amounts of pesticides, not only farmers, but also wider environments and human health will be benefitted to a large extent; however, the evidences for its impact and pesticide use or on yield improvements remain very less. They evaluated the data from 85 IPM projects from 24 countries of Asia and Africa that have been applied for more than past twenty years. Their analysis has shown that IPM approaches had a yield increase of 40.9% and brought down the pesticide use by 30.7% compared to the baseline. Further, 35 out of 115 (30%) crop combinations have shown to be in a transition to zero pesticide use. They assessed the success of IPM and observed that 50% of the agroecosystems does not require the usage of pesticides with the adoption of IPM approaches. However, the policy support for IPM is comparatively very rare. The counter-interventions from pesticide industry always exist, to challenge the IPM as new pests, diseases, and weeds grow and move further always.

8.4 FACTORS INFLUENCING INSECT MANAGEMENT TECHNOLOGY

Harper et al. (1990) studied factors affecting the insect management technology adoption. They undertook Logit analyses to evaluate survey data with respect to rice stink bug [*Oebalus pugnax* (Fabricius)] management by Texas rice producers. They investigated the impacts of rice production traits on the implementation of insect sweep nets associated with treatment thresholds, and the spraying of insecticides for the management of the rice stink bug. They mention that the proportion of neighboring land use in the pasture, the proportion of rice acreage planted to semi-dwarf varieties, and producers' presence at particular field days had a noteworthy influence on the possibility of accepting sweep nets and treatment thresholds. Acceptance of sweep nets and treatment thresholds improve the chances of spraying by 11.3%.

Way and Heong (1994) made a review on the role of biodiversity in the dynamics and management of insect pests of tropical irrigated rice and discussed about it relating to variation within rice plants, rice fields, groups of rice fields and rice related ecosystems. It is determined that, under the distinctive cropping conditions and constant factors like water supply of tropical irrigated rice, the use of a comparatively few controllable components of diversity can convey stability so that pests are mostly retained at levels which do not explain the usage of insecticides. Long-lasting rice plant resistance, together with moderate resistance as well as capacity to pay off the damage of definite key pests, are measured as basic fundamentals for accomplishing thriving biological control caused by the natural enemy complex. Besides, consistent natural enemy action is also measured to depend on all-year-round continuity of prey or hosts which are made probable by the moderately short fallow periods amid staggered two to three rice crops per year and by immediacy of certain non-rice habitats, mostly the occurrence of the vegetation-covered bunds around each field. On the contrary, synchronous cropping could affect stability by destroying the continuity required for natural enemy success. Their deductions were favored by the experiences of the farmers who used very little quantities of insecticides. A great extent of evidences on the obliteration of natural enemies by certain insecticides supports the hypothesis that insecticide use, particularly early on, in the crop season, upset normal enemy control of insects such as *Nilaparvata lugens*. Simultaneously, it also exerts profound selection pressure for strains of pests that can surmount earlier resistant rice cultivars. These situations cause outbreaks of secondary pests and affect biological control of some chief major pests such as stem borers. It is determined that pest management of much tropical irrigated rice must be based on natural controls seldom by insecticides. The achievement of this method depends especially on additional research on dynamics of natural enemy and pest communities present in rice ecosystems, particularly where climatic conditions and water supply are slightly steady. Besides, there is more to know about the nature and use of rice plant compensation for damage, chiefly by defoliators and stem borers. There is no proof that a natural control-based approach, as suggested in their review, is contrary with farmer practicability or with upcoming developments in rice production technology, excepting possibly, the likely mechanization-driven add to an increase in field size and may decline bund area. On the other hand, the insecticide-based method is not only harmful to natural controls but is expensive and generally demands unfeasible decision making by farmers on need-base use.

Heong et al. (2002) in his discussion on the usage of insecticides in Laos and the farmer's management practices, had mentioned that comparatively the usage of insecticides is less though some increase in usage is observed recently. By using a social psychology framework survey, they found that farmers are of the opinion that productions will be reduced by insects and insecticides can effectively control the damage of insects. Farmers' spray frequencies were interrelated with both the belief and norm indices suggesting that these two components have an influence on farmers spray frequencies of insecticides. Amongst these two, the norm influences are more on the perception of farmers. Agricultural technicians seem to be influenced more. The study discovered that Lao farmers are also probably going to become susceptible sufferers of insecticide misapplication alike that of many of the farmers in Asian countries who turned out the victims of insecticides in the rice intensification programs. Policies on the usage of insecticides and management and real strategic plans are to be developed and implemented parallelly. These should entail the establishment of wide-ranging pesticide policies, expansion of extension programs to teach specialists and farmers on the health hazards and ecological principles associated to pesticide use. Likewise, identical educational programs should be organized in schools simultaneously these should also be explored through approaches of radio and television dramas, etc., which would be effective as belief change panels.

Heinrichs (1986) reviewed on the status of breeding and commercial use of insect-resistant rice varieties. He developed proficient approaches for screening the World collection of rice and used them in identification of resistance sources against 32 insect pests of rice. They mentioned that identified insect resistance sources are to be incorporated within the commercial rice varieties to increase the number of varieties exhibiting multiple resistances. They discussed strategies to overcome restrictions to the breeding and use of insect-resistant rice varieties.

The quick improvement of transgenic biotechnology has significantly boosted the breeding of genetically engineered (GE) rice in China. Xia et al. (2011) reported enlarged yield performance of two insect-resistant *Bt* rice lines under target-insect attacks and its effects for field insect management. No noticeable underlying cost of the *Bt* transgene was detected in the two insect-resistant GE hybrid rice lines. The yield performance of these transgenic rice lines was high in environmental conditions with non-target insect control compared with no insect control. It is determined that insect-resistant *Bt* GE rice, especially the hybrid line, has greater potentiality in maintenance of high yields even when there is a high ambient insect pressure also. Furthermore, appropriate use of insecticides to control non-target insects will assure the optimum performance of insect-resistant *Bt* GE rice.

8.5 DISEASES

The endemic plant pathogens of rice cause lot of damage to the crop. The identification of these plant pathogens is essential. It is also essential to find out causes, symptoms, and eradications of diseases. Sayedaparveen-qureshi Sayeda et al. (2018) undertook the survey of disease in fifteen rice fields of Bhandara district. The different diseases were evaluated on rice crop. Then, farmer's behavior on rice cultivation was determined to increase awareness about diseases and promote efficient and environmentally sound disease management. They stated that the fungi, bacteria, and virus causes disease pressure from seedling stage to harvest and are the major causes of low production in rice crop.

8.5.1 BLAST

Rice blast is a major disease in rice. Miah et al. (2017) reviewed the basics of rice blast disease to find the various methods of rice blast disease control. At present, more than 100 blast resistance genes have been identified in rice, which can be transferred into a susceptible variety by means of marker-aided backcrossing. The primary inoculums for the spread of disease are infested residues and seeds. In view of the significance of this disease, different management methods have been followed to control blast disease, of which utilization of resistant varieties is an important measure. In this review complete information on the pathogen, its epidemiology, evaluation of resistance genes and effective control measure of rice blast disease was provided. This up-to-date information would be useful and direct rice breeders to develop rice varieties resistant to blast (Figure 8.8).

FIGURE 8.8 Blast disease of rice.

Rice blast is a major menace to the sustainable rice production. In developing countries, the utilization of resistant cultivars is the only feasible way to manage the disease, where most farmers cannot afford fungicides. But the resistance often breaks down. To achieve durable resistance, quantitative and qualitative resistance alleles should be combined in breeding programs by exploring rice genetic diversity. Genome-wide association studies (GWAS) facilitated the high resolution exploration of rice genetic diversity. In this study, GWAS of resistance to rice blast was undertaken by Louis-Marie et al. (2016). They genotyped 150 accessions of tropical japonica type with 10,937 markers, and 190 accessions of *indica* type with 14,187 markers. In the field, the *indica* and tropical *japonica* groups exhibited contrasting distribution of blast disease scores indicating a higher level of quantitative resistance in the *japonica* type than the *indica* type. In *japonica* type, two different loci one on chromosome 1 with cloned resistance genes Pi37 and Pish/Pi35 and another on chromosome 12 with partial resistance to blast were identified. In the *indica* type, only one locus on chromosome 8 with Pi33 gene was detected. Further, the haplotype diversity analysis in the target region unveiled two different haplotypes, both associated with Pi33 resistance. Therefore, in this study using GWAS three chromosomal regions related with blast resistance can be identified. Future research studies should focus on specific *indica* markers targeting the known insertion in the Pi33 zone and to explore quantitative resistance among tropical *japonica* varieties specific experimental designs should be applied.

8.5.2 BROWN SPOT, BACTERIAL SEEDLING ROT, AND BACTERIAL GRAIN ROT

The optimal temperature of around 30°C is favorable for the growth of diseases such as brown spot, bacterial seedling rot and bacterial grain rot in rice. Therefore, to ensure sustainable rice production the varieties resistant to these diseases must be developed. To date, the genes for complete resistance to these diseases were not identified. Hence, rice breeders should utilize partial resistance, which is mostly affected by environmental conditions. The detection of QTLs linked with resistance has been facilitated by recent advancements in molecular genetics and development of evaluation methods for disease resistance. Ritsuko Mizobuchi et al. (2016) reviewed the outcomes of global screening for varieties with brown spot, bacterial seedling rot and bacterial grain rot resistance and conferred the detection of QTLs related with resistance to these diseases, so as to make available beneficial information for further rice breeding programs.

8.5.3 BROWN SPOT

Brown spot disease caused by *Bipolaris oryzae* is one of the major foliar diseases affecting rice Worldwide (Figure 8.9). Marco Martin Mwendo et al. (2017) carried out a study to identify the mode of inheritance for resistance to brown spot disease. They crossed nine rice genotypes with variable resistance levels in a full diallel mating design. Then parents, reciprocals, and F_2 progenies were evaluated in an alpha lattice design in the screen house and field trials. Among the tested genotypes, significant differences were observed for brown spot resistance. The low estimates of baker's ratio (0.29) and narrow sense coefficient of genetic determination (0.24) revealed the dominance of non-additive genetic effects in the brown

FIGURE 8.9 Brown spot symptom.

spot resistance genetic control. This shows that progeny performance was better only in specific cross combinations and cannot be projected from parents GCA effects. Further, the genetic control of brown spot resistance by dominant genes was proved from segregation patterns. The significantly different reciprocal effects for the crosses, suggested the cytoplasmic genetic effects modified the resistance expression. Therefore, care should be taken while choosing female parents during hybridization and family-based breeding programs would be effective for enhancing brown spot resistance in rice varieties.

8.5.4 SHEATH ROT

A rice disease-causing rotting of sheaths was first reported in Taiwan. *Acrocylindrium oryzae,* presently known as *Sarocladium oryzae* was recognized as causal agent of this disease (Figure 8.10). Later, it has become clear that other organisms such as *Fusarium* sp. and fluorescent *pseudomonas* can cause similar disease symptoms. These organisms

FIGURE 8.10 Sheath rot.

produce a range of phytotoxins that cause necrosis in plants. All these organisms are seed-transmitted and cause grain discoloration, chaffiness, and sterility in rice. The symptoms of rice sheath rot disease are noticed in all rice-growing areas. The disease is now gaining impetus and is deliberated as a significant evolving menace to rice production. The disease can bring about varying yield shortfalls, which can be as high as 85%. Bigirimana et al. (2015) conducted a review on this disease to improve the understanding of the disease etiology, causal agents, pathogenicity factors, interactions among the various pathogens, epidemiology, and geographical distribution and discussed the feasible disease management options.

8.5.5 SHEATH BLIGHT

Rice sheath blight has been a most destructive disease resulting in heavy yield loss every year. Singh et al. (2016) performed an experiment to study the nature of gene action regulating sheath blight resistance in rice. From the cross between susceptible rice variety HUR 105 and resistant variety Tetep six generations viz., P_1, P_2, F_1, F_2, B_1 and B_2 were used. For sheath, blight the number of effective factors ranged from 1.31 to 4.14 indicating that 1 to 4 genes are involved in the inheritance of resistance. The area under disease progress curve showed significant positive association with days to 50% flowering (FL), days to maturity, number of productive tillers per plant, flag leaf length, fertile spikelet per panicle, spikelet fertility (SF) percentage, and test weight and grain yield (GY) per plant and significant negative association with plant height. On area under disease progress curve the highest positive direct effect was contributed by the trait days to 50% FL, followed by SF percentage, GY per plant, plant height, test weight, fertile spikelet per panicle and number of productive tillers per plant. Thus, the information on genetics of different contributing traits of resistance would assist researchers in selecting suitable breeding methods for the enhancement of sheath blight resistance in rice.

In rice, the sheath blight (ShB) disease is caused by a soil-borne pathogen *Rhizoctonia solani*. Many studies have concentrated on assessing individual mapping populations for QTLs for ShB resistance, but so far no study has employed association mapping to study hundreds of lines for potentially new QTLs. Limeng Jia et al. (2012) conducted association mapping to identify ShB QTLs in rice. From the USDA rice core collection,

217 sub-core entries were phenotyped with a micro-chamber screening method and genotyped with 155 genome-wide markers. For association mapping of ShB, PCA5 with genomic control was found to be the best model. A total of ten marker loci located on seven chromosomes exhibited significant correlation with the ShB pathogen and the allele contributing the greatest effect to ShB resistance was termed as the putative resistant allele. Out of 217 entries, maximum putative alleles were located in the entry GSOR 310389. Most of the resistant entries having putative resistant alleles were of *indica origin.* Thus, through their research findings they demonstrated the efficiency of marker-assisted selection (MAS) to pyramid putative resistant alleles from different loci in a cultivar for improved ShB resistance in rice.

8.5.6 TUNGRO VIRUS DISEASE

Tungro virus disease of rice has been a severe menace in rice production. During 1977–1981, the annual population flux pattern of the vectors *Nephotettix virescens* and *N. nigropictus* assessed by light trapping peaked at the end of the wet and the dry crop seasons. Mean tungro incidence in monthly planted fields was higher in plots planted in the months when vector catches were greater. To solve tungro problems, an integrated management scheme was developed in 1983, with four components. With the execution of four components there was a decline in the tungro incidence and the insecticide use, this was more evident in those areas where there was the implementation of IPM approaches. Sama et al. (1991) assessed the significance of the practices followed in tungro management and studied its epidemiology at Sidrap and Maros, South Sulawesi. The information from their surveys conducted over a period of two years 1985 and 1986 has shown that at Sidrap 90% of rice fields were planted following the suggested planting schedule. There was a fallow period for one or more months between the dry and wet crop seasons and 78–85% of on an average the rice growers have taken up the recommended cultivars for planting. At Maros, they observed that one month a head there was the occurrence of peak periods in the annual green leafhopper populations to a pattern that was recorded in 1977–1981. The change was perhaps related with a shift of 1 month in the crop establishment for dry and solves wet seasons. In the 1987–1988 dry season in Sidrap, where cropping was scheduled for planting in November, four rice cultivars were planted in the first week of October, November, December, January, and February, and development of the vectors and tungro were compared. October planting had a lower vector density which later increased at a slower rate in planting during the November month. However, there was a rapid increase in their population in December and late plantings. The mean tungro incidence at 56 days after transplanting was insignificant in October, November, and December plantings, but it was as high as 60% in the February planting. In the January and February plantings, tungro incidence was also elevated in cultivars suggested for that dry season in the area. In the related trial in the same season at Maros, with wet weather, the tungro incidence was negligible in all plantings but the vector density was high in the February and March plantings. These findings indicated that the vector source was inadequate and the tungro source was insufficient in the two locations at the time when planting was arranged. Tungro infection could occur, but only at the maturing stage or when planting was delayed. Low tungro incidence under the management system was mostly due to slow rate of vector population upsurge and lack or shortage of tungro source in the area. With reference to the component practices involved, the utilization of cultivar rotation seemed to have an inadequate role in decreasing tungro incidence.

Silicon is deliberated as a beneficial element and in many crops, it is not considered as an essential element. However, in many of the graminaceous and non-graminaceous crop species it is involved in bringing out plant development, growth, and yield improvements. Silicon is also known to lessen plant diseases particularly in rice. Its fertilization is routinely practiced in rice production in Florida. Datnoff et al. (1997) provided an outline of the silicon its history, application, and ability in disease suppression in rice genotypes along with its relation of interaction with other fungicides. Although their focus was on rice and organic soils, the information provided is of much interest to all those researchers working on grass crops in organic soils and the rice production that is taken up on low silicon mineral soils or on weathered soils. They presented a viewpoint and its potential requisites in the future.

Well-timed diagnosis of crop diseases in fields is vital for exactitude on-farm disease management. Remote sensing technology can be used as a successful and low-cost technique to find diseased plants in a field scale.

Conversely, due to variety of crops and their related diseases, appliance of the technology to agriculture is yet in research stage. Thus, there is a need for its elaborative investigation for development of algorithm and standard images of the processing procedures that are followed. Qin and Zhang (2005) studied the applicability of broadband high spatial resolution ADAR (Airborne data acquisition and registration) remote sensing data to identify rice sheath blight. Further, they developed a method to promote and explore it's the applicability. The findings revealed that the broadband remote sensing imagery has the potential to identify the disease. Some image indices such as RI_{14}, SDI_{14}, and SDI_{24} worked better than others. A correlation coefficient above 0.62 was recorded. It specified that these indices are more valuable in utilizing them for identification of the rice disease. In the validation analysis, they obtained a small root mean square error (RMS = 9.1), which has confirmed the applicability of their developed method. Though their results were hopeful, they could not easily detect the variation that existed between the lightly infected plants and the sound health plants, owing to their spectral similarities. Therefore, it was clear that detection precision increases when infection attains medium-to-severe levels (DI > 35). This phenomenon demonstrated that remote sensing images with higher spectral resolution (more bands and narrower bandwidth) were essential so as to further assess the ability of separating the light diseased plants from healthy plants.

8.6 RESEARCH ADVANCES IN BIOTIC STRESS MANAGEMENT IN RICE

Peng et al. (2009) discussed about the challenges and position of rice production in China. They mentioned that rice production there has tripled in the last five decades. This is mostly affected by the improved GYs that occurred without any increase in its area of production. Many high yielding varieties were taken up for planting and the farmers adopted the effective management practices. However, it faced the yield stagnation in the last ten years. To meet the domestic need of increased population it needs to produce at least 20% higher yields by 2030, if rice consumption per capita stays at the current level. However, it's not that easy to produce 20% higher yields as several constraints limit the sustainable increase in rice productivities. Some of these include a reduction in arable land, increasing water shortage, global climate change, labor scarcities, and increasing consumer demand for high-quality rice. The main problems opposing rice production in China are constricted genetic background, misuse of fertilizers and pesticides, breakdown of irrigation infrastructure, oversimplified crop management, and a weak extension system. In spite of these challenges, good research approaches can drive to realize an increase in rice production in China. These comprise the development of new rice varieties with high yield potential, enhancement of resistances to major diseases and insects, and to major abiotic stresses such as drought and heat, and the setting up of integrated crop management. They believe that a sustainable growth in rice production is attainable in China with the development of new technology through rice research.

8.7 CONCLUSION

Under field conditions, rice crop confront different biotic stresses such as insects and diseases. These stresses and their impact on crop yield have been discussed in this chapter. Further, in this chapter, an overview of pest and disease management strategies and the latest achievements made in biotic stress management in rice has been presented.

KEYWORDS

- **airborne data acquisition and registration**
- **biotic stress**
- **brown planthopper**
- **gall midge biotype**
- **quantitative trait loci**
- **RNA interference**

REFERENCES

Adesina, A. A., Johnson, D. E., & Heinrichs, E. A., (2008). Rice pests in the Ivory Coast, West Africa: Farmers' perceptions and management strategies. *International Journal of Pest Management*, *40*, 293–299.

Alam, S.N., & Cohen, M. B., (1998). Detection and analysis of QTLs for resistance to the brown plant hopper, *Nilaparvata lugens*, in a doubled-haploid rice population. *Theoretical and Applied Genetics*, *97*, 1370–1379.

Begum, S., Reddy, V. R., Srinivas, B., & Kumari, C. A., (2018). Combining ability and gall midge resistance for yield and quality traits in hybrid rice (*Oryza sativa* L.). *International Journal of Pure and Applied Bioscience*, *6*, 712–724.

Berg, H., (2001). Pesticide use in rice and rice–fish farms in the Mekong Delta, Vietnam. *Crop Protection*, *20*, 897–905.

Datnoff, L. E., Deren, C. W., & Snyder, G. H., (1997). Silicon fertilization for disease management of rice in Florida. *Crop Protection*, *16*, 525.

Grigarick, A. A., (1984). General problems with rice invertebrate pests and their control in the United States. *Protection Ecology*, *7*, 105–114.

Grist, D. H., & Lever, R. J. A. W., (1969). *Pests of Rice* (pp. 520). London & Harlow, Longmans.

Gurpreet, S. M., & Bentur, J. S., (2017). *Breeding For Stem Borer and Gall Midge Resistance in Rice* (pp. 323–352). Springer Nature Singapore Pvt. Ltd.

Harper, J. K., Rister, M. E., Mjelde, J. W., Drees, B. M., & Way, M. O., (1990). Factors influencing the adoption of insect management technology. *American Journal of Agricultural Economics*, *72*, 997–1005.

Heinrichs, E. A., (1986). Perspectives and directions for the continued development of insect-resistant rice varieties. *Agriculture, Ecosystems and Environment*, *18*, 9–36.

Heong, K. L., Escalada, M. M., Sengsoulivong, V., & Schiller, J., (2002). Insect management beliefs and practices of rice farmers in Laos. *Agriculture, Ecosystems and Environment*, *92*, 137–145.

Himabindu, K., Suneetha, K., Sama, V. S. A. K., & Bentur, J. S., (2010). A new rice gall midge resistance gene in the breeding line CR57-MR1523, mapping with flanking markers and development of NILs. *Euphytica*, *174*, 179–187.

Jia, L., Yan, W., Zhu, C., Agrama, H. A., Jackson, A., Yeater, K., Li, X., Huang, B., Hu, B., McClung, A., & Wu, D., (2012). Allelic analysis of sheath blight resistance with association mapping in rice. *PLoS One*, *7*, e32703.

Louis-Marie, R., Elsa, B., Didier, T., Alain, R., Julien, F., Brigitte, C., & Nourollah, A., (2016). Association mapping of resistance to rice blast in upland field conditions. *Rice*, *9*, 1–12.

Mangan, J., & Mangan, M. S., (1998). A comparison of two IPM training strategies in China: The importance of concepts of the rice ecosystem for sustainable insect pest management. *Agriculture and Human Values*, *15*, 209–221.

Matteson, P. C., (2000). Insect pest management in tropical Asian irrigated rice. *Annual Review of Entomology*, *45*, 549–574.

Miah, G., Rafii, M. Y., Ismail, M. R., Sahebi, M., Hashemi, F. S. G., Yusuff, O., & Usman, M. G., (2017). Blast disease intimidation towards rice cultivation: A review of pathogen and strategies to control. *The Journal of Animal and Plant Sciences*, *27*, 1058–1066.

Mwendo, M. M., Ochwo-Ssmakula, M., Mwales, S. E., Lamo, J., Gibson, P., & Edema, R., (2017). Inheritance of resistance to brown spot disease in upland rice in Uganda. *Journal of Plant Breeding and Crop Science*, *9*, 37–44.

Panda, N., & Heinrichs, E. A., (1983). Levels of tolerance and antibiosis in rice varieties having moderate resistance to the brown planthopper, *Nilaparvata lugens* (Stål) (Hemiptera: Delphacidae). *Environmental Entomology*, *12*, 1204–1214.

Peng, S., Tang, Q., & Zou, Y., (2009). Current status and challenges of rice production in China. *Plant Production Science*, *12*, 3–8.

Powell, K. S., Spence, J., Bharathi, M., Gatehouse, J. A., & Gatehouseb, A. M. R., (1998). Immunohistochemical and developmental studies to elucidate the mechanism of action of the snowdrop lectin on the rice brown planthopper, *Nilaparvata lugens* (Stal). *Journal of Insect Physiology*, *44*, 529–539.

Prasad, S. S., Gupta, P. K., Singh, R. V., Prakash, N., & Mishra, J. P., (2013). Screening of semi deep water rice genotypes against yellow stem borer, *Scirpophaga incertulas*. *International Journal of Current Microbiology and Applied Science*, *2*, 62–65.

Preetinder, S., Gurpreet, K., Kumari, N., Gurjit, S., Bhaskar, C., & Kuldeep, S., (2016). Donors for resistance to brown plant hopper *Nilaparvata lugens* from wild rice species. *Rice Science*, *23*, 219–224.

Pretty, J., & Bharucha, Z. P., (2015). Integrated pest management for sustainable intensification of agriculture in Asia and Africa. *Insects*, *6*, 152–182.

Qin, Z., & Zhang, M., (2005). Detection of rice sheath blight for in-season disease management using multispectral remote sensing. *International Journal of Applied Earth Observation and Geo Information*, *7*, 115–128.

Renuka, P., Madhav, M. S., Padmakumari, A. P., Barbadikar, K. M., Mangrauthia, S. K., Rao, K. V. S., Marla, S. S., & Babu, V. R., (2017). RNA-seq of rice yellow stem borer *Scirpophaga incertulas* reveals molecular insights during four larval developmental stages. *G3 Genes/Genome/Genetics*, *7*, 3032–3045.

Ritsuko, M., Shuichi, F., Seiya, T., Masahiro, Y., & Hiroyuki, S., (2016). QTLs for resistance to major rice diseases exacerbated by global warming: Brown spot, bacterial seedling rot and bacterial grain rot. *Rice*, *9*, 1–12.

Roling, N., & Fliert, E., (1994). Transforming extension for sustainable agriculture: The case of integrated pest management in rice in Indonesia. *Agriculture and Human Values*, *11*, 96–108.

Sama, S., Hasanuddin, A., Manwan, I., Cabunagan, R. C., & Hibino, H., (1991). Integrated management of rice tungro disease in South Sulawesi, Indonesia. *Crop Protection*, *10*, 34–40.

Savary, S., Horgan, F., Willocquet, L., & Heong, K. L., (2012). A review of principles for sustainable pest management in rice. *Crop Protection*, *32*, 54–63.

Sayeda, P. Q., Belurkar, Y., Mehar, P., Kodape, D., & Selokar, M., (2018). Study of diseases on rice (*Oryza sativa*) in major growing field of Bhandara district. *International Journal of Agriculture Sciences*, *10*, 5573–5575.

Settle, W. H., Ariawan, H., Astuti, E. T., & Cahyana, W., (1996). Managing tropical rice pests through conservation of generalist natural enemies and alternative prey. *Ecology*, *77*, 1975–1988.

Singh, A. K., Singh, P. K. Ram, M., Kumar, S., Arya, M., & Singh, N. K., (2016). Gene interactions and association analysis for sheath blight resistance in rice (*Oryza sativa* L.). *SABRAO Journal of Breeding and Genetics*, *48*, 42–52.

Teng, P. S., (1994). Integrated pest management in rice. *Experimental Agriculture*, *30*, 115–137.

Bigirimana, V. P., Hua, G. K. H., Nyamangyoku, O., & Höfte, M., (2015). Rice sheath rot: An emerging ubiquitous destructive disease complex. *Frontiers in Plant Science, 6*, 1066.

Way, M. J., & Heong, K. L., (1994). The role of biodiversity in the dynamics and management of insect pests of tropical irrigated rice a review. *Bulletin of Entomological Research, 84*, 567–587.

Xia, H., Lu, B. R., & Xu, K., (2011). Enhanced yield performance of *BT* rice under target-insect attacks: Implications for field insect management. *Transgenic Research, 20*, 655–664.

CHAPTER 9

Methods of Cultivation

ABSTRACT

This chapter deals with different rice cultivation methods. The methods involved in rice cultivation include the method of land preparation, nursery management, sowing, planting methods, how fertilizer is managed, and methods of protection of the rice crop. Besides, it discusses the management of greenhouse gas emissions from rice fields and climate change effects on rice productivity.

9.1 CROPPING SYSTEMS

In the Mekong Delta, the most common method of the spatial distribution of heading date and rice-cropping system estimation in relation to seasonal variations in water resources was a multi-temporal moderate resolution imaging spectroradiometer (MODIS). Sakamoto et al. (2006) developed a wavelet-based filter to determine crop phenology (WFCP) and to assess the spatial distribution of cropping systems (WFCS). The results were interpreted with MODIS time-series data. The results were in correspondence with the physical characteristics of the cropping system in the Mekong Delta, which have altered over time in response to local and seasonal variations in water resources. For instance, in the upper Mekong Delta due to the sequential floods, the double-irrigated rice-cropping system is usually employed in the dry season. The unavailability of appropriate irrigation water and intrusion of saline water in the coastal regions during the dry season has restricted the use of dry-season cropping and the double- and single-rainfed rice-cropping systems are used only in the rainy season. While, in the central part of the Mekong Delta which is situated in between the flood-prone and salinity intrusion areas a triple-irrigated rice-cropping system is employed. Analysis of yearly variations in the rice cropping systems revealed the expansion of triple-cropped rice to the flood- and salinity-intrusion areas. This extension designates the employment of methods to reduce the magnitude of flooding and salinity intrusion by advanced farming technologies and advances in land management. In 2003, the heading dates in the upper Mekong Delta were 20 to 30 days earlier than in 2002. This may be due to reduced flood runoff in 2002 and execution of government policies regarding early sowing of dry-season. Further, the MODIS data analysis confirmed the close association between the spatial distribution of rice-cropping systems and seasonal variations in river runoff regime in the Mekong Delta.

9.1.1 RICE-WHEAT SYSTEM OF THE INDO-GANGETIC PLAINS (IGP)

The raised coverage of high-yielding varieties (HYVs) in 1960s and 1970s, facilitated rice (*Oryza sativa* L.)-wheat (*Triticum aestivum* L. and *T. durum* Desf.) system to arise as a major cropping system in the Indo-Gangetic plains (IGP), escorting the green revolution (GR). In this paper, Gupta et al. (2003) described different choices accessible to farmers for crop establishment and diversification in rice-wheat system ecology and the prospective of advanced technologies across the IGP. They stated that the GR technologies are the base of the South Asian scheme for food security, rural development, conservation of natural resources, and poverty mitigation. But, the currently appearing evidences indicate that the rice-wheat system has destroyed the natural resource base. Therefore, to exploit new sources of productivity growth an infusion of technologies is needed in agriculture. On the basis of driving variables for agricultural development, the IGP has been demarcated into five comparatively homogeneous areas to deal with the location-specific rice-wheat system ecology challenges. The integration of crop improvement; water management; nutrient management; weed, disease, and pest management; and policy and social science research associated to the establishment of crop within these areas permits constant production, diversification, and improved system productivity. Further, this paper highlighted new approaches and some Chinese agrotechniques applicable to the rice-wheat system of the IGP.

9.2 LAND PREPARATION

9.2.1 *TILLAGE AND CROP ESTABLISHMENT OPTIONS*

Rice and wheat are very important cereal crops and sustain the food necessities of more than half of the World population. At present, rice, and wheat contributes more than 80% of the total cereal production and significant to employment and food security for millions of rural families. It is expected that the demand for these crops may increase from 2% to 2.5% per annum till 2020, which requires sustained attempts to improve productivity while ensuring sustainability. The extension of area and major increases in rice-wheat productions system were realized with the widespread adoption of "GR" technology in the 1960s. GR technology focused on the use of HYVs, fertilizers, and irrigation resulting in improved production and productivity of these crops. But, in recent years the intensive use of GR technologies lead to lower marginal returns and, in some areas to salinization, overexploitation of groundwater, physical, and chemical deterioration of the soil, and pest problems. In this article, Gupta and Seth (2007) reviewed the results of current research on resource conservation technologies including tillage and crop establishment options which enables farmers to maintain productivity of intensive rice-wheat systems. Field outcomes unveiled that the resource conserving technologies, for example of conservation agriculture, increases yields, decreases water usage, and reduces negative influences on the environmental quality. The paper deliberates contributions of innovative inter-institutional collaboration in international agricultural research and socio-economic changes in the IGP countries that resulted in quick improvement and implementation of these technologies by farmers.

9.2.2 *EFFECTS OF STRAW MULCHING AND SHALLOW SURFACE TILLAGE*

During the land preparation of soils for rice production, high water loss occurs from bypass flow through cracks. It was assumed that the processes that reduce crack development during the soil drying period can decrease these losses or obstruct the water flow through these cracks. Cabangon and Tuong (2000) conducted a field experiment to study the straw mulching and shallow surface tillage effect on formation of crack during the fallow period, and on components of water flow during land preparation. During land preparation of the control plots heavy water loss occurred because, the rewetting did not completely close the cracks. Though straw mulching did not significantly decrease mean crack depth and the quantity of water utilized in land preparation, it helped to conserve moisture in the soil profile by reducing the average crack width. The small soil aggregates formed from shallow tillage makes the crack water flow discontinuous and obstructs the groundwater recharge from the water flow through cracks. This decreases the total water input for land preparation by 31–34%, which is equal to around 120 mm of water. When compared to the control, the mean surface irrigation water flow was faster and less time was required for land preparation in the shallow tillage plots. Thus, the study proposed that the shallow tillage is an effective method for increasing water-use efficiency of irrigation systems.

Simultaneously, Wopereis et al. (1994) studied the bypass flow processes in a cracked, formerly puddled rice soil. By means of a morphological staining technique vertical continuity and depth (0.65 m) of soil cracks (10–10 mm in width) in the field was ascertained. An infiltration experiment revealed that water was mostly absorbed in the subsoil between 0.2 and 0.5 m depth. A laboratory experiment was carried out by applying simulated rainfall to large undisturbed soil columns acquired from the field. For low (10 mm h^{-1}) and high (30 mm h^{-1}) rain intensities, high bypassing ratios (greater than 0.9) were noted. In the field, shallow surface tillage (0–50 mm) reduced the bypass flow, and laboratory experiments revealed that the formation of discontinuous cracks saves 45–60% water. Therefore, in rainfed areas, shallow tillage may help in early crop establishment, reducing the probability of late-season drought and this may widen the scope for a second crop. In irrigated rice systems, water savings during land preparation may permit an extension of irrigation system command area.

9.2.3 TILLAGE PRACTICES

Soil tillage is a key factor that affects soil properties and crop yield through its interaction with timely planting and good plant stand. In the South Asian rice-wheat systems of the IGP, puddling (wet tillage) is the common method of rice establishment. The field is generally plowed dry, flooded, and then puddled (plowed when flooded) to form a soil with poor soil physical properties and low water infiltration. The wheat and other non-rice crops grown in this puddled soil after rice receive normal dry plowing before broadcasting the seeds. For rice and wheat cultivation the land is prepared by multiple plowings and planking to level the fields and break the clods. When animal power is used, the land preparation can take many days leading to late planting of the crop. With tractor power, also, land preparation can take several weeks; further low accessibility to tractor power can delay planting. Delayed planting is a main cause of low wheat yields. So as to overwhelmed the difficulties of late planting, poor plant stands and to decrease the costs of production researchers has been assessing the use of zero- and reduced-tillage systems for wheat in the rice-wheat system. Since it is presumed that the non-puddled soils have better soil physical conditions for the growth of the next upland crop, the research on system that can avoid puddling in rice is initiated. In this review, Hobbs (2001) described the existing and new practices of tillage and crop establishment for rice and wheat cropping system.

9.3 SOWING

9.3.1 PRE-SOWING SEED TREATMENTS EFFECTS

Basra et al. (2005) studied the pre-sowing seed treatments effects on the germination and emergence of fine rice, on reducing, non-reducing total sugars and α-amylase activity. For this study, the fine rice seeds were soaked in tap water using a traditional method or hardened for 18 or 24 h (2 cycles) or osmoconditioned (−1.1 MPa KNO_3) for 24 or 48 h. The traditional soaking treatments and seed hardening (24 h) lead to a higher germination percentage, germination index (GI), and energy of germination. Further, it shortened the mean germination time and mean emergence time. Owing to the lower T_{50} and non-reducing sugars plus higher reducing total sugars and α-amylase activity of hardened seeds, Their performance of was better than the seeds soaked by traditional method. The performance of seeds exposed to osmoconditioning for 48 h was impaired probably due to the KNO_3 toxicity.

9.3.2 DETERMINATION OF SAMPLE SIZE

To get the reliable assessments of agroecosystem variables, prior to the collection of experimental data the sample size should be determined. So as to study the experimental factors (artificially-induced variability) effects on rice sample size, Stroppiana et al. (2006) conducted an experiment in northern Italy. They determined the sample sizes for different levels of fertilizer, varieties (*Indica* and *Japonica* type), development stages, sowing methods and typologies of the sampling unit. To study the effect of each factor the acquired sample sizes were compared by keeping the other factors constant. For instance, the sample sizes calculated for different levels of fertilizers were compared within the same variety, phenological stage, and sampling unit. Usually, the original data is not normally distributed and the variances of the original samples are non-homogeneous, so a new method of sample size distribution which is based on a visual evolution of the jackknife was chosen over classical methods. The results revealed that the sample sizes calculated in an early phenological stage (between 21 and 27) were higher than those for later stages (15–21). They noticed that the unfertilized plots (21–27) required a larger sample size than fertilized plots (15–27) because the fertilization hides the variability of soil N content. Further, they noted that the *Indica* type variety needed a larger sample size (always 27) for the early sampling when compared to the *Japonica* type variety (21–24). Thus, these outcomes have given an insight of the effect of experimental factors and development stage on within-plot variability, and highlighted the significance of sample size determination in field experiments.

9.3.3 SOWING TIME

Numerous researches have shown that the sowing time is determined by the climatic conditions and crop-specific temperature requirements. On the basis of these characteristics, Waha et al. (2012) simulated the sowing dates of 11 major annual crops worldwide at high spatial resolution. Depending on the crop- and climate-specific, characteristics a set of rules were determined, and then on the basis of these rules sowing dates were simulated for rainfed conditions. They assumed that farmers decide sowing time of the crop based on their past precipitation experiences and temperature conditions. Further, the initiation of the growing period is estimated to be reliant on the commencement of the wet season or on the exceeding of a crop-specific temperature threshold for emergence. This procedure is validated by using a worldwide data set of noted monthly growing periods (MIRCA2000). For all the major annual crops except for rapeseed and cassava, the variance between simulated and observed sowing dates was less than 1 month. This difference was noted in at least 50% of the grid cells and on at least 60% of the cultivated area. However, in tropical regions characterized by multiple-cropping systems and in countries with large climatic gradients a deviation of more than 5 months was observed. Therefore, their present study demonstrated that the sowing dates for annual crops can be effectually assessed from climatic conditions for large parts of the earth.

9.4 NURSERY MANAGEMENT

9.4.1 NURSERY MANAGEMENT METHODS

Adhikari et al. (2013) undertook a study to analyze various nursery management methods and their influence on grain yield (GY) and yield traits under rainfed conditions. Using the rainfed lowland rice variety Radha-4, three management factors viz., and fertilizer management in the nursery, seeding density, and seedling age at transplanting were tested. Two levels of fertilization (0 and 20:20:0:13 kg NPKS ha^{-1} at 15 DAS), two levels of seeding density (607 and 303 $g \cdot m^{-2}$) and two seedling ages (20 and 40 days old) were applied as treatments in the study. The results revealed that the top-dressed fertilizer in the nursery had no reliable influence on GY. The lower seeding density (303 $g \cdot m^{-2}$) produced taller plants with more effective tillers per m^2, less sterility and higher GY. Furthermore, older seedlings (40 days) bring about taller plants, more effective tillers, more filled grains, and a higher grain and straw yield. The economic analysis proved that the combination of these two management factors gives highest net returns and B:C ratio in both seasons, irrespective of the fertilizer treatment. Thus, their study findings indicated that this combination of low seeding density and 40 days old seedlings is economically feasible and cost-effective, and can be suggested to farmers. But, the consequences should be verified for different varieties employed by farmers (Figure 9.1).

FIGURE 9.1 Wet sowing of paddy on nursery beds and seedling stage.

9.4.2 AGE OF TRANSPLANTED SEEDLINGS

Transplanting rice seedlings is very common practice for major rice growing areas and proved best for maximum GY due to its higher tiller production. Pasuquin et al. (2008) undertook an experiment at the IRRI farm during the dry and wet seasons. The main aims of the research were to determine the result of using even younger seedlings and contrasting nursery management on GY in different plant types and to find plant characteristics associated with high performance under a given establishment technique. For an elite line, a new plant type (NPT) and hybrid rice the seedlings transplanted at age varying from 7 to 21 days and different nursery types such as seedling tray, dapog, mat nursery, and traditional wet-bed seeding were assessed. The results revealed that there was a delay in the onset of linear dry matter accumulation and tiller emergence when older seedlings were transplanted. This delay also reduced nitrogen content in the seedlings. However, the plants recovered immediately after transplanting. Further, the delay prolonged crop duration with late maximum tillering, flowering (FL), and maturity. The younger seedlings exhibited higher GY consistently, with a variance as large as 1 t ha^{-1} between 7- and 21-day transplanting in some cases. This outcome was applicable for the four genotypes evaluated, with a higher influence during the dry season. However, the nursery type had no effect on the tiller emergence time and on GY. Though high variances were noted in seedling vigor (plant dry weight, specific leaf area, N content), in dapog and wet bed method, and maximum tillering in the case of the seedling tray method, these variances did not have a significant effect on the delayed increase in crop dry matter and on panicle number at maturity. Significant interactions were not observed between seedling age and nursery management for all genotypes and parameters measured. In spite of the associated reduction in tillering efficiency, transplanting young seedlings found to increase GY GY in all cases. Prolonged growth inside the nursery, but not transplanting shock per se, seemed to be the major cause for delayed tiller emergence in late transplanting (Figure 9.2).

FIGURE 9.2 Nursery beds.

9.5 METHOD OF PLANTING

9.5.1 EFFECTS OF IMPLEMENTING ALTERNATE SEEDING METHODS

The effects of implementing alternate seeding methods for rice and wheat establishment were investigated by Singh et al. (2011) at three geographically separate sites in the rice-wheat system of the IGP. The rice yields cultivated by wet or dry seeding methods were compared with those from zero-tillage plots and under conventional transplanting methods. In the same experiments, the influences of wheat crop establishment methods on both wheat yields and rice yields were evaluated. Rice crop establishment methods showed significant effects on the emergence of weed flora and probable yields were calculated with respect to intensity of weed management. In the absence of weed competition, the highest average rice GYs were noted from wet seeded rice. These yields were Akinto those from transplanted rice and dry seeded rice after dry soil tillage. The lowest yields were realized from dry seeded rice sown without tillage. The yield shortfalls due to uncontrolled weed growth were minimum in transplanted rice and highest in dry seeded rice sown without soil tillage. The significantly higher wheat GYs were observed from crops sown into tilled soil (3.89 t ha^{-1}) compared to crops sown without tillage (3.51 t ha^{-1}). The wheat yields were improved by 5% on the soil prepared by dry method for the preceding rice crops rather than puddled soils. The wheat cultivation method had no influence on rice yield. In the wheat season, soil infiltration rates were lowest in the soil prepared by puddled method for rice and greater in the dry-tilled soil and greatest in zero-tilled soils. Thus, through their research findings they suggested that direct seeding of rice (DSR) could offer a substitute to the traditional practice of transplanting, and help to deal with increasing costs and threats to sustainability in the rice-wheat rotation. While demonstrating the strength, viability, and significant potential of direct seeded rice, they also highlighted the critical nature of efficient weed management in effective application of direct seeding systems for rice (Figure 9.3).

9.5.2 *EFFECTS OF DIFFERENT PLANTING METHODS*

Prasad and Prasad (1980) conducted a field trial to study the effects of different planting methods, duration of varieties and nitrogen sources on GY and nitrogen uptake of rice. Significantly higher yield and nitrogen uptake were recorded for medium duration (135 days) variety Improved Sabarmati compared with the short duration (105 days) variety Pusa-33. The transplanted rice plants had higher GY and nitrogen uptake than the direct-seeded (drilled in moist soil) rice. The yield and nitrogen uptake of rice cultivated by broadcasting sprouted seeds on puddled seed bed was in between transplanting and direct-seeding and offers a practically suitable method of planting. About 160–170 kg N ha^{-1} was noted as an economically optimum dose of nitrogen for rice. Nitrogen applied in the form of urea briquettes resulted in highest yield and nitrogen uptake by rice. Further, urea briquettes were superior to sulfur-coated urea or neem-cake-coated urea pertaining to N-uptake. Therefore, these new nitrogen fertilizers which hold substantial promise should be included in rice culture to improve yield potential.

FIGURE 9.3 Dry sowing of paddy in soil.

9.5.3 EFFECTS OF DRY SEEDING AND TRANSPLANTING METHODS

Sharma et al. (1987) studied the consequences of two planting methods (dry seeding and transplanting) on the root growth and GY of rainfed lowland rice. They performed field experiments on two soils (clay loam and sandy loam) with variable texture and water table depth. The top 10 cm of soil was found to hold 75% of the roots. Owing to the greater moisture stress in dry-seeded rice between late tillering and harvest, the root length density (RLD) of dry-seeded rice at the FL stage and GY was less than the transplanted rice in the sandy loam soil, with a water table deeper than 1.5 m. In the clay loam soil, where the water table is mostly at the soil surface, the significant differences were not noticed between the GYs of dry-seeded and transplanted rice (Figure 9.4).

FIGURE 9.4 Rice seedlings transplanted in main field.

9.5.4 CAUSES OF THE HIGHER DRY MATTER PRODUCTION BY BROADCASTED RICE PLANTS

Previous studies reported that the broadcasted rice plants produces more dry matter and GY when compared with transplanted plants (TP). So as to investigate the probable reasons for the greater production dry matter by broadcasted rice plants in a submerged paddy field a study has been undertaken by San et al. (2004). They compared the characteristics of plants cultivated with various methods viz., the direct-sown plants (DSP) and TP and with different planting patterns viz., 51.3 hills m^{-2} and one plant per hill (planting pattern I) and 17.5 hills m^{-2} and three plants per hill (planting pattern III). With the same planting pattern, the DSP had higher aboveground parts dry weight and GY than TP. With the same cultivation method, the plants in planting pattern I provided higher dry matter and

GY. The GY differences were reflected in dry matter production differences. The factors contributing to the differences in dry matter production were found to be depended on the growth stage. At the tillering stage, the number of tillers and the leaf area index (LAI) increases quickly and the interception of solar radiation by the canopy was higher in DSP. These characteristics were accountable for higher crop growth rate (CGR) in the DSP compared to the TP, irrespective of the planting pattern. The number of crown roots produced by the plants in planting pattern I were more than those in pattern III. Further, during the maturity stage plants in planting pattern I were found to be characterized with the reduced exudation rate, high amount of nitrogen accumulation and higher levels of chlorophyll and ribulose-1,5-bisphosphate carboxylase/oxygenase (Rubisco). These features may support the higher dry matter and GY production by the plants in planting pattern I.

9.6 WATER MANAGEMENT

9.6.1 *COMPONENTS OF WATER USE IN RICE-BASED PRODUCTION SYSTEMS*

Tuong and Bhuiyan (1999) described the components of water use in rice-based production systems and identified water utilized during land preparation, and seepage and percolation of water during crop growth as main causes of water 'loss' from the rice-based production system. Further, they discussed the approaches for improving farm-level water-use efficiency, together with the constraints of up-scaling from on-farm to system-level water savings.

9.6.2 *WATER-SAVING IRRIGATION SYSTEMS*

Higher water requirement is a main problem in traditional transplanted rice production system. The transplanted rice method is an intensive water user and the increasing water scarcity demands some alternative establishment methods that need less water. Tabbal et al. (2002) reported the outcomes of on-farm experiments in the Philippines to reduce water input through water-saving irrigation systems and alternative crop establishment methods, like wet and dry seeding. Under water-saving conditions, the direct wet-seeded rice out yielded transplanted rice by 3–17%. The direct wet-seeded rice required 19% less water than the transplanted rice. This resulted in 25–48% higher water productivity in direct wet-seeded rice. Though the yields of direct dry-seeded rice were comparable with the transplanted and wet-seeded rice, it can use early season rainfall more efficiently in the wet season and conserve irrigation water for the following dry season. When compared with flooded conditions, the transplanted, and wet-seeded rice in which soil is retained continuously around saturation had 5% reduced yields and 35% less water inputs and 45% higher water productivity. Under water-saving irrigation, the yield of wet-seeded rice was 6–36% higher than transplanted rice. Thus, the wet-seeded rice is appropriate establishment method to save water and maintain high yields.

9.6.3 AEROBIC RICE SYSTEM

Decreasing water availability has intensified the search for alternative rice irrigated systems with less water input. In aerobic rice systems, rice grows in nonflooded and nonsaturated soil by providing supplemental irrigation. This irrigation system is initially developed by identifying promising varieties and quantifying the yield potential, water use, field water outflows, and water productivity. Here, Bouman et al. (2005) studied the productivity of tropical upland and lowland rice varieties under irrigated aerobic conditions. Under aerobic conditions, significantly higher yields were recorded for the upland variety, Apo (5.7 t ha^{-1}) and the lowland hybrid rice Magat (6 t ha^{-1}) in the dry season. The similar yields were attained in moderately wet soil with 10–12 kPa seasonal-average soil moisture tensions in the root zone. Total 790–1430 mm water was applied for aerobic fields 1,240–1,880 mm for flooded fields. This resulted in 190 mm water saving in aerobic rice, further it had 250–300 mm less seepage and percolation, 80 mm less evaporation, and 25 mm less transpiration than flooded fields. In comparison, average yield of all varieties under aerobic condition was 32% lower in the dry season and 22% lower in the wet season. However, under aerobic conditions the water productivity of rice (with respect to rainfall and irrigation water input) was 32–88% higher than that under flooded conditions. Thus, the study concludes that aerobic rice is a promising rice cultivation system under rising water shortage conditions and the aerobic rice management research should concentrate on the development of high yielding varieties, the optimization of crop and water management, and the sustainability of aerobic rice under continuous cropping system.

9.6.4 *MODELS OF WATER BALANCE*

Odhiambo and Murty (1996) developed a water balance model which is more appropriate to paddy fields of the lowlands. The model inputs were irrigation supply, climatic data, soil parameters, and layout dimensions. They have formulated the model in such a way so as to simulate the different processes such as evapotranspiration, seepage, and percolation, and surface runoff that occur commonly in the fields and form a part of the field water balance systems. It is effective in the prediction of the variations in water balance components under diverse land management and hydrological situations. This model is also applicable to layouts of both plot-to-plot or independent plots.

9.6.5 *UTILIZATION OF WATER SAVING IRRIGATION*

In Asian regions, rice is under irrigated cultivation. There is a requirement of its increased production to meet the food demands of the rapidly increasing population. However, the water that is available to irrigate is scarce. Therefore, several challenges exist in improving rice production such as water-saving, increasing water productivity and production of rice with little amounts of water. Bouman and Tuong (2001) analyzed a few ways of water-saving irrigation at the field level to cope up with the existing challenges. The data of their analysis has shown that alternate wetting/drying enables the reduction of water input by declining water depths to soil saturation. Under these conditions in saturated soil conditions, water savings would be on an average 23% (±14%). The reduction in yield maybe only 6% (±6%). Further, they observed that when the root zone water potentials were −100 to −300 mbar there was only 10–40% decline in yields. In clayey soil, this intermittent drying may show the way to shrinkage and cracking. It might increase the risk of increased soil water loss, the water requirements, and a decrease in water productivity. They detected that in India, the water productivity in continuous flooded rice was 0.2–0.4 g grain per kg water, while in the Philippines it was 0.3–1.1 g grain per kg water. Thus, they mentioned that an increase in a maximum water productivity of 1.9 g grain per kg water by the water-saving irrigation, but also results in a decrease of yield. A concomitant increase in both field-level water productivity and yield can be obtained with an improvement in the factor productivity or by a raise in yield potential. Thus, it is possible to raise the total rice production in a location by utilizing the water saved of a location to irrigate new land at other location.

Alternate wetting and drying (AWD) irrigation has been reported to save water in rice cultivation than continuous flooding (CF). However, there were large variations in the reported effects of AWD on rice yields. Further, there is a gap in the complete agro-hydrological characterization for making the generalizations. Moreover, it is not yet clearly identified how these nutrient use efficiencies are modified by AWD. Cabangon et al. (2004) measured the agro-hydrological conditions of the usually practiced AWD at two typical lowland rice sites in China. At two sites, Jinhua and Tuanlin significant variations were not observed in the interactions of water by nitrogen on GYs or productivity or nitrogen uptakes. Further, there was no difference in the N timings in yield or biomass between water treatments of AWD or CF. A 5–35% higher productivity was observed in the productivity of irrigation water in AWD. The variations occurred significantly when there was low rainfall or high evaporation conditions. An increase in number of splits of 4–6 times had resulted in an increase in total N uptake only rather than that of P or K uptakes. Further, with an increase in the number of splits there was an increase in the apparent nitrogen recovery (ANR). However, there was no variation among the alternate wetting drying or in CF. Further, they observed that water tables rarely went deeper than –20 cm of the soil during the cycles of AWD irrigation in the root zone and soil has remained moist during most periods. Thus, the experimental results unveiled that in the classical irrigated low lands of China, the AWD can effectively lessen the inputs of water without any influence on the yields of rice and without any extra requirement of the nitrogen management (Figure 9.5).

FIGURE 9.5 Rouging of rice field (removal of off types).

9.6.6 IRRIGATED RICE SYSTEMS AND IMPROVED NITROGEN USE EFFICIENCIES

Much emphasis is given on research and extension work to bring about an improvement in the nitrogen (N) management of irrigated rice. In Asiatic regions farmers rice yields are mostly under the influence of the application of the N fertilizers. Most of the research works were concentrated on placement, form, and timing of applied N for reducing the volatilization and denitrification losses, and less emphasized in devising new approaches to adjust N rates in relation to the amount of N supplied by indigenous soil resources. In general, most of the recommendations of N fertilizers were typically made for districts or regions with the implied assumption that soil N supply is moderately uniform within these domains. Latest studies, nevertheless, document remarkable differences in soil N supply amongst the lowland rice fields with alike soil types or in the similar field over time. In spite of these differences, many of the farmers growing rice do not adjust applied N rates in accordance to the soil N supply. This accounts for large scale imbalances in the N supplies and also contributes to low N-use efficiency. Cassman et al. (1998) proposed a model for calculating N-use efficiency, which can clearly account the contributions from both indigenous and applied N to plant uptake and yield. They emphasized strongly that elevated N-use efficiencies are dependent to a large extent on the tactics of field-specific N management that are receptive to soil N supply and plant N status. Thus, they considered the N fertilizer losses as a sign of incongruence amid N supply and crop demand relatively than to a driving force of N efficiency. They concluded that wetland rice systems have intrinsic capacities in conserving nitrogen and the rice plants have immense potential of fast uptake of nitrogen. Thus, they provide an opportunity in increasing the N efficiency through proper monitoring and improved management of indigenous N resources, N fertilizers, and plant N status, and straw residues.

9.7 FERTILIZER MANAGEMENT

9.7.1 FERTILIZER MANAGEMENT

In Asia, the yield increment of irrigated rice have reduced in recent years, which can be increased with adequate management of the crop. So, to assess a new method for site-specific nutrient management (SSNM) Dobermanna et al. (2002) conducted on-farm trials at 179 locations in eight main irrigated rice fields of Asia. Among the eight intensive rice fields and among farms within each field large variation in primary soil fertility characteristics and indigenous supply of N, P, and K was noted. NPK applications specific to field and season were estimated by accounting for the local nutrient supply, yield targets, and nutrient requirement as a function of N, P, and K interactions. Based on the season and field-specific crop N status monitoring data the nitrogen applications were adjusted. The SSNM performance was investigated for four consecutive rice crops. In the same cropping seasons the average GY of SSNM plots was 0.36 Mg ha^{-1} (7%) more than the plots with the present farmers' fertilizer practices (FFP). When compared with the baseline data the mean nutrient uptake under SSNM was increased by 13% (N) and 21% (P, K). The yield increments were found to be related with a 4%

reduction in the average N rate and high amount K fertilizer application at K deficient sites. The practice of improved in-season N management systems increased the average N use efficiencies by 30–40% which in turn increased the profitability by US$ 46 ha^{-1} per crop or total average net return by 12%. But the SSNM performance did not show any significant difference between high-yielding and low-yielding climatic seasons. In the first year, the improved SSNM method and re-capitalization of P and K applied increased the average profitability from US$ 32 ha^{-1} per crop in the first year to US$ 61 ha^{-1} per crop in the second year. Though SSNM needed little extra credit for financing, it remains profitable even if rice price are slightly lower than existing levels. SSNM may probably close the current yield gaps by improving yields and nutrient efficiency in irrigated rice. The foremost challenge for SSNM will be to maintain the success of the method while decreasing the complication of the technology as it is disseminated to farmers. The SSNM method specific to conditions in different countries should be designed. In some areas, SSNM may be field or farm specific, but in many areas, it is likely to be just region and season-specific. Thorie et al. (2013) conducted experiment to study the effect of different organic sources of nutrients on upland rice. They found that FYM @ 12 t/ha+Azotobacter @ 20 g/kg seed treated plot produced highest yield contributing characters like number of panicles/m^2, grain yield, straw yield, grains/panicle, fertility percentage over forest litter @ 20 t/ha, farmyard manure (FYM) @ 12 t/ha, poultry manure @ 4 t/ha, forest litter @ 20 t/ha+Azotobacter @ 20 g/kg seed, poultry manure @ 4 t/ha+Azotobacter @ 20 g/kg seed. In another experiment, Ovung and Sarkar (2013) reporrted that Sesbania aculeata @ 6.25 t/ha+Azotobacter@ 20 g/kg seed signifcantly recorded maximum number of grains/panicle (173.14), number of panicles/ m^2 (163.43) and grain yield (30.88 q/ha), giving highest gross return and net return.

A drop in productivity in wetland rice has been noticed in some intensively-cultivated experimental farms in Asia since the early 1980s. Increased doses of fertilizer N are being applied in both experimental and farmers' fields to uphold the original yield levels. Much attention has not been paid to judicious management of native soil N, which is the main N source for rice, and to biological N_2 fixation (BNF), which basically replenishes the soil N concentration. Kundu et al. (1995) reviewed a series of effects of long-term flooding and puddling associated with intensive cultivation of wetland rice on soil N availability and BNF. Some strategies are recommended to efficiently manage these two N sources to uphold high productivity of the rice lands.

9.7.2 *APPLICATION OF N FERTILIZERS ON BASIS OF SPAD READINGS*

Nitrogenous fertilizers are applied to rice crop at various growth stages, varied seasons as wet and dry and also to rice varieties with diverse crop durations. Balasubramanian et al. (2000) has discussed about the Nitrogen management in rice on the basis of the relative chlorophyll contents in rice leaves through the use of a SPAD chlorophyll meter. They studied on the factors affecting SPAD values, such as radiation (season), plant density, varietal groups, soil types, nutrient status in soil and plant, and biotic and abiotic stresses that induce leaf discoloration. They strongly suggested that N dose (in kg ha^{-1}) is to be applied when measured SPAD values fall below established critical values with respect to different growth stages, crop duration and the dry and wet seasons. They mentioned in their review that when a reduction in the SPAD values below the critical values are noticed, with respect to varied crop growth stages, duration, and seasons of wet and dry, the N rates (in kg ha^{-1}) are to be applied.

9.7.3 *EFFECT OF FERTILIZER MANAGEMENT PRACTICES ON GLOBAL WARMING POTENTIAL (GWP)*

The effect of agricultural management practices on global warming potential (GWP) and greenhouse gas intensity (GHGI) is not well recognized. Shang et al. (2011) used a long-term fertilizer trial in Chinese double rice-cropping systems started in 1990 to understand the complete greenhouse gas effect responsible for GWP and GHGI. An experiment was planned using inorganic fertilizer [nitrogen and phosphorus fertilizer (NP), nitrogen, and potassium fertilizer (NK), and balanced inorganic fertilizer (NPK)], mixed inorganic/organic fertilizers at full and reduced rate (FOM and ROM), and no fertilizer application (control) as treatments. Then by means of static chamber method from November 2006 through October 2009 methane (CH_4) and nitrous oxide (N_2O) fluxes were evaluated. By observing the variations in topsoil (0–20 cm) organic carbon (SOC) density over the 10-year period, the net ecosystem carbon balance was assessed.

Annual topsoil SOC sequestration rate was projected to be 0.96 t C ha^{-1} yr^{-1} for the control and 1.01–1.43 t C ha^{-1} yr^{-1} for the fertilizer plots. They observed that long-term fertilizer application significantly increases the GYs. Further, the long-term inorganic fertilizer application significantly increased the CH_4 emissions and N_2O emissions from the drained soils during flooded rice season and the nonrice season, respectively. When compared with the control, the mixed inorganic/organic fertilizer application remarkably increased the net annual GWPs, whereas only slight increment was noted in inorganic fertilizer application plots. The GHGI was highest for the FOM and ROM plots and lowest for the NP and NPK plots. The study findings suggested that with the balanced fertilizer application both agricultural economic viability and GHGs mitigation could be achieved concurrently.

9.7.4 NITROGEN (N) DYNAMICS IN LOWLAND RICE FIELDS

Buresh and Datta (1991) undertook a review of the nitrogen (N) dynamics in lowland rice fields. They studied the influence of typical soil drying and wetting cycles on nitrogen dynamics, the effect of legumes on soil N transformations and N accretion in rice-based cropping systems, the N contribution of legumes to rice, and the integrated management of legume N and industrial fertilizer N for rice. In the tropics, most of the studies on N contributions from legumes have been concentrated on short-duration legumes grown and then incorporated solely as green manures before the monsoon rice crop. The nitrification from the non-flooded rice fields and denitrification from the flooded rice are the main causes of N loss. In lowland rice-legume sequences, the contribution of N by legumes, as in upland crop-legume sequences, depends on the amount of legume N resulting from N_2 fixation, the NHI, the quantity of legume N mineralized and the efficacy of use of this mineralized N by the following crop.

9.7.5 EFFECTS OF ORGANIC MATTER APPLICATION ON SOIL PROPERTIES

Datta and Hundal (1984) discussed the influences of organic matter application on soil properties. They mentioned that the incorporation of organic matter into the soil improves its structure by increasing aggregation. This increased aggregation favorably affects the tillage properties like crusting, water infiltration, moisture retention, drainage, aeration, temperature, and root penetration. The addition of organic matter may change the properties of soil shrinkage properties, improve tilth, and decrease draft required for making the seedbed of dryland crops.

9.7.6 MICRONUTRIENT DEFICIENCIES UNDER SUBMERGED CONDITIONS

Micronutrient deficiencies are widespread and are main causes of yield decline in many rice-growing countries. Savithri et al. (1998) reported that the submerged conditions created for cultivation of rice affects the electrochemical and biochemical reactions, and alters pH, pCO_2 and the concentration of certain ions. This condition raises the availability of Fe and Mn with simultaneous decrease in availability of Zn and Cu. The Zn deficiency is most common disorder in lowland ecosystems, whereas the Fe deficiency is predominant in sodic and upland soils and calcareous coarse-textured soils with low organic matter. The acid soils and the low-lying poorly drained alluvial and colluvial soils are susceptible to Fe toxicity. Rice cultivars generally do not encounter B and Mo deficiencies. Experiments conducted in diverse agroecological zones all over India indicated that Zn doses varying from 2.5 to 22 kg ha^{-1} are most suitable for correcting Zn deficiency, with a rice yield advantage of 4.8 t ha^{-1}. When compared with the foliar sprays of 0.5% $ZnSO_4$ or use of Zn-enriched seedlings, the soil amended with requisite amount of Zn before transplanting was found to be more effective in the lowland ecosystem. Both Hepta and monohydrated $ZnSO_4$ were superior to other sources of Zn (ZnO, $ZnCl_2$ and Zn frits). The organic manures such as farmyard manure, green leaf manure, and coir pith compost enriched with Zn were found beneficial for the direct and residual crops. Zinc fertilization with an optimum dose of 25 kg $ZnSO_4$ ha^{-1} once a year provided high economic return. The foliar spray of 1–2% $FeSO_4$ solution or soil incorporation of Fe (50 kg $FeSO_4 ha^{-1}$) with bulky organic manure (12.5 t ha^{-1}) bring about maximum yield increase in rice up to up to 4.8 t ha^{-1}. The use of 12.5 kg $CuSO_4$ ha^{-1} corrected Cu deficiency and significantly improved rice production. Further, the management schemes such as liming and additional multinutrient supply

(P, K, Mg, Zn, Cu, and B) were found to enhance the rice productivity by correcting the micronutrient deficiency syndrome.

9.8 CROP PROTECTION

Many plants exhibit allelopathic properties. This characteristic feature is effectively exploited and utilized as a tool to bring down the populations of weed and pathogens. For more than a decade, in ecosystems of Japan and Southeast Asia opening surveys of few hundred allelopathic plants were done. In these surveys, they observed that there were more than 30 species including crops, exhibiting huge potentials of allelopathy. These plant species were chosen and studied for their effects on emergence of pathogens and weeds by Xuan et al. (2005). Their research has highlighted that application of these plant materials @ 1–2-tonne ha^{-1} effectively reduced the weed biomass by about 70%, and resulted in 20% increase in rice yields. Further, they observed that a strong inhibition on some key plant pathogens was exhibited by a few plant species. Thus, these plant species exhibiting high allelopathic effects might become efficient tools in decreasing plant pathogens and weeds. Conversely, application of 1–2-tonne ha^{-1} of plant material to the field makes weighty fieldwork. Abundant growth inhibitors recognized from these allelopathic plants are accountable for their allelopathic properties and may be helpful sources for the future development of bio-herbicides and pesticides.

Tu et al. (2000) described the improvement of transgenic elite rice lines expressing a *Bt* fusion gene obtained from *cryIA(b)* and *cryIA(c)* under the control of rice *actinI* promoter, in the lines of *indica* CMS restorer line of Minghui 63 and its resultant hybrid rice Shanyou 63. The level of *Bt* fusion protein CryIA(b)/CryIA(c) noticed in Minghui 63 (T51-1) plants was 20 ng/mg soluble protein. The *Bt* Shanyou 63 was field-tested in natural and recurrent heavy manual infestation of two lepidopteran insects, leaf folder, and yellow stem borer. The transgenic hybrid plants exhibited high protection against both insect pests without yield reduction.

The golden apple snail *Pomacea canaliculata* (Lamarck) in recent times is a newly introduced rice pest in Asia. Though its control is mostly based on preventative and corrective methods which emphasize the significance of cultural and mechanical practices last resort is the application of a molluscicide. Litsinger and Estano (1993) had mentioned that the period of susceptibility of the rice seedlings is mostly influenced by the planting methods. They observed that wet bed-transplanted 20-day-old seedlings were less damaged than dapog-transplanted 13-day-old seedlings or direct-seeded rice, similarly those seedlings ≥ 30 days old were more tolerant of snail damage. Immobilization of snails occurred with drainage. They specified that there is a prerequisite of adopting more preventative measures in high-risk (≥ 2 snails m^{-2}) than in low-risk (< 2 snails m^{-2}) fields. These in low-risk fields can be managed by employing one or two cultural methods. Among the various methods, the Dapog or direct-seeded methods are more effectively benefited from drainage and mechanical control (hand-picking). Under circumstances where it is not feasible to effectively, control the snail populations by cultural methods low dosages of molluscicide can be applied. However, in the high-risk fields, combination of cultural and mechanical methods is required for effective control of snail populations and also these can be substituted with low or high dosages of molluscicide. However, only molluscicide requires a dosage of 0.5 kg a.i. ha^{-1}, which most farmers cannot afford.

9.8.1 BLAST RESISTANCE

Kim et al. (2004) developed a sequential planting method to detect durable blast resistance in Korean rice cultivars. After inoculating with mixed 26 rice blast isolates the rice cultivars were tested using sequential planting method. Two rice cultivars viz., Daesanbyeo, and Gihobyeo exhibited high initial resistance in farmer's field with a rapid breakdown of the resistance in very short periods. These cultivars showed resistance till the third and fourth planting and afterwards exhibited higher than 40% DLA (diseased leaf area).

9.8.2 RICE WATER WEEVIL

In Louisiana, two field experiments were performed by Thompson et al. (1994) to assess the influence of planting date on the damaging effects of rice water weevil, *Lissorhoptrus oryzophilus* Kuschel, larvae on rice yield indicators. They observed that weevil infestation did not show any effect on yields of rice planted before mid-April. The data analysis from both studies revealed that unlike later-planted rice, the early planted rice did not escape harmful populations of rice water weevils; however, it can tolerate such infestations without losing yield. The economic analysis of the data unveiled that planting date has more effect on yield than does carbofuran treatment. Thus, this study specified that the use early rice seeding as a cultural control method in southwest Louisiana may decrease the harmful effects of rice water weevil.

9.9 MANAGEMENT OF GREENHOUSE GAS EMISSION FROM RICE FIELDS

9.9.1 *EFFECT OF DIFFERENT WATER MANAGEMENT SCHEMES ON THE EMISSION OF METHANE (CH_4)*

Yagi et al. (1996) studied the effect of different water management systems on the emission of methane (CH_4) from rice paddies to the atmosphere. With the help of an automated sampling and analyzing system, the experiment site was divided into two plots: a continuously flooded plot, maintained flooded by continuous irrigation from May to August, and an intermittently drained plot in which short-term draining practices were performed several times during the flooding period. The CH_4 emission was found to be influenced with draining practice and large flush of CH_4 emission was observed in the intermittently drained plot. During the cultivation period the total CH_4 emission was 14.8 g m^{-2} from continuously flooded and 8.63 g m^{-2} from intermittently drained plots. The companion N_2O flux measurements exhibited that nearly no N_2O was released from either plot till the final drainage. These findings indicated that intermittent draining practices can effectively decrease the CH_4 emission from rice paddy fields, and that improvement in water management can be one of the major mitigation approaches for CH_4 emission from rice paddy fields.

9.9.2 *REDUCTION OF METHANE EMISSIONS FROM PADDY FIELDS*

Methane is one of the greenhouse gases being released from the paddy fields. Yagi et al. (1997) researches have focused on measures for reducing these CH_4 emissions from paddy fields. They mentioned that some of the options available for mitigating its emissions are management of water, soil organic matter, and addition of soil amendments, adoption of various tillage methods and crop rotation and selection of suitable cultivars. A key promising method is altering the water management, particularly by promotion of midseason aeration through a short-term drainage. However, might pose limitations particularly in those paddy fields with well-irrigated systems. Similarly, organic matter can be efficiently managed with the incorporation of organic matter into the soil during off-season drained period or by composting in order to enhance aerobic degradation. There are many intricate obstacles to take on the mitigation options into local rice farming, including restricted applicability to diverse types of rice fields, rising cost, and labor, negative effects on rice yield and soil fertility, etc.

Many researchers have shown that if the methane emissions from the paddy fields are decreased these would also result in a decline of the global methane concentrations in the atmosphere. Over the past 20 years, there was an upsurge in the global atmospheric methane concentrations. It is attributed that a mid-season drainage can effectively reduce the methane fluxes emitted from the paddy fields. This was for the first time implemented in China in early 1980s. They adopted midseason paddy drainage and slowly substituted the CF in much of the paddy areas of China. By usage of a DNDC biogeochemical model, Li et al. (2002) had made a regional prediction for China's rice paddy methane emissions. They combined all the results of CF and midseason drainage simulations obtained from all the paddy fields of China with regional scenarios. This was done to time the transition from CF to predominantly mid-season drainage. They generated the estimates of total methane flux for 1980–2000. They found out that methane emissions from China's paddy fields were decreased by ~5 Tg CH_4.

9.9.3 EFFECT OF WATER MANAGEMENT ON METHANE EMISSION

Many research studies have highlighted the significance of water management in bring about a decline in the emissions of methane from the paddy fields. Minamikawa et al. (2006) recommended the water management based on the soil redox potential (Eh) named 'Eh control' in rice. Their results have indicated that there was a decline in the total methane emissions by 36% under Eh control on an average of two years of CF. Though straw application was effective in hastening the decrease in soil Eh, there was no effect on the total methane emissions. Their experimental factors had no impact on the yields of rice grain or the straw, though there was an increase in the number of panicles per hill in brown rice in Eh control. Therefore, their study demonstrated the practicability of Eh control under the field conditions and suggested that Eh control is effective in minimizing the methane emissions to a low level in accordance to the soil Eh and it is also feasible to bring about a change in range of soil Eh to more positive values to have decreases in the methane emissions and increases rice GY.

9.9.4 EFFECTS OF UREA PLUS $NH_4H_2PO_4$ APPLICATION RATES ON THE CH_4 EMISSIONS

The ammonium-based, non-sulfate fertilizers, such as urea and/or ammonium phosphate ($NH_4H_2PO_4$), are being used more and more for rice cultivation. Dong et al. (2011) conducted a four-year field campaign was conducted in the Yangtze River Delta to assess the effects of different application rates of urea plus $NH_4H_2PO_4$ on the CH_4 emissions from a paddy rice field. The experimental field was under a typical Chinese water regime that follows a flooding-midseason drainage-reflooding-moist irrigation mode. Over the course of four years, the mean cumulative CH_4 emissions during the rice seasons were 221, 136 and 112 kg C ha^{-1} for nitrogen addition rates of 0, 150 and 250 kg N ha^{-1}, respectively. Compared to the treatment without nitrogen amendments, the 150 kg N ha^{-1} decreased the CH_4 emissions by 6–59%. When the addition rate was further increased to 250 kg N ha^{-1}, the CH_4 emissions were significantly reduced by 35–53% compared to the no-nitrogen treatment. Thus, an addition rate of 250 kg N ha^{-1}, which has been commonly adopted in the delta region in the past two decades, can be regarded as an effective management measure as regards increasing rice yields while reducing CH_4 emissions. Considering that doses of ammonium-based, non-sulfate fertilizers higher than 250 kg N ha^{-1} currently are, and most likely will continue to be, commonly applied for paddy rice cultivation in the Yangtze River Delta and other parts of China.

9.9.5 IRRIGATION EFFECTS ON THE EMISSIONS OF NITROUS OXIDES AND METHANE

Methane and nitrous oxide emissions are common from the paddy fields. An important practice to bring down these emissions is the effective water management. Hou et al. (2012) designed a field trial for studying the controlled irrigation effects on the emissions of nitrous oxides and methane from the fields of paddy. In controlled irrigations a peak in the emissions of methane were observed after 1–2 days of disappearance of the water layer. Later, there was a quick decline in these emissions and these remained small till the re-flooding of the soil. On immediate re-flooding, they observed a minor enlargement in the methane emissions within a short period of re-flooding. They also found increases in the emissions of N_2O from controlled irrigated paddies after 8–10 d after fertilizations. Though soil drying resulted in substantial N_2O emissions, these were not so prominent after rewetting of the dry phase. They noticed a reduction of 81.8% in controlled irrigations in the methane emissions while there was a concomitant increase of 135.4% increase in the N_2O emissions. Sarkar (2005) studied the effect of neem formulations viz., Azadirachtin rich neem coated urea (ARNCU), Pusa neem golded urea, neem cake coated urea and prilled urea (PU) and levels of N on the release of nitrous oxide emission from the incubated soil. The laboratory study revealed that at field capacity, ARNCU treated soil resulted in lowest N_2O-N flux (218.1 g/kg) with only 0.34% loss of N, which accounted about 78% less loss than PU. On the other hand, a lowest N2O-N flux of 126.1 g/kg soil was registered from soil treated with ARNCU under submergence condition indicating about 32% less loss than PU with negligible loss (0.01%) of fertilizer-N. Thus, their research results showed that controlled irrigations can be more effective in reducing the nitrous oxide and methane emissions from the paddy fields.

9.9.6 EMISSIONS OF GREENHOUSE GASES AND RELATED SOIL MICROBIAL PROPERTIES

Hadi et al. (2010) research elucidated the emissions of greenhouse gases and associated soil microbial properties in paddy soils of Japan (rice variety Nipponbare) and Indonesia (Siam Pandak) which were continuously flooded or intermittently drained. The findings unveiled that there was a decline in the emission of greenhouse gases from the intermittent drainage in both the paddy soils of Japan and Indonesia without any noteworthy variations in soil microbial population. The declines of greenhouse emission from Japanese peaty and alluvial paddy soil as a result of intermittent drained were correspondingly about 32 and 37%.

However, the declines in greenhouse gas emission from alluvial soil of Indonesia due to intermittent drainage were extremely alike to that of in Japan, i.e., average about 37%. The research has unveiled that intermittent drainage can be one of the suitable technologies of choice to bring about a decline in the emissions of greenhouse gases from paddy soils in Japan and Indonesia.

9.9.7 MODELS TO ASSESS METHANE EMISSIONS

A widespread biogeochemistry model, DNDC, was studied by Fumoto et al. (2008) to simulate crop growth and soil processes more openly and advance its facility to assess methane (CH_4) emission from rice paddy fields under a broad array of climatic and agronomic conditions. The revised model simulates rice growth by tracking photosynthesis, respiration, C allocation, tillering, and release of organic C and O_2 from roots. For anaerobic soil processes, it quantifies the production of electron donors [H_2 and dissolved organic carbon (DOC)] by decomposition and rice root exudation, and simulates CH_4 production and other reductive reactions based on the availability of electron donors and acceptors (NO_3^-, Mn^{4+}, Fe^{3+}, and SO_4^{2-}). The model further revealed methane emission through rice is simulated by a diffusion routine based on the conductance of tillers and the CH_4 concentration in soil water. The revised DNDC was tested against observations at three rice paddy sites in Japan and China with varying rice residue management and fertilization, and produced estimates consistent with observations for the variation in CH_4 emission as a function of residue management. It also successfully predicted the negative effect of $(NH_4)_2SO_4$ on CH_4 emission, which the current model missed. Predicted CH_4 emission was highly sensitive to the content of reducible soil Fe^{3+}, which is the dominant electron acceptor in anaerobic soils. The revised DNDC generally gave acceptable predictions of seasonal CH_4 emission, but not of daily CH_4 fluxes, suggesting the model's immaturity in describing soil heterogeneity or rice cultivar-specific characteristics of CH_4 transport. Further, the model also overestimated CH_4 emission at one site in a year with low temperatures, suggesting uncertainty in root biomass estimates due to the model's failure to consider the temperature dependence of leaf area development. However, the revised DNDC clearly reveals the influences of soil electron donors and acceptors. It can be used to quantitatively estimate CH_4 emissions from rice fields under a range of conditions.

9.9.8 *IMPACT OF NITROGENOUS FERTILIZERS ON THE EMISSIONS OF METHANE AND N_2O EMISSIONS*

Research studies have indicated that application of nitrogenous fertilizers have an impact on the emissions of Methane and N_2O emissions. Under intermittent irrigation in 1994, in rice fields these emissions as affected by nitrogen fertilizers were measured by Cai et al. (1997). They have applied Ammonium sulfate and urea at rates of 0 (control), 100 and 300 kg N ha^{-1}. Their research has indicated that there was a decline in methane emissions in both ammonium sulfate and urea; however, a drastic decline of 60% in methane emission was seen with the application of ammonium sulfate at 300 kg N ha^{-1}. But, there was an increase in the N_2O emissions with the increasing application rate of nitrogen, where higher rates of its emission occurred from the application of ammonium sulfate rather than urea applications at identical application rates. They observed a clear trade-off effect between the emissions of CH_4 and N_2O. Further, it was observed that in the rice paddy fields on flooding there was very meager fluxes of N_2O, however, these fluxes attained peak values at the commencement of the disappearance of the floodwater, in contrast, during flooding there was a peak of CH_4 flux which was later reduced during the mid-season aeration. Therefore, the research findings has advocated that it is imperative to evaluate the integrative influence of water management and fertilizer application for alleviating greenhouse gas emissions so as to assuage the greenhouse effect of the rice paddy fields contribution.

9.9.9 *DETERMINATION OF CH_4 EMISSION RATES*

CH_4 emission rates have been measured by Schutz et al. (1989) in an Italian rice paddy between 1984 and 1986, in 3 vegetation periods by usage of a fully automated, computerized sampling and analyzing system. This system that was developed enabled in the concurrent determination of CH_4 emission rates at 16 different field plots. There was a strong variation within the diurnal and seasonal variation emission rates of methane. There was a close association between the changes of the soil temperatures and the diurnal changes of methane emissions. During the season, a peak in methane emission rate was evidenced before tillering in May-June and later a second peak during the reproductive stage of rice plants in July. Two peak maximas in summer were noticed in 1985 and 1986, in addition to a first maximum peak in the rate of CH_4 emission that was observed during spring. Based on the type, rate, and mode of fertilizer application of either mineral or organic there were variations in the rate of methane emissions. Deep incorporation of urea and ammonium sulfate @ 200 kg N ha^{-1} correspondingly resulted in a decline in the methane emission rates by 40 and 60%. Though the first maximum of decline in methane emission occurred with the application of 200 kg N ha^{-1} calcium cyanamide, there was a subsequent elevation in the CH_4 emission. Further, there was an enhancement of methane emission by a factor of two with the rice straw application @ of 12 t ha^{-1}, though higher rates of straw application failed to bring about subsequent increases in methane emissions. The average methane emissions during the season from unfertilized fields were in the range of 0.16–0.38 g CH_4 m^{-2} d^{-1}. On the basis of their experimental values and also on the temperature dependence of methane emissions, they estimated the global methane emission rates to be in the range of 50–150 Tg, with a probable average of 100 Tg, which is between 19% and 25% of the global CH_4 emission, signifying that rice paddies are one of the most significant sources of atmospheric CH_4.

9.9.10 ANTHROPOGENIC SOURCE OF METHANE AND NITROUS OXIDE

Akiyama et al. (2005) mentioned that a key anthropogenic source of methane and nitrous oxide emissions in the atmosphere is rice cultivation. They collected and examined data on N_2O emissions from rice fields described in peer-reviewed journals. The projected whole year background emission was 1820 g N ha^{-1} yr^{-1}. A large ambiguity remains, specifically for background emission due to inadequate data availability. Although midseason drainage usually lessens CH_4 and increases N_2O emissions, it may be an effectual substitute for extenuating the net GWP of rice fields.

9.9.11 EFFECTS OF BIOCHAR ON EMISSIONS OF METHANE AND NITROUS OXIDES

Zhang et al. (2010) explored the biochar effects on emissions of methane and nitrous oxides in rice (*Oryza sativa* L., cv. Wuyunjing 7) grown at Tai Lake Plain China. They observed that Biochar application at 40 t ha^{-1} resulted in increased total soil CH_4-C emissions 41% in soils without N fertilization, though there was a sharp decline in the emissions of total N_2O by 40–51% in soils amended with biochar and N fertilization. Similarly, it also resulted in reduced emission factor of 0.0013 kg N_2O-N kg^{-1}N fertilized with biochar at 40 t ha^{-1}. Thus, their research highlights revealed that biochar is capable of increasing not only rice yields but also effective in declining the N_2O emission, though it resulted in an increase in total methane emissions.

9.9.12 NITROUS OXIDE (N_2O) FLUXES FROM RICE PADDY FIELDS

Hua et al. (1997) measured the Nitrous oxide (N_2O) fluxes from rice paddy fields with the help of closed chamber method. The outcomes revealed that these N_2O fluxes varied temporally, spatially, and geographic regionally, with the total amounts of N_2O emissions during the rice growth period ranged from 13.66 to 98.11 mg m^{-2} in Nanjing, 1.73 to 3.65 mg m^{-2} in Yingtan and 178.04 to 472.26 mg m^{-2} in Fengqiu of China, respectively. Further, these fluxes were significantly affected by soil water regime and soil texture of rice paddy fields. The average N_2O fluxes from sandy, loamy, and clayey rice paddy fields were 182.2, 82.8, and 68.7 μg N_2O-N m^{-2} h^{-1}, respectively. They observed that alternation of irrigation and drainage resulted in the occurrence of high N_2O fluxes from the paddy fields rather than from continuously submerged paddy fields. Similarly, those paddy fields which were applied with ammonium sulfate had more emissions of N_2O than with urea. The N_2O-N losses of applied ammonium sulfate and urea were in the range of 0.038 to 0.28% and 0.033 to 0.16%, respectively.

9.10 CLIMATE CHANGE AND ITS EFFECTS ON RICE PRODUCTIVITY

9.10.1 RICE GROWING ENVIRONMENTS

Wassmann et al. (2009) reviewed about the two rice growing environments and their importance in supply of rice under climate changes. Rice is the main staple crop of Asia. Deterioration of rice production systems in the course of climate change would acutely weaken food security. This review assessed the spatial and temporal susceptibilities of diverse rice production systems to climate alteration effects in Asia. They suggested that probable adaptation options for heat stress are resultant from regions where the rice crop is by now subjected to very high temperatures including Iran and Australia. Drought stress is also likely to exacerbate via climate change; a map superimposing the sharing of rainfed rice and precipitation irregularities in Asia highlights particularly susceptible areas in east India/Bangladesh and Myanmar/Thailand. It mentioned that mega-deltas in Vietnam, Myanmar, and Bangladesh which seem to be the backbone of the rice economy in the individual country are likely to encounter particular climate change shocks because of sea-level rise. Noteworthy progress of the rice production systems, i.e., advanced resilience to flooding and salinity, are fundamental for upholding or even rising yield levels in these very productive deltaic regions. The additional 'hotspot' particularly high climate change threats in Asia are the IGP which will be exaggerated by the melting of the Himalayan glaciers. They finally concluded that geospatial susceptibility assessments may turn out to be decisive for planning besieged adaptation programs, but that policy outlines are desirable for their accomplishment.

9.10.2 GENOTYPE-BY-ENVIRONMENT (G × E) INTERACTION

Though the genotype-by-environment (G × E) interaction component of the variance for GY has been broadly described for rainfed lowland rice, the causes for such interactions are not well documented. Ouk et al. (2007) examined the extent and nature of G × E interactions for GY of 34 genotypes collected from the Cambodian rice improvement program. They evaluated the genotypes in a multi-environment trial (MET) by repeating the experiment for 3 years and on eight locations in the rainfed lowlands. The interaction was divided into factors accredited to genotype-by-location (G × L), genotype-by-year (G × Y) and genotype-by-location-by-year (G × L × Y) interactions. For GY, the G × L × Y interaction was recorded as the main variance component. A key factor promoting these large G × L × Y interactions was late maturing genotypes being influenced greatly when the availability of soil water at FL was lowered. The genotypic component (G) and G × L interaction were significant and comparable in size. The significant G component was partially described using a group of four genotypes that were broadly adjusted to diverse environmental conditions characterized by three environmental groups. The three environmental groups were recognized from pattern analysis, and grouped on the basis of sowing time, and availability of water at FL. Though large G × E interaction for GY was elucidated by the different genotypic responses to the water availability environment, other non-water related environmental conditions also appeared to be a part of the interaction. In Cambodia, three target environments which were potential for the breeding programs were identified and four putative genotypes with high yield and wide adaptation in the rainfed lowland were selected.

9.10.3 MULTIPART REGIONAL PATTERN OF CLIMATE VARIABLES

The statistical results of the crop yields explicate the multipart regional pattern of predictable climate variables, CO_2 effects, and agricultural systems. These have a say to aggregations of global crop production. The A1FI scenario, as predictable with its big upsurge in global temperatures, shows the maximum declines of temperature in global as well as regional crop yields especially by 2080s. There is a wide dissimilarity in the yield variations within the developed and developing countries under A2a–c scenarios, however, these exhibit less dissimilarities in yield changes of crops, specifically in B1 and B2 scenarios. They suggested that these changes in crop yields would be much favorable in B2 scenarios. On combined analysis of crop yields with BLS World food trade system model and combined model, it is observed that many parts of the World would be able to feed the population in the next centuries un SRES scenarios. Nevertheless, this result is achievable when the production in the developed countries is able to compensate the declines in yield that in most of the developing countries as per the climate change projections. The analysis revealed that though the global production seems to remain stable yield dissimilarities may be seed across the regional production areas.

9.10.4 PHOTO-THERMAL MODELS OF RICE GROWTH DURATION

The comprehensive photo-thermal models of rice growth duration from sowing to heading were set up by Gao et al. (1987). By means of multiple regression technique the models were developed for 3 major varietal types i.e., early, medium, and late rice. The crop data was collected from the 12 typical varieties grown at diverse locations all over China and the meteorological data, was acquired from the same locations with the same periods. Then by comparing the modeling method with the traditional temperature summation method, the errors in the models were reduced by 3 days for early rice, 6–9 days for medium rice and 18–20 days for late rice varieties.

9.10.5 HadCN3 MODEL

Many climate models were developed to understand the effects of likely changes of climate in the forthcoming years, these impacts were assessed for the forthcoming climate change scenarios using HadCN3 model of global climate change Intergovernmental Panel on Climate Change IPCC, special report on emissions scenarios (SRES) A1FI, A2, B1, and B2. Parry et al. (2004) assessed the projected changes in yield through calculations by the use of some transfer functions that were obtained from crop model simulations with experimental climate data and predictable climate change scenarios. They employed the basic linked system (BLS) to assess the subsequent changes in global cereal production, cereal price, and the number of people at risk from hunger.

9.10.6 CERES-RICE V3 CROP SIMULATION MODEL

Saseendran et al. (2000, 2008) used the CERES-Rice v3 crop simulation model and validated for its suitability to simulate rice production in the tropical humid climate of Kerala State of India. They calibrated and applied it for analysis of climate change effect on rice productivity in Kerala state. The likely climate change picture for the Indian subcontinent as anticipated by the middle of the next century, in view of the predicted emissions of greenhouse gases and sulfate aerosols, in a coupled atmosphere-ocean model experiment conducted at Deutsches Klimarechenzentrum, Germany, is adopted for their study. The adopted scenario showed that over the state of Kerala in the decade 2040–2049 there might be an increase in monsoon seasonal mean surface temperature by about 1.5°C, and an increase in rainfall of approximately 2 mm per day, with respect the climatic conditions that existed in 1980s. The crop model simulation of IPPC inclusive of plant usable concentration of CO_2 about 460 PPM has indicated that under the changed climate scenarios studied there would be a decrease in the length of the rice maturity period by 8% and a yield increase may be to the extent of 12%. Taking into consideration only the increase in temperatures in the crop simulation model has shown that there would be a decrease in crop maturity period to the extent of 8% and also the yield by 6%. This shows that the rise in yields may be due to fertilization effects of elevated CO_2 and augmented rainfall over the Kerala state as anticipated in the climate change picture virtually makes up for the harmful negative impact on rice yield due to temperature rise. The experiments of the sensitivity of the rice model to CO_2 concentration alterations have shown that in Kerala state, an increase in CO_2 concentration leads to yield increase because of its fertilization effect and also due to improved water use efficiency (WUE) in paddy. Further, these experiments have also shown that for a positive change in temperature up to 5°C, there is likely to have a continuous decline in the yield and for every one-degree increment of temperature, this decline in yield might be about 6%. Other experimental results have shown that the physiological effect of ambient CO_2 at 425 ppm concentration may possibly be effective to compensate for the yield losses that would have arisen because of an increase of temperature up to 2°C. Similarly, the rainfall sensitivity experiments have indicated that a boost in rice yield due to an increase in rainfall above that of the observed values is near exponential, however, the decline in rainfall may bring about a constant loss of yield of about 8% per 2 mm/day of decrease in rainfall up to 16 mm/day.

9.10.7 *EFFECTS OF CLIMATE CHANGES*

Lobell et al. (2008) tried to predict about the effects on food availabilities from the climate changes. They mentioned that it becomes easier to measure the benefits only when one has a clear perception of the impact of climate changes. They detected that except in the United States, the temperature drifts from 1980 to 2008 have gone above one standard deviation of remarkable year-to-year changeability in several countries in the cropping regions and seasons. Few models specified that major commodity crops like maize and wheat may exhibit a decrease in their yields by 3.8 and 5.5% respectively. However, the yield productions and yield declines due to climate change trends may be balanced in crops of soybeans and rice. Nevertheless, they predicted that improved technologies that were in use in some countries may be able to compensate the yield declines that are probable to be caused due to variable climate trends in the forthcoming years.

9.10.8 INFLUENCE OF CO_2 CONCENTRATIONS AND TROPOSPHERIC OZONE CONCENTRATIONS ON RICE YIELDS

Rice is the most essential food source on the planet. More than half of the World's populations consume rice. Therefore, to meet the feeding requirements of growing World population there is a substantial requirement of bringing increased yields over this century. Elizabeth (2008) synthesized the meta-analysis of the response yields as influence by rising CO_2 concentrations and tropospheric ozone concentrations. Their analysis revealed that average increased concentrations of 627 ppm may bring an increase of 23% in rice yields. This response might arise due to slightest increase in grain mass and much larger increases in grain and panicle number on the rice plants. This response of rice yields may exhibit variations with fumigation technique to the elevated increase of carbon dioxide. Experiments of free air concentration enrichment (FACE) have shown only an increase of 12% in rice yield. They mentioned that this rise in carbon dioxide in the atmosphere will also accompany an increase in tropospheric ozone and temperatures. Rice that has been exposed to 62 ppb of ozone had a decrease in yield by 14% in comparison to the one that is grown under charcoal filtered air. There was a decrease by increased ozone concentrations in many determinants of yield, including photosynthesis, biomass, and LAI, grain number, and grain mass. Elevated temperature treatments canceled any improvement in rice yield at elevated [CO_2], which suggests that identifying high-temperature tolerant germplasm will be key to realizing yield benefits in the future.

9.10.9 EFFECTS OF CO_2 CONCENTRATION AND AIR TEMPERATURE ON RICE GROWTH AND YIELD

Nakagawa et al. (2003) has worked out on responses of rice growth and yield to increases of CO_2 concentration and air temperature at International Rice Research Institute (IRRI) and summarized these for parameterization of these responses for growth simulation models and exploration of adaptive production technologies. These studies unveiled that rice biomass was increased by 25% with the season-long doubling of CO_2, and a decrease in rice yields with elevated temperatures, due to spikelet fertility (SF) decline. This might still worsen by high CO_2 levels. They have discussed about the main traits linked with rice adaptations and their genetic disparities in rice germplasm. They simulated the sound influences of climate change on rice production in Japan by the rice growth simulation model SIMRIW. These simulation studies have shown that an increment in T50 temperature of 1.5°C (the daily maximum temperature at which SF becomes 50% because of high-temperature damage) would extremely lessen the negative effects of climate change in several prefectures in central and southern Japan and add to total rice production in Japan by about 5%. Similarly, the model simulation under a variety of adaptive technologies indicated that a combination of advanced transplanting and the taking on of later-maturing cultivars may facilitate to make use of the gain of high CO_2 climates for rice production in Japan.

9.10.10 GLOBAL WARMING EFFECTS ON CROP YIELDS

Peng et al. (2004) evaluated the blow of predictable global warming on crop yields by indirect methods using simulation models by analysis of weather data at IRRI Farm from 1979 to 2003 and examined the temperature patterns and the correlation between rice yield and temperature. They reported that during the period of 1979–2003, there was an increase in the annual mean maximum and minimum temperatures by 0.35°C and 1.13°C, respectively, and there was a close up of association between rice GY and average minimum temperature throughout the dry cropping season (January to April). There was a reduction in GY by 10% for each increase of 1°C of minimum temperature in growing-season in the dry season. This report provides a straight confirmation of reduced rice yields from larger nighttime temperature related with global warming.

9.10.11 STRESSES AGGRAVATE UNDER CLIMATE CHANGE

Wassmann et al. (2009) reviewed about the probable adaptation strategies in rice production to various stresses as heat, drought, salinity, etc. that are likely to aggravate under climate change. They discussed about every stress of its present state of damage and probable developments to be taken in crop management as well as in germplasm technologies for overcoming the production losses in various crops. They mentioned that high temperatures along with elevated humidities result in more of spikelet sterilities and affect drastically the grain quality, further an increase in night temperatures may bring about a decline in rice assimilate accumulation rates. Both these would exhibit a drastic effect on rice yields. Further, they mentioned that the rice cultivars growing under severe hot environments are more useful in the improvement and development of rice germplasm by capturing of this germ pool for heat resistance. Similarly, significant progress was made in improvement of germplasm for drought, salinity, and submergence tolerance (SUB). It also made a comparative assessment of rice with other crops.

Fernando (1993) discussed about the fish culture and rice field ecology. Rice fields are primary part of the landscape all the way through in most part of the tropics. Rice is also grown extensively in higher latitudes. The majority of rice cultivation is carried out in flooded fields where a temporary aquatic fauna is produced. Rice cultivation has supported some of the ancient civilizations. The usage of the aquatic phase for raising a crop of fish has not been practiced extensively. Fragmented records indicated that rice and fish have been cultivated parallel, but infrequently over 2 or 3 millennia. More consistent records of rice and fish culture in rice fields are only found during the past 150 years. Though rice cultivation currently is extremely mechanized, it utilizes high fertilizer and pesticide inputs. Widespread irrigation facilities have been developed for increasing the area under rice cultivation and achieve improved yields. The rice fields with adequate waters generate a distinctive, temporary, and quickly changing habitat which is often very productive and can be used to raise fish on an artisanal or intensive scale.

Fish culture in rice fields has had a variable history during the past 150 years when records are presented. Its previous history is unclear. Enduring records of fish culture activities are not available from any part of the World though it looks like that successful enterprise might have occurred in Japan, Italy, USSR, and China. Efforts to culture fish in rice fields have been made on all continents except Australasia and Antarctica of course. Nowadays the emphasis of rice-cultivation seems to have moved to China, Indonesia, and Thailand. Whether this enterprise will sustain even in these countries or not, is not be projected with any degree of certainty.

9.11 CONCLUSION

This chapter presented the various aspects of rice cultivation. These included land preparation, nursery management, sowing, and methods of planting, water management, fertilizer management, and crop protection. The chapter also presented the management of greenhouse gas emissions from rice fields and climate change effects on rice productivity.

KEYWORDS

- **alternate wetting and drying irrigation**
- **free air concentration enrichment**
- **global warming potential**
- **green revolution**
- **greenhouse gas emission**
- **Indo-Gangetic plains**

REFERENCES

Adhikari, B. B., Mehera, B., & Haefele, S., (2013). Impact of rice nursery nutrient management, seeding density and seedling age on yield and yield attributes. *American Journal of Plant Sciences*, *4*, 146–155.

Akiyama, H., Yagi, K., & Yan, X., (2005). Direct N_2O emissions from rice paddy fields: Summary of available data. *Global Biogeochemical Cycles*, *19*, 1–10.

Balasubramanian, V., Morales, A. C., Cruz, R. T., Thiyagarajan, T. M., Nagarajan, R., Babu, M., Abdulrachman, S., & Hai, L. H., (2000). Adaptation of the chlorophyll meter (SPAD) technology for real-time N management in rice: A review. *International Rice Research*, *25*, 4–8.

Basra, S. M. A., Farooq, M., Tabassam, R., & Ahmad, N., (2005). Physiological and biochemical aspects of pre-sowing seed treatments in fine rice (*Oryza sativa* L.). *Seed Science and Technology*, *33*, 623–628.

Bouman, B. A. M., & Tuong, T. P., (2001). Field water management to save water and increase its productivity in irrigated lowland rice. *Agricultural Water Management, 49*, 11–30.
Bouman, B. A. M., Peng, S., Castañeda, A. R., & Visperas, R. M., (2005). Yield and water use of irrigated tropical aerobic rice systems. *Agricultural Water Management, 74*, 87–105.
Buresh, R. J., & Datta, S. K., (1991). Nitrogen dynamics and management in rice-legume cropping systems. *Advances in Agronomy, 45*, 1–59.
Cabangon, R. J., & Tuong, T. P., (2000). Management of cracked soils for water-saving during land preparation for rice cultivation. *Soil and Tillage Research, 56*, 105–116.
Cabangon, R. J., Tuong, T. P., & Castillo, E. G., (2004). Effect of irrigation method and N-fertilizer management on rice yield, water productivity, and nutrient-use efficiencies in typical lowland rice conditions in China. *Paddy and Water Environment, 2*, 195–206.
Cai, Z., Xing, G., Yan, X., Xu, H., Tsuruta, H., Yagi, K., & Minami, K., (1997). Methane and nitrous oxide emissions from rice paddy fields as affected by nitrogen fertilizers and water management. *Plant and Soil, 196*, 7–14.
Cassman, K. G., Peng, S., Olk, D. C., Ladha, J. K., Reichardt, W., Dobermann, A., & Singh, U., (1998). Opportunities for increased nitrogen-use efficiency from improved resource management in irrigated rice systems. *Field Crops Research, 56*, 7–39.
Datta, S. K., & Hundal, S. S., (1984). *Effects of Organic Matter Management on Land Preparation and Structural Regeneration in Rice-Based Cropping System*. Agricultural Information Bank for Asia, South-East Asian Regional Center for Graduate Study and Research in Agriculture.
Dobermanna, A., Witt, C., Dawe, D., Abdulrachman, S., Gines, H. C., Nagarajane, R., et al., (2002). Site-specific nutrient management for intensive rice cropping systems in Asia. *Field Crops Research, 74*, 37–66.
Dong, H., Yao, Z., Zheng, X., Mei, B., Xie, B., Wang, R., Deng, J., Cui, F., & Zhu, J., (2011). Effect of ammonium-based, non-sulfate fertilizers on CH4 emissions from a paddy field with a typical Chinese water management regime. *Atmospheric Environment, 45*, 1095–1101.
Elizabeth, A., (2008). Rice production in a changing climate: A meta-analysis of responses to elevated carbon dioxide and elevated ozone concentration. *Global Change Biology, 14*, 1642–1650.
Fernando, C. H., (1993). Rice field ecology and fish culture: An overview. *Hydrobiology, 259*, 91–113.
Fumoto, T., Kobayashi, K., Yagi, K., & Hasegawa, T., (2008). Revising a process-based biogeochemistry model (DNDC) to simulate methane emission from rice paddy fields under various residue management and fertilizer regime. *Global Change Biology, 14*, 382–402.
Gao, L. Z., Jin, Z. Q., & Li, L., (1987). Photo-thermal models of rice growth duration for various varietal types in China. *Agricultural and Forest Meteorology, 39*, 205–213.
Gupta, R. K., Naresh, R. K., Hobbs, P. R., Jiaguo, Z., & Ladha, J. K., (2003). Sustainability of post green evolution agriculture: the rice-wheat cropping systems of the Indo-Gangetic plains and China. In: *Book: Improving the Productivity and Sustainability of Rice–Wheat Systems*. Issues and Impact, 1–25.

Gupta, R., & Seth, A., (2007). A review of resource conserving technologies for sustainable management of the rice–wheat cropping systems of the Indo-Gangetic plains (IGP). *Crop Protection, 26*, 436–444.

Hadi, A., Inubushi, K., & Yagi, K., (2010). Effect of water management on greenhouse gas emissions and microbial properties of paddy soils in Japan and Indonesia. *Paddy and Water Environment, 8*, 319–324.

Hobbs, P. R., (2001). Present practices and future options. Tillage and crop establishment in South Asian Rice-Wheat systems. *Journal of Crop Production, 4*, 1–22.

Hou, H., Peng, S., Xu, J., Shihong, Y. S., & Mao, Z., (2012). Seasonal variations of CH_4 and N_2O emissions in response to water management of paddy fields located in Southeast China. *Chemosphere, 89*, 884–889.

Hua, X., Guangxi, X., Cai, Z. C., & Tsuruta, H., (1997). N fertilizer-induced emission factor of N_2O GHGs greenhouse gases GWP global warming potential SOC soil organic carbon WRGS whole rice growing season. *Nutrient Cycling in Agro Ecosystems, 49*, 23–28.

Kim, B. R., Ahn, S. W., & Roh, J. H., (2004). Durability of rice cultivars to blast in Korea by sequential planting method. *Korean Agricultural Science Digital Library,* 21, https://doi.org/10.5423/RPD.2015.21.1.020.

Kundu, D. K., & Ladha, J. K., (1995). Efficient management of soil and biologically fixed N_2 in intensively-cultivated rice fields. *Soil Biology and Biochemistry, 27*, 431–439.

Li, C., Qiu, J., Frolking, S., Xiao, X., Salas, W., Moore III, B., Boles, S., Huang, Y., & Sass, R., (2002). Reduced methane emissions from large-scale changes in water management of China's rice paddies during 1980–2000. *Geophysical Research Letters, 29*, 331–334.

Litsinger, J. A., & Estano, D. B., (1993). Management of the golden apple snail *Pomacea canaliculata* (Lamarck) in rice. *Crop Protection, 12*, 363–370.

Lobell, D. B., Schlenker, W., & Roberts, J. C., (2008). Climate trends and global crop production. *Science, 333*, 616–620.

Minamikawa, K., & Sakai, N., (2006). The practical use of water management based on soil redox potential for decreasing methane emission from a paddy field in Japan. *Agriculture, Ecosystems and Environment, 116*, 181–188.

Nakagawa, H., Horie, T., & Matsul, T., (2003). *Effects of Climate Change on Rice Production and Adaptive Technologies*. International Rice Research Institute. 635–658.

Odhiambo, L. O., & Murty, V. V. N., (1996). Modeling water balance components in relation to field layout in low land paddy fields: I. Model development. *Agricultural Water Management, 30*, 185–199.

Ouk, M., Basnayake, J., Tsubo, M., Fukai, S., Fischer, K. S., Kang, S., Men, S., Thun, V., & Cooper, M., (2007). Genotype-by-environment interactions for grain yield associated with water availability at flowering in rain fed lowland rice. *Field Crops Research, 101*, 145–154.

Ovung, Z., & Sarkar, N. C., (2013). Effect of Biofertlizer on Productvity of Upland Rice (Oryza satva L.) in Nagaland, India. *International Journal of Bio-resource and Stress Management, 4*(1), 23–27.

Parry, M. L., Rosenzweig, C., Iglesias, A., Livermore, M., & Fischer, G., (2004). Effects of climate change on global food production under SRES emissions and socio-economic scenarios. *Global Environmental Change, 14*, 53–67.

Pasuquin, E., Lafarge, T., & Tubana, B., (2008). Transplanting young seedlings in irrigated rice fields: Early and high tiller production enhanced grain yield. *Field Crops Research, 105*, 141–155.

Peng, S., Huang, J., Sheehy, J. E., Laza, R. C., Visperas, R. M., Zhong, X., Centeno, G. S., Khush, G. S., & Cassman, K. G., (2004). Rice yields decline with higher night temperature from global warming. *PNAS, 101*, 9971–9975.

Prasad, M., & Prasad, R., (1980). Yield and nitrogen uptake by rice as affected by variety, method of planting and new nitrogen fertilizers. *Fertilizer Research, 1*, 207–213.

Sakamoto, T., Nguyen, V. N., Ohno, H., Ishitsuka, N., & Yokozawa, M., (2006). Spatio-temporal distribution of rice phenology and cropping systems in the Mekong Delta with special reference to the seasonal water flow of the Mekong and Bassac rivers. *Remote Sensing of Environment, 100*, 1–16.

San, Y., Mano, Y., Ookawa, T., & Hirasawa, T., (2004). Comparison of dry matter production and associated characteristics between direct-sown and transplanted rice plants in a submerged paddy field and relationships to planting patterns. *Field Crops Research, 87*, 43–58.

Sarkar, N. C., (2005). Efficacy of neem formulations and levels of N on productivity of rice-mustard cropping system. PhD. Thesis. Division of Agronomy, Indian Agricultural Research Institute, New Delhi, India.

Saseendran, S. A., Singh, K. K., Rathore, L. S., Singh, S. V., & Sinha, S. K., (2008). Rice production in a changing climate: A meta-analysis of responses to elevated carbon dioxide and elevated ozone concentration. *Global Change Biology, 14*, 1642–1650.

Saseendran, S. A., Singh, K. K., Rathore, L. S., Singh, S. V., & Sinha, S. K., (2000). Effects of climate change on rice production in the tropical humid climate of Kerala, India. *Climatic Change, 44*, 495–514.

Savithri, P., Perumal, R., & Nagarajan, R., (1998). Soil and crop management technologies for enhancing rice production under micronutrient constraints. *Resource Management in Rice Systems: Nutrients, 81*, 121–135.

Schütz, H., Pschorn, A. H., Conrad, R., Rennenberg, H., & Seiler, W., (1989). A 3-year continuous record on the influence of daytime, season, and fertilizer treatment on methane emission rates from an Italian rice paddy. *Journal of Geophysical Research Atmosphere, 94*, 16405–16416.

Shang, Q. Y., Yang, X., & Gao, C., (2011). Net annual global warming potential and greenhouse gas intensity in Chinese double rice cropping systems: A 3-year field measurement in long-term fertilizer experiments. *Global Change Biology, 17*, 2196–2210.

Sharma, P. K., Datta, S. K., & Redulla, C. A., (1987). Root growth and yield response of rainfed lowland rice to planting method. *Experimental Agriculture, 23*, 305–313.

Singh, Y., Singh, V. P., Singh, G., Yadav, D. S., Sinha, R. K. P., Johnson, D. E., & Mortimer, A. M., (2011). The implications of land preparation, crop establishment method and weed management on rice yield variation in the rice-wheat system in the Indo-Gangetic plains. *Field Crops Research, 121*, 64–74.

Stroppiana, D., Boschetti, M., Gusberti, D., Bocchi, S., & Acutis, M., (2006). Analysis of rice sample size variability due to development stage, nitrogen fertilization, sowing technique and variety using the visual jackknife. *Field Crops Research, 97*, 135–141.

Tabbal, D. F., Bouman, B. A. M., Bhuiyan, S. I., Sibayan, E. B., & Sattar, M. A., (2002). On-farm strategies for reducing water input in irrigated rice, case studies in the Philippines. *Agricultural Water Management*, *56*, 93–112.

Thompson, R. A., Quisenberry, S. S., Nguessan, F. K., Heagler, A. M., & Giesler, G., (1994). Planting date as a potential cultural method for managing the rice water weevil (Coleoptera: Curculionidae) in water-seeded rice in Southwest Louisiana. *Journal of Economic Entomology*, *87*, 1318–1324.

Thorie, M., Sarkar, N. C., & Kharutso, A., (2013). Effect of Biofertlizer on the Productvity of Terraced Upland Rice (Oryza Satva L.). *International Journal of Bio-resource and Stress Management*, *4*(3), 400–403.

Tu, J., Zhang, G., Datta, K., Xu, C., Yuqing, H. Y., Zhang, Q., Singh, G. K., Swapan, & Kumar, D. S., (2000). Field performance of transgenic elite commercial hybrid rice expressing *Bacillus thuringiensis* δ-endotoxin. *Nature Biotechnology*, *18*, 1101–1104.

Tuong, T. P., & Bhuiyan, S. I., (1999). Increasing water-use efficiency in rice production: Farm-level perspectives. *Agricultural Water Management*, *44*, 117–122.

Waha, K., Bussel, L. G. J., Müller, C., & Bondeau, A., (2012). Climate-driven simulation of global crop sowing dates. *Global Ecology and Biogeography*, *21*, 247–259.

Wassmann, R., Jagadish, S. V. K., Heuer, S., Ismail, A., Redona, E., Serraj, R., et al., (2009). Climate change affecting rice production: The physiological and agronomic basis for possible adaptation strategies. *Advances in Agronomy*, *101*, 59–122.

Wassmann, R., Jagadish, S. V. K., Sumfleth, K., Pathak, H., Howell, G., Ismail, A., Serraj, R., Redona, E., Singh, R. K., & Heuer, S., (2009). Regional vulnerability of climate change impacts on Asian rice production and scope for adaptation. *Advances in Agronomy*, *102*, 91–133.

Wopereis, M. C. S., Bouma, J., Kropff, M. J., & Sanidad, W., (1994). Reducing bypass flow through a dry cracked and previously puddle rice soil. *Soil and Tillage Research*, *29*, 1–11.

Xuan, T. D., Shinkichi, T., Khanh, D. T., & Min, C., (2005). Biological control of weeds and plant pathogens in paddy rice by exploiting plant allelopathy: An overview. *Crop Protection*, *24*, 197–206.

Yagi, K., Tsuruta, H., & Minami, K., (1997). Possible options for mitigating methane emission from rice cultivation. *Nutrient Cycling in Agro Ecosystems*, *49*, 213–220.

Yagi, K., Tsuruta, H., Kanda, K., & Minami, K., (1996). Effect of water management on methane emission from a Japanese rice paddy field: Automated methane monitoring. *Global Biochemical Cycle*, *10*, 255–326.

Zhang, A., Cui, L., Pan, G., Li, L., Hussain, Q., Zhang, X., Zheng, J., & Crowley, D., (2010). Effect of biochar amendment on yield and methane and nitrous oxide emissions from a rice paddy from Tai Lake plain, China. *Agriculture, Ecosystems and Environment*, *139*, 469–475.

CHAPTER 10

Rice Grain Quality Analysis, Food Quality, Chemistry, and Food Processing

ABSTRACT

Research work on improvement of rice grain quality is being done through various approaches and many advancements have been achieved that need to be assembled at one place therefore in this chapter recent approaches and their efficiency in the determination of grain characteristics and food quality is being presented. A brief overview of the recent progress made in rice grain quality research is provided which can be reconsidered in novel ways.

10.1 RICE GRAIN QUALITY ANALYSIS

In cereals, grain quality is controlled by epistasis, maternal, and cytoplasmic effects and endosperm are triploid in nature. This makes its inheritance more complex than other agronomic characters. He et al. (1999) undertook a genetic analysis of rice grain quality using the Double Haploid population obtained from anther culture of an *indica/japonica* hybrid. For the DH lines and the parent varieties the parameters such as amylose content (AC), alkali-spreading score (ASS), gel consistency (GC), percentage of grain with a white core (PGWC) and the square of the white core (SWC), were assessed. All the five parameters were moderately stable in three locations for each parent and exhibited significant differences between the parents. AC exhibited a bimodal distribution, and the ASS distribution was skewed toward the JX17 value, whereas the remaining three parameters showed continuous distributions among the DH lines with partly transgressive segregations. For AC, a minor and a major gene were identified on chromosomes 5 and 6. For GC, two QTLs were detected on chromosomes 2 and 7 correspondingly. For ASS, both minor and major genes were detected on chromosome 6. For PGWC, two QTLs were found on chromosomes 8 and 12. For SWC, one minor QTL was identified on chromosome 3. Further, in QTL mapping a genetic linkage was found between AC major allele *wx* and alkali degeneration the gene *alk*.

The edible rice grain constituents and the current progresses made in rice grain quality research are defined by Huang et al. (1998). In the grain quality research the traditional methods followed and the problems faced while improving the rice grain quality are analyzed. They mostly comprised: (1) the conflict between the rice grain quality dynamic forming and static research methods on rice grain quality features; (2) the associations among amylose and protein content, constituent, and structure; (3) the effect of cytoplasm on rice grain quality. For the research on rice grain quality forming uniformity novel approaches and measures are put forth.

At present, wild rice species, *Oryza rufipogon*, *O. officinalis*, and floating rice, *O. sativa*, have become rare in the Mekong Delta. This necessitates the rigorous survey of diversity and urgent conservation of these wild species. So as to assess and to preserve genetic resources of two wild rice species distributed in the Mekong Delta. Thanh and Hirata (2001) compared their morphological features with cultivated rice. Further, they analyzed protein variations by using SDS-PAGE analysis method. Results unveiled that the morphological materials comprising *O. officinalis* specimens were diverse from cultivated rice in two distinct features such as thick rhizome, green grain color. The molecular weight of some proteins like glutelin and prolamin was high in *O. officinalis*. In comparison, protein profiles of *O. officinalis* populations displayed a complete absence of prolamin 16kDa, less 13kDa and low waxy content. High waxy protein content was noticed in two F_6 generation RIL lines signifying that waxy content is still a complex trait, but it may be suitable for enhancement of AC in rice.

10.2 INTRODUCTION TO GRAIN QUALITY

To study the effect of various nitrogenous (N) fertilizers on growth, yield, and quality of hybrid rice variety 'Proagro 6207' Chaturvedi (2006) conducted a field trial. From 10 different treatments, the data on all the growth characters and yield parameters was recorded. The results unveiled the significant increase in growth characters, yield parameters, and grain nitrogen (N) with the sulfur-containing nitrogen fertilizer-super net application. Similar results were recorded in treatment T4, where ammonium sulfate nitrate was applied. In this experiment significantly lowest grain yield (GY) and grain nitrogen (N) content were found in treatment applied with non-sulfur-containing nitrogen fertilizer, urea, and these results were consistent in all the experiments.

In rice plants, the 'active' *indica*-type and 'inactive' *japonica*-type soluble starch synthase Ha (SSIIa) differs for four amino acids. Nakamura et al. (2005) carried out *SSIIa* gene fragments shuffling experiments, where they have replaced Glu-88 and Gly-604 in *SSIIa* of *indica*-cultivars, IR36 and Kasalath with Asp-88 and Ser-604, respectively. In *japonica* cultivars, Nipponbare and Kinmaze SSIIa, Met-737 and Phe-781 were replaced by Val-737 and Leu-781, respectively. The results unveiled that Val-737 and Leu-781 are vital for the optimum activity of SSIIa and involved in *indica*-type amylopectin synthesis. The introduction of the 'active' *indica*-type SSIIa gene facilitated synthesis of *indica*-type amylopectin by *japonica*-type cv. Kinmaze. The gelatinization-resistant properties which are specific to *indica*-rice starch were observed in transformed *japonica* rice plants. Variety of starches with amylopectin chain-length distribution patterns were produced in transformed lines expressing various levels of the IR36 SSIIa protein. Unexpectedly, a combination of Phe-781 and Gly-604 which can restore approximately 44% of the SSIIa activity provided that Val-737 was conserved. The findings of this research established that the SSIIa activity performs an important function in the synthesis of the long B_1 chains by extending short A and B_1 chains and determines the rice starch amylopectin structure type to be either the typical *indica*-type or *japonica*-type.

Owing to their reduced harvest index (HI), vulnerability to lodging and increased susceptibility to foliar diseases conventional basmati rice varieties are poor yielders. To meet the current demand, the new varieties with grain quality characteristics of basmati and high yield potential are essential. Genetic control of grain and cooking quality characters of basmati is somewhat complicated. But, the use of molecular markers strongly linked to these characters eased the breeding work. Amarawathi et al. (2008) undertook quantitative trait loci (QTLs) mapping for seven key quality traits such as grain length (GL), grain breadth (GB), grain length to breadth ratio (LBR), cooked kernel elongation ratio (ELR), AC, alkali-spreading value (ASV) and aroma. For QTL mapping 209 recombinant inbred lines (RILs) derived from a cross between basmati quality variety Pusa 1121 and a complimentary quality breeding line Pusa 1342 were used. Using 110 polymorphic simple sequence repeat (SSR) markers dispersed across the 12 rice chromosomes a framework of molecular linkage map was created. On seven different chromosomes several QTLs, involving three for GL, two for GB, two for LBR, three for aroma, and one each for ELR, AC, and ASV were mapped. The position of most of these QTLs was in conformity with the previous reports. The QTLs for GL, ELR, and aroma on chromosomes 1, 11 and 3, respectively, were reported for the first time in this research. In Pusa 1121 the aroma was found to be controlled by three genes present on chromosomes 3, 4, and 8 which is in contrast to the former reports of monogenic recessive inheritance and alike to the informed correlation of *badh2* gene with aroma QTL on chromosome 8. In the RILs with high aroma, a discontinuous 5 + 3 bp deletion was reported in the seventh exon of *badh2* gene. But this deletion was not enough to impart this trait to the rice grains and the RILs having this deletion exhibited only slight or no aroma expression.

10.2.1 GRAIN QUALITY CHARACTERISTICS

Rice varieties grown in the cultivated globally possess different cooking, sensory, and processing qualities. Rice preferred by one group may not be chosen by another group. Researchers started observing these cultivars diversity early in the twentieth century. Warth and Darabsett (1914) observed the cultivar divergence in rice kernel response to dilute alkali. This was the primary documented study on the subject. Further research was irregular and was concentrated specifically on the "cooking" (i.e., hydration property) of rice in India. However, the scope of this study was extended to a greater magnitude in the 1950s. Then, on the basis of specific dimensional and cooking-processing characteristics, rice was classified into three grain type's viz. long, medium, and short in the United States (Adair et al., 1973). The unintended release of the cultivar Century Patna 231 in the early 1950s, which possessed the features of characteristic U.S. long-grain rice but diverse cooking and processing traits, was an economic disaster for the rice industry. This experience brings about the necessity to know the source of rice end-use quality and to develop simple tests for it. A research work, coordinated by the United States Department of Agriculture (USDA), elucidated the basics of rice quality, which remain considerably effective even today. Beachell and Halick (1957) documented the initial phases of this work and Beachell and Stansel (1963) its ultimate result. Since the 1950s, many research in Germany (Federal Research Center for Cereal and Potato Processing), Spain (Institute of Agronomy and Food Technology), Japan (National Food Research Institute and Niigata University), and especially India (Central Food Technological Research Institute) and the Philippines (International Rice Research Institute (IRRI)) has taken our knowledge of rice end-use quality to new levels and currently, research endeavors on this subject exist on almost every continent.

Throughout the World, rice occurs in many combinations of length-width ratio, grain weight, bran color, and endosperm chemical characteristics. The cooking and processing qualities of rice are not related with the physical forms of kernels. End-use quality characteristics preferred by one section of the World may be entirely undesirable to another. Maximum rice is utilized nearby to where it was produced and does not reach the export market. Therefore, consumers are frequently subjected only to native cultivars and are unacquainted to the abundant variation in the physical and chemical traits of the World's rice germplasm. Rice that reaches the export market is usually restricted to a few combinations of kernel physical and chemical characteristics. This enables trade by simplifying the language necessary to recognize rice quality-types and is suggested to meet the quality necessities of rice-importing nations. For instance, in the export market the rice with a long, thin grain normally possess a firm, nonsticky cooked texture and is termed as "long-grain rice." But, in the World's rice germplasm, the rice with a long grain may not be firm cooking and similarly firm-cooking rice may not possess a long grain character.

On the basis of *indica* hybrid rice, *indica* conventional rice, *japonica* hybrid rice and *japonica* conventional rice classification, the correlation analysis and multiple comparisons of grain quality characters were undertaken by Wang et al. (2005). Results revealed the significant positive association between brown rice rate, milled rice rate and head rice rate in various rice types. In appearance quality, the length/width ratio of various rice types was significantly and negatively associated to milled rice rate, head rice rate, chalkiness grain rate, chalkiness degree. In *indica* rice, significantly positive association was observed between the AC and ASV, while AC was negatively associated with GC. In *japonica* rice, significant correlations were not found between AC, ASV, and GC. In multiple comparison, different types of rice were alienated into three groups (high, middle, and low) according to the values of head rice rate, chalkiness, and AC, which were significantly different respectively.

The genetic factors affecting cooking and eating quality of rice were studied by Bao et al. (2002). From an *indica* (IR64) and a *japonica* (Azucena) rice with a related apparent amylose content (AAC) a doubled haploid (DH) population comprising of 135 lines were obtained. Then the parameters such as AAC, gelatinization temperature (GT), GC and six starch pasting viscosity were measured and using 193 molecular markers the QTLs for these parameters were mapped on the DH population. A total of 17 QTLs were identified for the 9 traits. Because of the related AAC between the parents, QTLs for the parameters were not noticed at the *wx* locus. In this study, based on populations obtained from parents with different AAC and *wx* gene alleles, many QTLs with significant effects on the differences in the measured parameters were identified. For GT hindrance and consistency viscosity a QTL flanked by Amy2A and RG433 was detected on the end of the long arm of chromosome 6, which might cover the gene encoding starch branching enzyme I. Likewise, for hot paste viscosity and breakdown viscosity a QTL flanked by RG139 and RZ58 was identified on chromosome 2, which might cover the gene encoding starch branching enzyme III. It accentuates the requirement of additional elucidation of fine molecular mechanisms underlying these traits to improve the eating and cooking quality of rice.

10.2.2 QUALITY CHARACTERISTICS OF PADDY OR RICE

Using a laboratory gas-fired infrared (GIR), wet paddy was dried under different process conditions. Then, by changing the peak wavelength of infrared (IR) emitted and initial moisture content of paddy studied, the drying behavior was studied (Laohavanich and Wongpichet 2009). The drying data was fitted with the available thin-layer drying models. By comparing the root mean square deviation (RMSD) values, average percentage of error (%E) and chi-square (χ^2) between the observed and expected moisture ratios best-fitted over other models and the modified Page model with higher coefficient of determination (R^2 = 0.9985 – 0.9695) were selected. Then, the head rice yield (HRY) and whiteness qualities of GIR and temper-dried wet paddy were estimated. Response surface methodology unveiled that the increment in GIR drying time (DT) may increase HRY. But, the whiteness decreases with the increase in both GIR DT and tempering time. Then, based on the industrial or user requirement, the suitable drying conditions could be determined using contour overlay graph.

IR drying technology seemed to be the possible substitutes to the conventional heating methods for acquiring high-quality dried agro-products. For long-grain paddy drying, at present there is a shortage of reports investigating IR drying kinetics, and in such reports, milling quality was the central point of investigation with regard to suitable drying process conditions. This study has described the modeling determination and milling qualities of paddy dried with GIR dryer. Further, it provided a guideline on basic information to construct a commercial drying system. Finally, these findings would have practical uses in upholding or improving the paddy milling qualities in the present industry.

During milling process of rice paddy, husking operation is the most significant stage which efficiently reveals the quantitative and qualitative losses of rice. Mohammad (2011) studied the effect of four levels of husked ratio (HR) viz. 0.6, 0.7, 0.8, and 0.9 on broken brown rice (BBR), broken milled rice (BMR) and rice whiteness (RW). Three Iranian rice varieties, viz. Binam, Khazar, and Sepidroud, and a commercial rice milling system comprising rubber rolls husker and blade-type whitener were used in the experiment. The findings revealed that with the increase of HR from 0.6 to 0.9 the BBR increased significantly ($P<0.01$) from 7.42 to 10.28% for Binam, 9.17 to 13.39% for Khazar and 15.17 to 21.82% for Sepidroud varieties of rice. At the HR of 0.8, the lowest BMR values were found for all the three varieties. While, with the increase of HR from 0.6 to 0.9, the RW reduced from 36.1 to 30.8, 36.5 to 30.1, and 35.4 to 29.8 for Bianm, Khazar, and Sepidroud varieties, respectively.

10.2.3 QUALITY CHARACTERISTICS OF MILLED RICE

The primary agronomic factors affecting the rice yield and milling quality are nitrogen fertilization and water management. To date, much data is not available on the effect of nitrogen and time of drainage on the uniformity of rice grain development. Jongkaewwattana and Geng (1991) undertook a research to study the effect of nitrogen application on the uniformity of grain development on a panicle as well as to determine the combined effect of drainage time and nitrogen application on yield and milling quality of rice. Results unveiled the significant effect of nitrogen on grain filling characteristics. They found that nitrogen decreases the grain filling rate and grain weight, while it increases the duration of grain filling and number of tillers. The time of drainage is critical to the head rice recovery. In general, late drainage increases the head rice recovery, and its influence on milling quality of rice increases with an increasing rate of nitrogen application.

To evaluate the cooking and eating qualities of milled rice a survey was conducted by Juliano in 1982. This report represents the response of 41 scientists in 27 countries. For evaluation of cooked rice, texture thirty-six respondents were selected; of those, 20 applied sensory evaluation, 11 applied instruments, and 5 applied both. Details of the approaches are examined and compared. Out of 41 respondents; amylose was tested by 30 respondents, ASV was evaluated by 27 respondents, and 17 assessed GC. Most respondents were interested in cooperative testing of cooking methods. The relative hardness of rice cooked to 75% water content by cooker and excess-water methods, single-grain vs. bulk cooked rice for assessment of hardness by instruments, grain elongation during cooking and amylography were the probable areas for cooperative testing.

Generally, the rice varieties with large and heavy grains are chosen by insects inhabiting in grains. Degree of milling, tight hull, grain hardness, high AC and high GT of endosperm starch, and low moisture content are the important physicochemical factors contributing to resistance against stored-grain insects in rice. At relative humidity of 75% and above, high AC and high GT contribute to low grain equilibrium moisture content. This also contributes to poorer digestibility of raw starch granules by insects (Juliano, 1981).

10.2.4 IMPROVING OR MAINTAINING RICE QUALITY

In many rice-producing regions, one of the most serious problems are cooking and eating quality of rice. The main constituents of the cooking and eating quality of rice grains are AC, GC, and GT. Tan et al. (1999) carried out genetic analysis of these traits with the help of molecular markers. F_2

seeds, an $F_{2:3}$ populations, and an F_9 recombinant inbred-line population obtained from a cross between the parents of 'Shanyou 63' were used in the analysis. Segregation analyses of these three generations exhibited that each of the three traits was under the control of a single Mendelian locus. The QTL (quantitative trait locus) analyses, both by one-way analysis of variance with single marker genotypes and by whole-genome scanning using MAPMAKER/QTL, unveiled that all three traits were under the control of the *Wx* locus or a genomic region strongly linked to this locus present on the short arm of chromosome 6. Thus, these outcomes have given an insight of the molecular bases of GC and GT which can be applied in future studies.

Ample World rice supply has led to renewed interest in enhancing grain quality of modem rice varieties. Unnevehr (1986) projected implied price of grain characteristics for Thailand, Indonesia, and the Philippines. Then using these price rice-breeding goals were evaluated and returns to research for quality improvement were estimated. Their findings showed that the former emphasis on improvement of physical quality was suitable, and future well-being gains are probable from improvement of chemical quality. Returns to quality improvement are large and indicate underinvestment in this type of research.

10.2.5 MEASURING PHYSICAL PROPERTIES OF RICE

The improperly designed machinery and operations may cause cracking and breakage of rice kernel and subsequently a low marketing price. The effect of the rice processing operations on physical and mechanical properties of various rice varieties was ascertained by Correa et al. (2007). They used three rice varieties viz. rough, brown, and milled in the study. It was observed that bulk densities of all varieties increases with processing up to 51% and there were variances among the varieties. Further, the processing affected the porosity of the bulk rice grains by reducing the external static and dynamic friction coefficients. The higher values of friction coefficient were noted on wood surface and the lowest on steel surface; the compression force required to stimulate the rice kernel breakdown was influenced significantly by the processing.

Toyoshima and Ohtsubo (1999) measured physical properties of cooked rice grains using 55 rice samples which include both waxy genotypes and genotypes with high AC. Then with the help of single apparatus (Tensipresser) multiple measurements were applied. These measurements include low, high, and continuous progressive compression tests (LC, HC, and CPC). The HC and LC tests determined the total hardness (H_2) and the surface hardness (H_1), which was used as indices to categorize the samples into the various groups corresponding to AC. The surface hardness was more appropriate than the total hardness for distinguishing the influence of protein contents. Using LC test, from the surface adhesion distance (L_3) the variance of stickiness among the cooked rice samples could be detected. The back pressure curve on the CPC test provides the estimates of elastic limit length ratio, which increases with AC. Thus, these three tests were important in describing the physical properties of cooked rice samples with waxy and high AC genotypes.

Lu et al. (2005) studied the influences of whole polished rice grains fermentation on the physical properties of rice flour and the rheological characteristics of rice noodles. The X-ray diffraction revealed the insignificant effect of natural fermentation on the crystalline structure of rice starch. However, the fermentation increased the ratio of the crystalline to the amorphous regions. Then differential scanning calorimetry and a rapid viscosity analyzer (RVA) were utilized to find the thermal properties of rice flours. It was observed that the GT, T_p, and the RVA peak viscosity of rice flour decreases, while the gelatinization enthalpy, ΔH, increases after fermentation. The FTIR spectra of fermented and control rice flours were comparable. Small superficial corrosion was noticed on fermented rice starch granules when examined using scanning electron microscopy (SEM). Thus, fermentation may change the chemical components and amorphous region of the starch granule, thereby modifying the physical properties of rice flour and rheological characteristics of rice noodles.

Rice husk is the most commonly available lignocellulosic material and a major by-product of the rice milling industry. Through various thermochemical conversion processes, rice husk can be transformed to various types of fuels and chemical feedstocks. For designing the thermochemical, conversion systems the physical and thermochemical properties of rice husk need to be properly understood. In this paper, Mansaray and Ghaly (1997) provided information on moisture content, bulk density, particle size, heating values, proximate analysis, ultimate analysis, ash composition, and ash fusibility characteristics of rice husk from six rice varieties. The results displayed excessive volatile release of over 60%, high ash content varying from 15·30 to 24·60% (dry weight basis), and high silica content of the ash varying from 90 to 97%. The lower heating values ranged from 13·24 to 16·20 MJ kg^{-1} (dry weight basis). The average moisture content and bulk density were 10·44% and 114 kg m^{-3}, respectively. All the varieties maintained the ash fusion temperature of 1,600°C. The dissimilarities in varietal traits have significant influences on the chemical properties of rice husk.

10.2.6 MEASURING PHYSICAL PROPERTIES OF MILLED RICE

The cooking and processing characteristics of rice can be determined with the help of various physical and chemical tests. Plant breeders normally test the new rice lines for AC, ASV, protein content, viscosity properties of the flour-water paste, and the appearance of milled gains (whiteness, transparency, and degree of milling). A study has been conducted by Delwiche et al. (1996) to find out the amount to which near-infrared reflectance (NIR) spectroscopy could be used on whole-grain milled rice to quantify such characteristics. They milled the samples of U.S. rice (n = 196) from advanced breeders' lines and commercial releases and later scanned the samples in the visible and near-IR regions (400–2,498 nm). Further, reference chemical and physical assessments were carried out on each sample. Partial least squares modeling results showed that moderately precise models were achieved for apparent amylose, coefficient of determination, protein content, whiteness, transparency, and milling degree through NIR. The ASV could be modeled to a lesser extent by NIR, however this precision maybe sufficient for initial screening in breeding programs. On the other hand, models for the five flour paste viscosity properties attained by a rapid visco analyzer (RVA) were not satisfactorily accurate ($r^2 < 0.75$) to permit replacement of the RVA technique with an NIR model. Reduction of NIR scanning sample size roughly from 100 to 8 g did not show any significant effect on the model performance of constituents.

10.2.7 MEASUREMENT OF CHEMICAL CHARACTERISTICS OF MILLED RICE

The process of aging in rice is very complex, which includes modifications in physical and chemical properties of the rice grain. Zhou et al. (2002)

reviewed the changes in physical and chemical properties of the rice grain that occur during storage and discussed the effects of these changes on functionality of rice. The foremost rice grain factors having an effect on the cooking and eating quality are starch, protein, and lipids. Though, the total starch, protein, and lipid contents in the rice grain remain unchanged during storage, structural modifications do take place. These variations influence the pasting and gel properties, flavor, and texture of cooked rice.

Though AC is deliberated as the primary determining factor of cooked rice texture, this component cannot be used as a predictor, because cultivars with comparable ACs may vary in textural properties. Therefore, to get better differentiation of cultivars amylography can be used as one of a series of tests, besides AC measurement. Champagne et al. (1999) carried out a study to ascertain how well amylography performed with a rapid visco analyzer (RVA) acts as a predictor of cooked rice texture, alone as well as in combination, with AC. Using descriptive sensory and instrumental texture profile (TPA) analyses textural properties of 87 samples containing short-, medium-, and long-grain rice cultivars were measured and related to RVA measurements. None of the cooked rice textural characteristics, whether assessed by descriptive analysis or TPA, were modeled by RVA with high precision (i.e., high r^2). Sensory texture characteristics like mass cohesiveness, stickiness, and initial starchy coating and TPA characteristic adhesiveness exhibited intense associations with measurements of RVA. Addition of amylose and protein contents in regression analyses did not improved the models. However, the elimination of samples that cook peculiarly on the basis of AC or GT could strengthen the precision of RVA measurements to some extent for predicting cooked rice texture.

In comparison, the Australian rice industry is small and to compete with other industries it has to uphold the quality product. The understanding of quality traits is essential to maintain quality. Viscosity of the flour (RVA) is a valuable tool for assessing quality. Martin and Fitzgerald (2002) determined the proteins effect on formation of viscosity curve, so as to know how proteins influence the quality of cooked rice. The results clearly demonstrated that proteins affect viscosity curves both through binding water, which increases the dispersed and viscous phase concentration of gelatinized starch and through the activity of a network connected by disulfide bonds. Increased N nutrition rates found to decrease the peak height, which was constant with less disulfide bonds representing a smaller amount of contribution from the linked network. Storage decreased peak height by increasing disulfide bonds. Generally, proteins affect the quantity of water absorbed by rice early in cooking. The water availability early in cooking will reveal the hydration of the protein and the concentration of the dispersed and viscous phases of the starch, which will decide the cooked rice texture.

Rousset and Pons (1995) studied the raw-milled rice and four differently parboiled rice samples to relate sensory perception to instrumental measurements. Variance analysis has shown that some physico-chemical traits such as thickness of cooked grain, length/width ratio, water uptake, elastic recovery, and white core rate and amylose and protein contents varied greatly among rice samples. The most discriminating sensory characteristics were: elasticity, stickiness, pastiness, mealiness, length of grain, firmness, crunchiness, time in mouth, brittle texture, and juiciness. From the principal component analysis (PCA) high correlation was detected between some sensory traits and instrumental measurements. Water uptake had positive correlation with the melting texture, surface moistness, and juiciness. White core presence exhibited significant effect on granular texture, crunchiness, brittleness, and mealiness. Length of cooked grain showed a correlation with the length/width ratio of raw grain. Pastiness, compactness, stickiness were slightly affected by the raw grain thickness. The firmness measured by extrusion force with an Ottawa cell was found to be correlated with the sensory firmness to some extent. Elasticity was reliant on elastic recovery and firmness measured by the Viscoelastograph. Out of the four parboiled rice samples, two were very elastic, firm, granular, crunchy, and mealy. The other two were moister, more melting, and cooked longer. Among the three raw-milled rice samples, variances in GL and melting-granular-brittle characteristics were observed.

Krishnana et al. (2004) studied the rheological properties of two rice varieties using a farinograph and a rheometer. The hydroxy propyl methyl cellulose (HPMC) was added as gluten substitute to the rice dough and compared with wheat dough to find its suitability for making rice bread. From the farinogram, the data on water absorption and dough development time was obtained. The findings unveiled that rice flour added with HPMC require more time to reach a consistency of 500 BU than that of standard wheat dough. In the rheometer, oscillation measurements, shear measurement and creep tests with a direct loading of 50 Pa for 60 S were conducted. The rheological measurements from the oscillation tests and creep tests showed that the rheological properties of rice dough with 1.5%

and 3.0% HPMC are similar to that of wheat flour dough and were appropriate for making rice bread. Further, with all the dough samples baking tests were performed, in which the loaf volume and moisture loss of bread were measured. It was observed that the long grain rice sample gives bread with better crumb texture.

The rice varieties with high physicochemical and gel textural characteristics are suitable for making rice noodles. Bhattacharya et al. (1999) evaluated the eleven rice genotypes with different rapid visco analyzer (RVA) pasting characteristics for their physicochemical and gel textural traits. AAC was strongly correlated with swelling power, flour swelling volume (FSV), noodle hardness, gumminess, chewiness, and tensile strength (TS). Solubility was negatively associated with the pasting parameters and noodle rehydration and positively associated with cooking loss, noodle hardness, and gumminess. Noodle hardness exhibited a negative association with the FSV and most of the pasting parameters. RVA parameters and textural parameters of gels produced in the RVA canister were positively associated with the actual noodle texture. Therefore, the RVA parameters could be applied to predict the quality of rice noodles during initial screening of genotypes in breeding program.

Five rice cultivars (PR-106, PR-114, IR-8, PR-103, and PR-113) with varied amylose content (AC) were collected and analyzed for their morphological, thermal, and rheological properties of starch by Singh and Singh (2003). ACs of starches took from PR-103, IR-8, PR-106, PR-114, and PR-113 were 7.83, 15.62, 16.05, 16.13, and 18.86%, respectively. Using a Scanning Electron Microscope, the granular size was measured, which ranged from 2.4 to 5.4 μm in all rice starches. The lowest granular size, AC and solubility were observed in PR-103 starch, while, the highest granular size and AC were noted in PR-113 starch. The highest transition temperatures, enthalpies of gelatinization, peak height index, range, and enthalpies of retrogradation were noted in PR-103 starch. The retrogradation (%) and swelling power were highest in PR-113 starch and lowest in PR-103 starch. In all rice starches, turbidity value of gelatinized pastes was increased gradually up to the 3rd day during refrigerated storage. The lowest turbidity value was observed in PR-103 starch paste and highest value was noted in PR-113. Further, during the storage of rice cultivars at 4°C, the syneresis (%) of starch pastes was measured. Excluding PR-103, all rice cultivars showed increased syneresis of starch pastes with storage.

10.2.8 INTERNATIONAL STANDARDS (ISO 7301)

The International Organization for Standardization (ISO, 1988) recently adopted standard rice specifications and the meeting was conducted in Washington, DC, in October 1990 to consider international standards for rice.

Recently, rice grain quality has been given an increasing importance in most of the national rice breeding programs. The rice grain quality comprises the traditional physical and visual properties of the rice grain, cooked rice texture as indexed by AAC, ASV (index of GT), GC, actual texture measurement of cooked grain, and protein content (nutritional value). Texture measurement is performed mainly to validate amylose classification, because instrument methods are not sensitive enough to distinguish cooked rice with the same amylose type. The studies on consumer demand revealed the wide diversity in preference of grain quality characteristics. They are helpful in deciding priorities in a grain quality breeding program. A 1990 survey confirmed that grain quality is one of the foremost objectives in national breeding programs. Several national programs have described the grain quality of common varieties and their germplasm collection entries. Asian food scientists have been characterizing the properties of varieties mostly suitable for conventional rice products, to enhance product quality and shelf life and to offer value added products and employment in the rural areas. For getting the probable variation of grain quality, the mutants and transgenic rice acquired from biotechnology need to be screened. To perform a major role in research associated to evaluation of grain quality strong national programs are desirable (Juliano and Duff, 1991).

10.2.9 DETERMINATION OF QUALITY OF MILLED RICE

The total lipids and free fatty acid (FFA) determination is extensively utilized in the food industry to assess the quality of milled rice. To ascertain milled rice surface lipid, an improved rapid ambient temperature isopropanol (IPA) extraction method was found to be more efficient than Soxhlet solvent extraction. Another method explained for the determination of FFA is colorimetric method. This method needs only 30 μL of sample, has better precision, and is more objective. The results acquired from both new and conventional acid-base titration methods are similar. The novel methods are mostly appropriate for industrial application in offering quick results for large number of samples (Lam and Proctor, 2001).

The polished edible rice is produce by subjecting the rough rice to dehusking and milling. The removal of hulls is known as dehusking and the whitening or milling is the removal of brownish outer bran layer. In order to reduce the economic loss, the control of degree of milling and percentage of broken kernels in milled rice is essential. Yadav and Jindal (2001) conducted a study by subjecting ten varieties of Thai rice to varying degrees of milling and the test duration was adjusted from 0.5 to 2.5 min. To find out the HRY, which represent the proportion by weight of milled kernels with three quarters or more of their original length, and the whiteness of milled rice, digital image analysis was used. Then, from the images of single kernels in a milled sample, three-dimensional features viz., length, perimeter, and projected area were obtained. These images were used to estimate a characteristic dimension ratio (CDR) described as the ratio of the sum of a specific dimensional characteristic of all head rice kernels to that of all kernels containing head and broken rice in the sample. On the basis of projected area of kernels in their natural rest position, the CDR presented the best estimate of the HRY with the least root mean square error of 1.1% amongst all dimensional characteristics studied. For the whiteness of milled samples, a correlation with R^2 value of 0.99 was observed between the commercial whiteness meter values and the mean of gray level distribution revealed by image analysis. Thus, their studies have indicated that for the improved control of rice milling operation, the quantitative estimation of HRY and degree of milling can be performed effectively by using two-dimensional imaging of milled rice kernels.

Singh et al. (2005) evaluated milled rice of 23 varieties for their physicochemical, cooking, and textural properties. Using Pearson correlation the relationship between different properties was ascertained. Among the varieties, the thousand kernel weight varied from 13.3–19.9 g, bulk density from 0.77–0.88 g ml^{-1}, length-breadth (L/B) ratio from 2.62–4.55 and AC from 2.3–15.4%. Minimum cooking time, water uptake ratio, gruel solid loss, and ELR varied between 13.3–24.0 min, 2.37–4.45, 1.88–8.53%, and 1.29–1.74, respectively. Further, using Instron Universal Testing Machine, textural properties such as maximum force, cohesiveness, packability, hardness, and chewiness were estimated. Cooking time was negatively associated with AC and positively associated with bulk density of milled rice. Gruel solid was positively associated with the AC and negatively associated with cooking time. The rice varieties with higher cooking time exhibited lower gruel solid loss and vice versa. All textural parameters exhibited a significant association with each other and were positively associated with AC and negative associated with cooking time.

Jiang et al. (2007) conducted an investigation on the associations among potassium (K), calcium (Ca), sodium (Na), magnesium (Mg), iron (Fe), zinc (Zn), copper (Cu), and manganese (Mn) contents and between these mineral element contents and other rice quality traits in milled rice. The findings unveiled the significant associations among most of mineral element contents. Mg, Fe, and Mn contents exhibited significant correlations with most of the other mineral element contents, whereas Cu content was significantly and negatively correlated with the K and Mg contents of rice. The associations between mineral element contents and cooking quality traits revealed the significant associations between gel consistency (GC) and K, Cu, and Mn contents of rice. AC had significant correlation with K, Na, Mg, Cu, and Mn contents. The ASV was positively correlated with Ca, Mg, and Mn contents. Furthermore, eight mineral element contents showed clear associations with various amino acid contents. Protein content was significantly associated with Na, Mg, Zn, Cu, or Mn content of rice.

10.2.10 DETERMINATION OF QUALITY OF PADDY OR ROUGH RICE

A study has been undertaken by Bonazzil et al. (1997) to evaluate the influence of drying conditions on paddy mechanical behavior. They measured the percentage of broken kernels during drying. The findings revealed that the observed rice quality degradation during drying cannot be elucidated by air temperature alone. The temperature of air primarily affects the drying rate and time. At low temperature, the evaporating capacity of air also affects the drying rate and time to some extent. With the evaporating capacity of drying air, a very rapid increase in broken kernels was observed.

To find out the optimal combination of drying air temperature (X_1), air speed (X_2) and tempering time (X_3) for intermittent drying of rough rice an experiment was conducted by Madamba and Yabes (2005). Drying air temperatures varying from 35°C to 55°C were applied in the study. This temperature ranged bring about high milling recovery (MR), high head rice ratio (HRR), short DT, high grain hardness (H), high germination ratio (GR), high degree of whiteness (W) and low cracked grains ratio (CGR). However, airspeed exhibited significant effect on high grain hardness (H) and tempering time influenced DT at the 95% significance level. Finally, for the intermittent drying of rough rice, the drying conditions of 45°C drying air temperature and a tempering time of 2 hours were found to be optimum.

Many researchers reported the effect of fluidized bed paddy drying with drying air temperatures of over 100°C on head rice yield (HRY) and whiteness of dried rice. But, only few studies reported the effect of fluidized bed paddy drying using drying air temperatures below 100°C. Therefore, Tirawanichakul et al. (2004) has worked out on effects of fluidized bed drying air temperature on different quality parameters of Suphanburi 1 and Pathumthani 1 *Indica* rice. Using inlet drying air temperatures between 40 and 150°C at 10°C/step, they dried the paddy from the original moisture contents (MCs) of 25.0, 28.8, and 32.5% dry basis to 22.5 ± 1.2% dry basis. Next to fluidized bed drying, paddy was tempered and after that, ambient air aeration was applied till its final moisture content was lowered to 16.3 ± 0.5% dry basis. The outcomes unveiled the significant association between the HRY, the inlet drying temperature and initial moisture content for Suphanburi 1, while such significant association was not noticed for Pathumthani 1. With the increment in drying air temperature and initial moisture, content a slight decrease in whiteness of the two rice varieties was observed. Further, the thermal analysis by DSC has shown the partial gelatinization during drying at higher temperatures. Therefore, the study specified that the inlet drying air temperatures in the range of 40–150°C do not have any effect on the quality of cooked rice and maintains the milling quality of paddy.

Das et al. (2004) evaluated the infrared (IR) dried parboiled rice for their HRY, color, percent-gelatinized kernel and specific energy consumption. They used five levels of radiation intensity (5514, 4520, 3510, 2520, and 1509 W m^{-2}) and four levels of grain bed depths (3, 6, 12, and 25 mm) in a vibratory IR dryer. Then the influence of these parameters on quality variables and specific energy were predicted with response surface methodology. For all the grain bed depths, about 6–8% decline in head yield was observed with the increment of IR intensity from 1509 to 5514 W m^{-2}.

The color of the milled rice was found to be affected with both grain bed depth and radiation intensity, with the Yellowness Index varying between 30.54 and 52.44. The gelatinized kernels percent increased about 8% at maximum bed depth, and 4% at lower bed depths. With the increase in radiation intensity and bed depth, the specific energy consumption showed the variation between 14.7 and 73.4 MJ kg^{-1}.

Pan et al. (2008) examined the IR radiation heating effect on the drying characteristics, milling quality, and disinfestations efficiency of rough rice. They collected freshly harvested medium grain rice (M202) samples with low (20.6%) and high (25.0%) moisture contents (MCs) for the study. Then, using a catalytic IR emitter, single-layer of both non-infested rough rice and rough rice infested with the adults and eggs of lesser grain borers (*Rhizopertha dominica*) and Angoumois grain moths (*Sitotroga cerealella*)] were heated for various durations. The impacts of the tempering treatment and natural and forced air cooling methods on moisture removal, milling quality and disinfestation were ascertained.

A high heating rate and equivalent high moisture exclusion were attained with IR heating. After heating, they observed increased removal of moisture during cooling and enhanced milling quality of the rice samples. When the rice sample with 20.6% MC was heated by IR for 60 s, at temperature of 61.2°C removal of 1.7% MC during the heating period, and a further removal of 1.4% MC after tempering and natural cooling were observed. Moreover, heating, and tempering treatment entirely killed the tested insects. The researchers determined that by heating rough rice with catalytic IR emitter, followed by tempering and slow cooling along with simultaneous drying and disinfestation, the high quality milled rice can be realized.

Ondier et al. (2010) examined the impact of low air temperatures (26–34°C) and relative humidities (19–68%) applied to dry thin-layer samples of rough rice to the preferred 12.5% moisture content. In long- and medium-grain rice varieties harvested at 19.6% and 17.5% MCs, the effects of different drying rates and durations on the quality parameters like head-rice yield, color, and pasting viscosity were determined. Results revealed at 26°C drying air dehumidification increases drying rates with greater potential than at 30 and 34°C. Low drying air temperatures and relative humidities did not show any harmful effects on head-rice yield or color. The Peak and final viscosities of low-temperature and low-relative humidity dried samples were comparable to controls.

10.2.11 AROMA TESTING

Wilkie et al. (2004) conducted a sequence of triangle tests by collecting the aromatic rice samples from retail outlets. A large consumer group of varied age, genders, and cultural background participated in the test. Based on aroma, the group could differentiate between Australian fragrant (AF) and non-fragrant (NF) rice and between AF and imported fragrant (IF) rice. The capability to distinguish between the samples differed with age, gender, and cultural background. Furthermore, from the descriptions of the aromas by participants, the aroma of AF rice was found to be more preferable than IF and Australian NF rice. Compared to the fragrant rice, twice the amounts of 2-hexenal [E], nonanal, 2-pentylfuran, and 2-octenal [E], and significantly more 2-nonenal [E] and hexanal were found in NF rice. Compared to the IF rice twice the amount of 2-heptenal [E] and three times the amount of 2-decenal [E] and 2,4-decadienal [E,E] were present in AF rice. Both fragrant rice contained 2-Acetyl-1-pyrroline (2AP). A significant association was noted between the concentrations of aroma volatiles and aroma descriptions of the rice.

So far, more than a dozen diverse aromas and flavors have been detected in rice with descriptive sensory analysis and above 200 volatile compounds have found in rice with instrumental analyses. But, after over 30 years of research, much information is not available on the relationship between the many volatile compounds and aroma/flavor. Numerous oxidation products probably causing stale flavor have been identified. But, the amounts of oxidation products, alone or together needed to produce stale or rancid flavor in rice have not been ascertained. 2-acetyl-1-pyrroline (2-AP; popcorn aroma) is the only volatile compound which has been proved to add a characteristic aroma. Moreover, it is the only compound in which the association between its concentration in rice and sensory intensity has been recognized. Champagne (2008) reviewed literature on the genetic, preharvest, and postharvest factors effects on volatile compound profiles and the aroma and flavor of cooked rice. Further, the problems associated with quantification of aroma and flavor, instrumentally, and by human sensory panels were discussed.

Mahatheeranont et al. (2001) extracted volatile components of uncooked Khao Dawk Mali 105 brown rice using indirect steam distillation under condensed pressure and regulated temperature so as to prevent cooking. Then with capillary gas chromatography–mass spectrometry the fresh extract was analyzed, which unveiled the presence of >140 volatile components. Among these, 70 volatiles were recognized, which include 2-acetyl-1-pyrroline (2AP), an important aroma compound of cooked rice. Further, in uncooked brown rice 2AP was detected by an improved method. In this method, 2AP was quantified using a capillary gas chromatographic system and a more selective column, CP-Wax 51, for amines. This improved chromatographic system had outstanding detection sensitivity for 2AP in the rice extracts, so that 2AP in only 0.5 g of uncooked Khao Dawk Mali 105 brown rice extract could be detected.

Compared to the most prominent varieties, aromatic rice is low-yielding, susceptible to lodging, and prone to pests and diseases. Owing to these, different agricultural chemicals like fertilizers, pesticides, and growth regulators have been utilized for its cultivation. Many researchers have assessed the effect of these chemicals on aroma and flavor of rice. Goufo et al. (2011) examined the effect of gibberellic acid (GA), paclobutrazol, 3-indole acetic acid, and a regulator mixture containing paclobutrazol, proline, and zinc chloride on rice aroma. After the emergence of 25% of panicles, these chemicals were applied on plants of two aromatic rice cultivars. Then applying static headspace together with gas chromatography, 12 odor-active compounds were extracted and identified. At the tested concentrations, all treatments applied with growth regulators exhibited low aroma content that influenced overall flavor. In a smelling assessment, higher intensity of smell was found in control samples than treated samples. The variance between the control and treated samples aromas was mostly associated to 2-acetyl-1-pyrroline, the main rice aroma compound, and lipid oxidation volatiles. For example, the cultivar Guixiangzhan, when treated with GA showed reduced content of 2-acetyl-1-pyrroline by 19%, 3-indole acetic acid by 9%, paclobutrazol by 22%, and the regulator mixture by 21% when compared with the control. Thus, the research findings have indicated that use of growth regulators inhibits the metabolic processes related with the formation of volatile compounds.

10.3 FOOD QUALITY

In Dhaka, the share of the low-priced coarse rice was found to be rapidly declining in rice markets and it thus appears that the position of rice as only a cheap staple food is being redefined. The rising demand for the high-priced varieties are apparently related with a more significant off-farm food sector—especially, milling, retailing, and branding, along with a transformed milling industry. Further, Minten et al. (2013) observed that the labor repays for growing various, rice varieties are not significantly different, and the farmers do not benefit directly from consumers' increased willingness to pay for rice.

In the boro and aman seasons, Duxbury et al. (2003) determined the total arsenic content of 150 paddy samples using hydride generation-inductively coupled plasma emission spectroscopy (ICP). At moisture content of 14%, about 10 to 420 μg kg^{-1} of variation was observed in arsenic concentrations. Rice yields and grain arsenic concentrations were 1.5 times higher in the boro (winter) than the summer (monsoon) season with the average values of 183 and 117 μg kg^{-1} for the boro and aman season rice, respectively. The variation in rice arsenic concentrations were consistent with the excessive use of groundwater for irrigation in the boro season, while it was only partly consistent with the arsenic concentrations pattern in drinking water tube wells. The evidence of arsenic toxicity on rice yield and panicle sterility is not available. In processed rice, (parboiled and milled) 19% reduced arsenic concentrations were recorded. At rice, intake of 400g and water intake levels of 4 L^{-1} cap^{-1} day, human acquaintance to arsenic via rice would be equal to half of that in water containing 50 μg k^{-1}.

In recent times, near-infrared reflectance spectroscopy (NIRS) has been increasingly applied for the assessment of food quality. This technique is non-destructive, rapid, and provides more information to study the quality of food products. In this paper, Cen and He (2007) has provided an overview of the form of information that can be obtained from various established theories and food research of NIRS. It contains the principle of NIRS technique, the specific techniques with chemometrics for data pre-processing methods, qualitative, and quantitative analysis and model transfer, and the varied applications of NIRS in food science. Additionally, the potential of NIRS technique for assessment of food quality and the problems related with it are discussed.

Another technique that has been extensively applied for the assessment of food quality is image processing techniques. Here, Du and Sun (2004) reviewed the current developments in image processing techniques for food quality evaluation. These developments include charge-coupled device camera, ultrasound, magnetic resonance imaging, computed tomography, and electrical tomography for image acquisition; pixel and local pre-processing approaches for image pre-processing; thresholding-based, gradient-based, region-based, and classification-based methods for image segmentation; size, shape, color, and texture characteristics for object measurement; and statistical, fuzzy logic, and neural network systems for classification. They demonstrated the potential of image processing techniques for food quality evaluation. Further, various problems that need to be solved to increase the application of image processing technologies for food quality assessment were discussed.

10.4 FOOD CHEMISTRY

Previous studies proved that increased intake of dietary fiber can have beneficial effects in both human and experimental animals. These beneficial effects comprise prevention or alleviation of diseases such as cardiovascular disease, diabetes, and diverticulosis and colon cancer. Studies have repeatedly shown that rice bran gives remarkable health benefits. Compositional analysis unveils that rice bran contains nearly 27% dietary fiber and has been stated to have positive effects, for example, laxative and cholesterol-lowering ability. This proposes that rice bran is a source of good fiber that can be included in different food products. In this paper, Hamid and Luan (2000) examined the usage of a dietary fiber preparation, resulting from defatted rice bran, as a functional ingredient in bakery products. The findings unveiled that the water-binding capacity of rice bran-dietary fiber is similar to that of FIBERX, which is a commercial fiber from sugar beet. The fat binding and emulsifying capacity of rice bran dietary fiber was higher than FIBERX. But, the viscosity of rice bran fiber was less than FIBERX. Addition of 5 and 10% dietary fiber preparation significantly reduced the volume of loaf and improved the firmness of the breads. Sensory evaluations showed that breads with 5 and 10% rice bran fiber were similar to high-fiber bread present in the market. This proves that the dietary fiber made from defatted rice bran has greatpotential in food applications, particularly in improvement of functional foods.

Five local rice bran varieties, i.e., rice bran-super kernel (RB-kr), Rice bran-Super 2000 (RB-s2), Rice bran-Super Basmati (RB-bm), Rice bran-Super-386 (RB-86) and Rice bran-Super fine (RB-sf) were evaluated for their antioxidant activity by Iqbala et al. (2005). Using DPPH radical and ABTS cation radical and conjugated dienes total phenolic content, antioxidant activity in linoleic acid system, reducing power, metal chelating ability, scavenging capacity were measured, to assess the order of antioxidant activity. Then with the help of reverse phase HPLC the main antioxidant components in rice bran, i.e., tocopherols, tocotrienols, and γ-oryzanol were determined. The general order of antioxidant activity was RB-kr > RB-s2 > RB-bm > RB-86 > RB-sf. However, conferring the chelating activity and conjugated dienes assays the antioxidant efficiency of RB-sf was more than RB-bm and RB-86. Antioxidant activity exhibited strong correlation with growth period and irrigation water requirement (IWR) by a specific variety. The growth period of RB-kr is longest and it requires less amount of water, thus shows highest antioxidant activity.

Parrado et al. (2006) described the production, stabilization, physico-chemical composition, and biological properties (including the anti-proliferative activity) of a water-soluble rice bran enzymatic extract (RBEE). The major constituent of RBEE is proteins (38.1%) present in the form of peptide and free amino acids having 6% sulfur amino acids. In addition to this fat (30.0%), with oleic and linoleic acids is present as the key component. γ-oryzanol 1.2 mg g^{-1} and slowly absorbed carbohydrates (14.2%) are minor components of RBEE. Previous studies on the anti-proliferative influence of RBEE on leukemia tumor cell growth *in vitro* were also stated. All these properties make RBEE suitable for the treatment and prevention of chronic pathological forms related with unusual multiplication of cells, as seen in the case of cancer.

Butsat et al. (2010) carried out an analysis of rice bran, rice husk, brown rice and milled rice fractions of a Thai rice variety (Khao Dawk Mali 105) to find out phenolic acid composition, γ-oryzanol and tocopherols content and their antioxidant capacity. They used the 2,2-diphenyl-1-picrylhydrazyl (DPPH) radical scavenging and ferric reducing ability power (FRAP) assays for the analysis. On the basis of these assays, higher antioxidant activity values were noted in bran and husk fractions. Furthermore, the bran fraction exhibited the maximum γ-oryzanol and tocopherols content. Conversely, the husk fraction had higher phenolic acids concentration than the other fractions. The three most important phenolic acids viz. ferulic, vanillic, and *p*-coumaric acids were observed in all fractions. Ferulic acid was high in the bran, while vanillic and *p*-coumaric acids were predominant in the husk. Therefore, the study revealed that rice bran and husk contains high antioxidant properties and can be deliberated as significant sources of bioactive components.

10.5 FOOD PROCESSING

It has been well-known that Americans have moved towards eating more processed food than cooking at home. But, much is not known about whether these trends are consistent among low-income individuals, who are increasingly the target of public health programs promoting home cooking. Smith et al. (2013) examined in what way the patterns of home cooking and home food consumption have transformed from 1965 to 2008 by means of socio-demographic groups. The data from 6 nationally representative US dietary surveys and 6 US time-use studies was cross analyzed. The percentage of daily energy consumed from home food sources and time spent in food preparation decreased significantly, for all socioeconomic groups between 1965–1966 and 2007–2008. The food supply from home was 65 to 72% of total daily energy. In comparison, low-income groups have shown the highest decrease in the percentage cooking, but consumed the majority of daily energy from home sources and spent more time in cooking. Throughout socioeconomic groups, it was found that people consume the more daily energy from the home food supply, but, only slightly more than half spend any time for cooking on a given day. Attempts to boost up the healthfulness of the US diet must give emphasis on encouraging the preparation of healthy foods at home while integrating restrictions on time available for cooking.

Subsidiary products resultant from the manufacturing or processing of plant-based foods: cereals, fruits, vegetables, and algae contain ample dietary fiber. The fortification of these fiber-rich by-products increases their dietary fiber content and brings about healthy products, low in calories, cholesterol, and fat. These products also act as functional constituents to enhance physical and structural properties of hydration, oil holding capacity, viscosity, texture, sensory characteristics, and shelf-life (Elleuch et al., 2011).

A study has been undertaken by Brand et al. (1985) where they compared the *in vitro* starch digestibility and postprandial blood glucose response of conventionally cooked foods with that of factory-processed foods. Carbohydrate portions from three unprocessed foods (boiled rice, sweet corn, and potato) and six processed foods (instant rice, rice bubbles, corn chips, cornflakes, instant potato, and potato crisps) were incubated for 3 hours along with human saliva and porcine pancreatin. In comparison, the processed forms of rice, corn, and potato had a higher proportion of starch digested. The six healthy volunteers who consumed 50 g carbohydrate portions of the above-processed foods gave a higher glycemic index. But, the glycemic response of potato crisps was similar to a boiled potato.

The red yeast rice is a traditional Chinese medicine well-known for centuries to improve blood circulation. Chang et al. (2000) undertook detailed analyses of the natural components of red yeast rice. The preparation of red yeast rice by fermenting the fungal strain *Monascus purpureus* on moist and sterile rice specified the presence of a group of metabolites belonging to the monacolin family of polyketides, in addition to fatty acids, and trace elements. The presence of these compounds may partially describe the cholesterol-lowering capacity related with this traditional Chinese food.

Biodegradable blend films were developed from rice starch-chitosan by casting film-solution on leveled trays. Bourtoom and Chinnan (2008) studied the starch and chitosan ratio (2:1, 1.5:1, 1:1, and 0.5:1) effect on the mechanical properties, water barrier properties, and miscibility of biodegradable blend films. An increase in tensile strength (TS), water vapor permeability (WVP), lighter color, and yellowness and a reduction in elongation at the break (*E*), and film solubility (FS) were noticed in biodegradable blend film with the addition of chitosan. Further, the chitosan addition improved the crystalline peak structure of starch film, but, too high concentration of chitosan produced phase separation between starch and chitosan. The amino group band of chitosan molecule in the FTIR spectrum moved from 1541.15 cm^{-1} in the chitosan film to 1621.96 cm^{-1} in the biodegradable blend films. These effects indicated the existence of molecular miscibility amongst these two components. The rice starch-chitosan biodegradable blend film properties were comparable to the other chitosan films. But, the water vapor permeability of rice starch-chitosan biodegradable blend film was less than chitosan films and more than polyolefin.

Ramezanzadeh et al. (1999) examined the effect of microwave heating, packaging, and storage temperature on the production of free fatty acids (FFA) in rice bran. An increase in total FFA over the 16-week period from the initial value of 2.5% in raw bran stored at 25°C to 54.9% in vacuum bags and 48.1% in zipper-top bags was noted though, total FFA of raw bran stored at 4–5°C increased at a slower rate from an initial value of 2.5 to 25.4% in vacuum bags and 19.5% in zipper-top bags. After 16 weeks of storage, total FFA of microwave-heated bran stored at 25°C increased from 2.8 to 6.9 (vacuum bags) and 5.2% (zipper-top bags). Total FFA of microwave-heated samples stored at 4–5°C did not vary considerably with storage time. Findings unveiled that hydrolytic rancidity of rice bran can be prohibited by microwave heating and that the proposed storage condition for microwaved rice bran is 4–5°C in zipper-top bags.

Rice is one of the main agricultural products in Japan. It is consumed in various forms, each of which uses dissimilar quantities of resources and energy. Roya et al. (2009) evaluated the life cycle of rice (well- and partially-milled, brown, and germinated brown rice) and compared it to the parboiled milled rice to find out if the environmental load from the life cycle of rice can be condensed. The life cycle inventory (CO_2 emission) rises steadily from partially-milled (milling 2%) to the parboiled rice. The paper-bag option appears to be more environmentally friendly than the polyethylene-bag option. It is valuable to note that the partially-milled rice not only decrease the environmental load but also keep hold of more food nutrients compared to the well-milled rice. It was found that around 2–16% of the negative environmental influence arising from the present life cycle of rice in Japan could be reduced by a variation in rice production and consumption patterns.

Park et al. (2012) investigated the variations in the physicochemical properties of rice that take place during storage at diverse temperatures. A higher fat acidity was found in the milled rice that was stored at high temperatures. The moisture content was maintained at adequate value in milled rice when it was stored at 4°C for 3 months (15.50%) and at 20°C for 2 months (15.53%). Similarly, it was observed that the white coloration in rice was retained when its storage was at low temperatures. Its viscosity had reached peak values during 4 months of storage and bigger variations were observed at higher storage temperatures. Further, with storage they observed a decline in breakdown and an increase in setback, regardless of the storage temperatures, while its storage at high temperatures enlarged the cohesiveness and hardness. Similarly, high temperatures also brought a decline in adhesiveness with age. High temperatures (30°C and 40°C) significantly reduced all sensory values even after 1 month of storage. The results of their study specified that storage temperature is an imperative factor affecting the physicochemical properties of rice. Short storage periods below room temperatures are recommended to maintain rice quality.

Randal et al. (1985) developed an extrusion cooking procedure for producing stable rice bran, without an increase in contents of free fatty acid for at least 30–60 days. In the best process, 500 kg hr^{-1} of 12–13% moisture bran was extruded at 130°C and held 3 min at 97–99°C before cooling. Stabilized bran contained 6–7% moisture. It was in the form of small flakes, where 88% of these flakes were much larger than 0.7 mm (25 meshes). Energy required to extrude the bran was 0.07–0.08 kW-hr/lcg bran, and wear on the extrusion surface indicated a life of 500 hr for the cone and 1000–2000 hr for other wearing parts.

10.6 CONCLUSION

To enhance rice grain quality, the grain quality characteristics determination is required. In this chapter, overviews of the various techniques that can be utilized to determine the physical and chemical characteristics of rice grain quality and the processing methods to improve grain quality and to make the best use of rice by-products are presented. Further, the current advancements made in rice grain quality research are discussed.

KEYWORDS

- **breeding techniques**
- **characteristic dimension ratio**
- **doubled haploid**
- **drying time**
- **food quality**
- **grain quality**

REFERENCES

Adair, C. R, Bollich, C. N., Joson, D. H., et al., (1973). Rice breeding and testing methods in the United Stales. In rice in the United Stales. Varieties and production. U. S. Dept. of Agriculture. Agriculture Handbook. 289

Amarawathi, Y., Singh, R., Singh, A. K., Singh, V. P., Mohapatra, T., Sharma, R., & Singh, N. K., (2008). Mapping of quantitative trait loci for basmati quality traits in rice (*Oryza sativa* L.). *Molecular Breeding*, *21*, 49–65.

Bao, J. S., Wu, Y. R., Hu, B., Wu, P., Cui, H. R., & Shu, Q. Y., (2002). QTL for rice grain quality based on a DH population derived from parents with similar apparent amylose content. *Euphytica*, *128*, 317–324.

Bergman, C. J. (2004). Rice end-use quality analysis. In: Rice: Chemistry and Technology. E.T. Champagne (ed). 3rd. ed. U.S. Department of Agriculture, Agricultural Research Service, Southern Regional Research Cenetr. New Orleans, Louisiana, U.S.A. pp. 415–472.

Bhattacharya, M., Zee, S. Y., & Corke, H., (1999). Physicochemical properties related to quality of rice noodles. *Cereal Chemistry*, *76*, 861–867.

Bonazzil, C., Peuty, M. A., & Themelin, A., (1997). Influence of drying conditions on the processing quality of rough rice: *Drying Technology: An International Journal*, *15*, 1141–1157.

Bourtoom, T., & Chinnan, M. S., (2008). Preparation and properties of rice starch-chitosan blend biodegradable film. *LWT-Food Science and Technology*, *41*, 1633–1641.

Brand, J. C., Nicholson, P. L., Thorburn, A. W., & Truswell, A. S., (1985). Food processing and the glycemic index. *The American Journal of Clinical Nutrition*, *42*, 1192–1196.

Butsat, S., & Siriamornpun, S., (2010). Antioxidant capacities and phenolic compounds of the husk, bran and endosperm of Thai rice. *Food Chemistry*, *119*, 606–613.

Cen, H., & He, Y., (2007). Theory and application of near-infrared reflectance spectroscopy in determination of food quality. *Trends in Food Science and Technology*, *18*, 72–83.

Champagne, E. T., (2008). Rice aroma and flavor: A literature review. *Cereal Chemistry*, *85*, 445–454.

Champagne, E. T., Bett, K. L., & Vinyard, B. T., (1999). Correlation between cooked rice texture and rapid viscoanalyzer measurements. *Cereal Chemistry*, *75*, 764–771.

Chang, M., Ma, J., Li, Y., Ye, Q., Li, J., Hua, Y., Ju, D., Zhang, D., Cooper, R., & Chang, M., (2000). Constituents of red yeast rice, a traditional Chinese food and medicine. *Journal of Agricultural and Food Chemistry*, *48*, 5220–5225.

Chaturvedi, I., (2006). Effect of nitrogen fertilizers on growth, yield, and quality of hybrid rice (*Oryza sativa*). *Journal of Central European Agriculture*, *6*, 611–618.

Correa, P. C., Schwanzda, F. S., Jaren, C., Alfonso, P. C., & Arana, I., (2007). Physical and mechanical properties in rice processing. *Journal of Food Engineering*, *79*, 137–142.

Das, I., Das, S. K., & Bal, S., (2004). Specific energy and quality aspects of infrared (IR) dried parboiled rice. *Journal of Food Engineering*, *62*, 9–14.

Delwiche, S. R., Kenzie, K. S., & Webb, B. D., (1996). Quality characteristics in rice by near-infrared reflectance analysis of whole-grain milled samples. *Cereal Chemistry*, *73*, 257–263.

Du, C. J., & Sun, D. W., (2004). Recent developments in the applications of image processing techniques for food quality evaluation. *Trends in Food Science and Technology*, *15*, 230–249.

Duxbury, J. M., Mayer, A. B., Lauren, J. G., & Hassan, N., (2003). Food chain aspects of arsenic contamination in Bangladesh: Effects on quality and productivity of rice. *Journal of Environmental Science and Health, Part A: Toxic/Hazardous Substances and Environmental Engineering*, *38*, 61–69.

Elleuch, M., Bedigian, D., Roiseux, O., Besbes, S., Blecker, C., & Attia, H., (2011). Dietary fiber and fiber-rich by-products of food processing: Characterization, technological functionality and commercial applications: A review. *Food Chemistry*, *124*, 411–421.

Goufo, P., Wongpornchai, S., & Tang, X., (2011). Decrease in rice aroma after application of growth regulators. *Agronomy for Sustainable Development*, *31*, 349–359.

Hamid, A. A., & Luan, Y. S., (2000). Functional properties of dietary fiber prepared from defatted rice bran. *Food Chemistry*, *68*, 15–19.

He, P., Li, S. G., Qian, Q., Ma, Y. Q., Li, J. Z., Wang, W. M., Chen, Y., & Zhu, L. H., (1999). Genetic analysis of rice grain quality. *Theoretical and Applied Genetics*, *98*, 502–508.

Huang, F., Sun, Z., Hu, P., & Tang, S., (1998). Present situations and prospects for the research on rice grain quality forming. *Chinese Journal of Rice Science*, *3*, 1–7.

Iqbala, S., Bhangera, M. I., & Anwarb, F., (2005). Antioxidant properties and components of some commercially available varieties of rice bran in Pakistan. *Food Chemistry*, *95*, 265–272.

Jiang, S. L., Wu, J. G., Feng, Y., Yang, X. E., & Shi, C. H., (2007). Correlation analysis of mineral element contents and quality traits in milled rice (*Oryza stavia* L.). *Journal of Agricultural and Food Chemistry, 55*, 9608–9613.

Jongkaewwattana, S., & Geng, S., (1991). Effect of nitrogen and water management on panicle development and milling quality of California rice (*Oryza sativa* L.). *Journal of Agronomy and Crop Science*, *167*, 43–52.

Juliano, B. O., & Duff, B., (1991). Rice grain quality as an emerging priority in National rice breeding programmes. Rice grain marketing and quality issues. P.O. Box 933. Manila, Philippines. Pp. 55–64.

Juliano, B. O., (1981). Rice grain properties and resistance to storage insects: A review. International Rice Research Institute, Los Banos, Laguna, Philippines.

Juliano, B. O., (1982). An international survey of methods used for evaluation of the cooking and eating qualities of milled rice. IRRI Research paper No. 77. 3.

Krishnana, H. S., Sengeb, B., & Chattopadhyay, P. K., (2004). Rheological properties of rice dough for making rice bread. *Journal of Food Engineering*, *62*, 37–45.

Lam, H. S., & Proctor, A., (2001). Rapid methods for milled rice surface total lipid and free fatty acid determination. *Cereal Chemistry*, *78*, 498–499.

Laohavanich, J., & Wongpichet, S., (2009). Drying characteristics and milling quality aspects of paddy dried with gas-fired infrared. *Food Processing Engineering*, *32*, 442–461.

Lu, Z. H., Li, L. T., Min, W. H., Wang, F., & Eizo, T. E., (2005). The effects of natural fermentation on the physical properties of rice flour and the rheological characteristics of rice noodles. *Food Science and Technology*, *40*, 985–992.

Madamba, P. S., & Yabes, R. P., (2005). Determination of the optimum intermittent drying conditions for rough rice (*Oryza sativa* L.). *LWT-Food Science and Technology*, *38*, 157–16.

Mahatheeranont, S., Keawsa, S., & Dumri, K., (2001). Quantification of the rice aroma compound, 2-acetyl-1-pyrroline, in uncooked khao dawk mali 105 brown rice. *Journal of Agricultural and Food Chemistry*, *49*, 773–779.

Mansaray, K. G., & Ghaly, A. E., (1996). Physical and thermochemical properties of rice husk. *Energy Sources*, *19*, 1997, 989–1004.

Martin, M., & Fitzgerald, M. A., (2002). Proteins in rice grains influence cooking properties. *Journal of Cereal Science*, *26*, 285–294.

Minten, B., Murshid, K. A. S., & Reardon, T., (2013). Food quality changes and implications: Evidence from the rice value chain of Bangladesh. *World Development*, *42*, 100–113.

Mohammad, A. R., (2011). Effect of paddy husked ratio on rice breakage and whiteness during milling process. *Australian Journal of Crop Science*, *5*, 562–565.

Nakamura, Y., Francisco, P. B., Hosaka, Y., Sato, A., Sawada, T., Kubo, A., & Fujita, N., (2005). Essential amino acids of starch synthase IIa differentiate amylopectin structure and starch quality between *Japonica* and *Indica* rice varieties. *Plant Molecular Biology*, *58*, 213–227.

Ondier, G. O., Siebenmorgen, T. J., & Mauromoustakos, A., (2010). Low-temperature, low-relative humidity drying of rough rice. *Journal of Food Engineering*, *100*, 545–550.

Pan, Z., Khir, R., Godfrey, D. L., Lewis, R., Thompson, J. F., & Salim, A., (2008). Feasibility of simultaneous rough rice drying and disinfestations by infrared radiation heating and rice milling quality. *Journal of Food Engineering*, *84*, 469–479.

Park, E. U., Sook, K. Y., Park, K. J., & Kim, B. M., (2012). Changes in physicochemical characteristics of rice during storage at different temperatures. *Journal of Stored Products Research*, *48*, 25–29.

Parrado, J., Miramontes, E., Jover, M., Gutierrez, J. F., Terán, L. C., & Bautista, J., (2006). Preparation of a rice bran enzymatic extract with potential use as functional food. *Food Chemistry*, *98*, 742–74.

Ramezanzadeh, F. M., Rao, R. M., Windhauser, M., Prinyawiwatkul, W., Tulley, R., & Marshall, W. E., (1999). Prevention of hydrolytic rancidity in rice bran during storage. *Journal of Agricultural and Food Chemistry*, *47*, 3050–3052.

Randal, J. M., Sayre, R. N., Schultz, W. G., Fong, R. Y., Mossman, A. P., Tribelhorn, R. E., & Saunders, R. M., (1985). Rice bran stabilization by extrusion cooking for extraction of edible oil. *Journal of Food Science*, *50*, 361–364.

Rousset, S., & Pons, B., (1995). Carine piland on sensory texture profile, grain physico-chemical characteristics and instrumental measurements of cooked rice. *Journal of Texture Studies*, *26*, 119–135.

Roya, P., Ijirib, T., Neia, D., Orikasaa, T., Okadomea, H., Nakamuraa, N., & Shiina, T., (2009). Life cycle inventory (LCI) of different forms of rice consumed in households in Japan. *Journal of Food Engineering*, *91*, 49–55.

Singh, N. S., & Singh, N., (2003). Morphological, thermal, and rheological properties of starches separated from rice cultivars grown in India. *Food Chemistry*, *80*, 99–108.

Singh, N., Kaur, L., Singh, N. S., & Singh, K. S., (2005). Physicochemical, cooking, and textural properties of milled rice from different Indian rice cultivars. *Food Chemistry*, *89*, 253–259.

Smith, L. P., Ng, S. W., & Popkin, B. M., (2013). Trends in US home food preparation and consumption: Analysis of national nutrition surveys and time use studies from 1965–1966 to 2007–2008. *Nutrition Journal*, 12, 45.

Tan, Y. F., Li, J. X., Yu, S. B., Xing, Y. Z., Xu, C. G., & Zhang, Q., (1999). The three important traits for cooking and eating quality of rice grains are controlled by a single locus in an elite rice hybrid, Shanyou 63. *Theoretical and Applied Genetics*, *99*, 642–648.

Thanh, C.V., & Hirata, Y., (2001). Seed storage protein diversity of three rice species in the Mekong Delta. *Biosphere Conservation: For Nutrition, 87*, 107–115.
Tirawanichakul, S., Prachayawarakorn, S., Varanyanond, W., Tungtrakul, P., & Soponronnarit, S., (2004). Effect of fluidized bed drying temperature on various qualities attributes of paddy. *Drying Technology: An International Journal, 22*, 1731–1754.
Toyoshima, H., & Ohtsubo, K., (1999). Multiple measurements of physical properties of individual cooked rice grains with a single apparatus. *Cereal Chemistry, 76*, 855–860.
Unnevehr, L. J., (1986). Consumer demands for rice grain quality and returns to research for quality improvement in Southeast Asia. *American Journal of Agricultural Economics, 68*, 634–641.
Wang, D., Zhang, X., & Zhu, Z., (2005). Correlation analysis of rice grain quality characteristics. *Europe PMC, 31*, 1086–1091.
Wilkie, K., Wootton, M., & Paton, J. E., (2004). Sensory testing of Australian fragrant, imported fragrant, and non-fragrant rice aroma. *International Journal of Food Properties, 7*, 27–36.
Yadav, B. K., & Jindal, V. K., (2001). Monitoring milling quality of rice by image analysis. *Computers and Electronics in Agriculture, 33*, 19–33.
Zhou, Z., Robards, K., Helliwell, S., & Blanchard, C., (2002). Aging of stored rice: Changes in chemical and physical attributes. *Journal of Cereal Science, 35*, 65–78.

CHAPTER 11

Improvement of Rice: Rice Varieties and Hybrid Rice Technology

ABSTRACT

Rice breeding has been successful in developing improved rice cultivars and hybrids, but it is now facing challenges to develop new cultivars with high yield potential, better grain quality, and resistance to biotic and abiotic stresses. This chapter reviews the most significant advances in rice breeding. Further, presents the current and future objectives and tends of rice breeding and crop improvement.

11.1 HISTORY OF RICE BREEDING

The history of rice breeding was discussed by Khush et al. (2001). He mentioned that a researcher's team at IRRI began to conceptualize a semi-dwarf rice plant and observed that the architecture of the tropical rice plant is a major constraint to yield increase. Typical farmer varieties were tall, with long, weak stems and susceptible to lodging. When the plants lodge the photosynthesis would stop, and grain would be lost to the water, or eaten by rodents. This clearly indicated the need for a shorter, non-lodging rice plant to get higher yields. By that time, the concept of dwarfism was founded in other crops. Dwarf sorghum was already available. In 1949, the International Rice Commission (IRC), based in Rome, Italy was established by the Food and Agriculture Organization (FAO) of the United Nations. During the 1950s, IRC sponsored an *indica-japonica* hybridization program at Cuttack, India. Its main objective was crossing the shorter *japonica* (temperate climate rice) with the taller *indica* (tropical climate rice) to get progenies with higher yield potential. In that program, the ADT27 and Mahsuri were selected and cultivated on significant areas during the 1960s. In the meantime, U.S. breeders Nelson Jodan and Hank irradiated the tall U.S. varieties, expecting to induce a short-statured mutant. But, the selections from those initial attempts were plagued with high levels of sterility. In 1957, the Rockefeller Foundation sent Peter Jennings to Arkansas, Texas, and Louisiana to get to know with rice so as to develop new varieties for Latin America. Later, in 1960 Jennings and Sterling Wortman, traveled across Asia observing rice varieties, meeting rice scientists, and interviewing potential trainees and staff. In India, they come across a Taiwanese variety, Taichung Native 1 (TN1), which was the first extensively grown semi-dwarf variety in the tropics outside of Mainland China. But TN1 was highly susceptible to major disease and insect pests. Then, the inheritance of the short stature was unidentified. In 1961, Jennings joined IRRI as head of the Varietal Improvement Department. Among the germplasm that had been collected at IRRI, he found Dee-geo-woo-gen, a parent of TN1 and identified it as a source of dwarfism. Afterward, Jennings and Akira Tanaka, the first plant physiologist from Japan studied the semi-dwarf rice plant and thoroughly examined the causes, and effects, of lodging for 3 years. Jennings compared the tall varieties such as Peta and MTU-15 by supporting them with bamboo sticks and lodging resistant varieties. The yields of tall varieties and lodging resistant varieties were found to be comparable. Furthermore, the lodging-susceptible varieties, when supported, responded well to nitrogen applications, while the unsupported plants exhibited an obvious negative response. This confirmed that lodging was the main cause of low yields when conventional tropical varieties were exposed to modern management methods.

To reach a socially optimum level of investment in research, it must contribute to both public good and support. However, the evidence of typical underinvestment in agricultural research is provided by the very high rate of returns to rice research in Japan. Funding rice research from government tax revenue is rationalized in terms of its contribution to the national goal of economic development (Akino and Hayami 1975).

Peng et al. (2008) discussed advancements in ideotype breeding to boost rice yield perspective. The ideotype method has been in use in breeding programs at the International Rice Research Institute (IRRI) and in China to advance rice yield potential. The first-generation new plant type (NPT) lines were developed from tropical *japonica* at IRRI. They were not good yielders. This was mostly due to their ability of production of inadequate biomass making ability and poor grain filling. Second-generation NPT lines were developed by crossing elite *indica* with improved tropical *japonica* and led to subsequent progress. These second-generation NPT lines gave more yield than the first-generation NPT lines and *indica* check varieties. China's "super" rice breeding project has developed many F_1 hybrid varieties. These were developed by handling of a grouping of the ideotype approach and intersubspecific heterosis. The grain yield (GY) of hybrid varieties was 12 t ha^{-1} which was 8–15% advanced than the hybrid check varieties. The accomplishment of China's "super" hybrid rice became unbeaten because of its partial development by accumulating the good components of IRRI's NPT design, besides the utilization of inter-subspecific heterosis. For instance, both designs have given importance to hefty panicle size, abridged tillering capacity, and superior lodging resistance. China's "super" hybrid rice has attained improvement, as they have given much emphasis to the top three leaves and panicle position within a canopy for meeting the requirement of weighty panicles for a huge source supply. The accomplishment of "super" hybrid rice breeding in China and advancement in NPT breeding at IRRI suggests that the ideotype advance is effectual for exceeding the yield upper limit of an irrigated rice crop.

Cai et al. (2001) discussed that the new approaches suggested for rice breeding are mostly based on the exploitation of wide cross and polyploidization, i.e., using heterosis of intergenomes, polyploid, and some special genes. The studies on rice breeding history revealed that earlier research approaches were based on sexual reproduction and diploid, in general, all utilized the superior genes combination of cultivated and wild rice in the same genome (AA). Rice is a diploid plant with smaller genome, DNA content and chromosome size. Increasing ploidy level of rice by increasing its genome number with the help of allopolyploid heterosis will be a novel pathway of rice breeding. The low percentage of filled seeds is the main problem in autotetraploid, which can be reduced by expanding their relative distance of parentages, decreasing the formation of polyvalent chromosome and using apomixis, wide compatibility, and Ph genes. A set of 209 recombinant inbred lines (RILs) developed from a cross between basmati quality variety Pusa 1121 and a contrasting quality breeding line Pusa 1342, were used to map the quantitative trait loci (QTLs). Contrary to the earlier reports of monogenic recessive inheritance, the aroma in Pusa 1121 is controlled by at least three genes located on chromosomes 3, 4 and 8, and similar to the reported association of *badh2* gene with aroma QTL on chromosome 8, location of badh1 gene was identified in the aroma QTL interval on chromosome 4. Some germplasms with special genes such as wide compatibility, apomixis, and Ph genes will play an important role in the cross. Wide compatibility will be useful to overwhelmed male sterility and female abortion of inter subspecies F_1. The absence of meiosis and fertilization in apomixes will be helpful to escape segregation of progeny. As in wheat, Ph gene can be used to avoid the pairing of partial homologous chromosomes and formation of multivalent, chromosomal bridge, lag chromosome in intergenomes of allopolyploid. These will ensure the high rate of seed set and good stability in progenies. Superior polyploid plant lines with powerful heterosis and up to 85% seed, the set had obtained by using apomictic lines as a parent to cross with *O. sativa* ssp. *indica* and *japonica* and cross between *O. sativa* ssp. Asian and American cultivars. Thus, it has offered a very good background to achieve the novel breeding strategy.

The domestication of Asian rice (*Oryza sativa*) was a complicated process interrupted by events of introgressive hybridization among and between subpopulations. Though the domestication of rice from at least two different wild populations was suggested from the profound genetic divergence between the two key varietal groups (*Indica* and *Japonica*), the genetic uniformity around main domestication genes through diverse subpopulations proposes a cultural exchange of genetic material among ancient farmers. Zhao et al. (2010) studied genome-wide patterns of polymorphism with a new 1,536 SNP panel genotyped through 395 different accessions of *O. sativa* to describe population structure, and to deduce the introgression history of domesticated Asian rice. The analysis of population structure has shown the presence of five major subpopulations (*indica*, *aus*, tropical *japonica*, temperate *japonica,* and Group V), which is in conformity with previous studies. The introgression analysis unveiled that most of the accessions have some amount of admixture, with many individuals sharing a similar introgressed segment because of artificial selection within a population. Admixture mapping and association analysis of amylose content (AC) and grain length (GL) demonstrated the possibility of analyzing the genetic basis of complicated characters in domesticated plant populations.

11.2 SUCCESS OF RICE BREEDING

In Thailand, the conventional rice-breeding program took 20 years to develop new cultivars to substitute Kao Dawk Mali 105 (KDML105) and

Kao Khor 6 (RD6) for the rainfed lowland rice conditions. The susceptibility to diseases and unacceptable grain qualities are the key reasons for the reduced adoption of new cultivars by farmers. Further, the traditional breeding program takes about 15 years from crossing to the release of new cultivars. Using marker-assisted selection (MAS), rapid generations advance (RGA), and early generation testing in multi-locations the period for cultivar improvement could be shortened. Fukai (2005) performed four generations of MAS backcross breeding to transfer genes and QTL for bacterial blight resistance (BLB), submergence tolerance (SUB), brown planthopper (BPH) resistance and blast resistance (BL) into KDML105. The backcross lines introgressed with target gene/QTL were found to be tolerant to SUB and resistant to BLB, BPH, and BL, with better agronomical characters and grain quality than KDML1.

Wünn et al. (1996) transferred the truncated form of a synthetic *crylA(b)* gene from *Bacillus thuringiensis* into the *Indica* rice breeding line IR58 using particle bombardment method. The CaMV 35S promoter regulates the expression of this gene and allows efficient production of the δ-endotoxin specific to lepidopteran. Ro, R_1 and R_2 generation plants exhibited a significant insecticidal effect against many lepidopterous insect pests. Feeding studies displayed 100% mortality rates for the yellow stem borer (*Scirpophaga incertulas*) and the striped stem borer (*Chilo suppressalis*), the most destructive insect pests of rice in Asia, and feeding inhibition of the two leaf folder species *Cnaphalocrocis medinalis* and *Marasmia patnalis*. The introgression of stem borer resistance gene into the *Indica* rice breeding line makes this agronomically significant trait accessible for conventional rice breeding program.

In India, the plant breeding and seed multiplication structure are highly centralized. Moreover, only a few new varieties are officially released each year. Therefore, the system seems to be inappropriate to meet the requirements of the large proportion of Indian farmers situated in risk-prone and highly diverse environments. Maurya et al. (1988) described an alternate strategy whose central feature is identical to the features of farmers' traditional rice varieties with those of advanced breeders' lines. A selection from these lines is then distributed in small quantities for on-farm trials managed by farmers themselves. If the success of these preliminary attempts is to be continued, a more decentralized method for breeding and multiplication will be needed.

Many agronomically significant traits are quantitative in nature viz. governed by several genes. Owing to their complex inheritance, the identification of QTLs controlling agricultural traits has been challenging. However, the completion of the rice genomic sequence has enabled the cloning and pyramiding of QTLs for breeding. In QTL, analysis of the utilization of a wide range of variation found in wild species is essential because QTLs are resultant of natural variation. Further, introgression lines (ILs) developed from wild species in combination with Marker Assisted Selection should enable effective gene identification. In this paper, Ashikari and Matsuoka (2006) described current advances in rice QTL analysis together with mapping, cloning, and pyramiding of QTLs.

11.3 EFFECT OF GREEN REVOLUTION (GR) IN INDIA AND WORLD

The green revolution (GR) has been announced as a political and scientific achievement – exceptional in human history. So far in the decades that have followed it, this apparently nonviolent revolution has left lands deteriorated by violence and ecological dearth. A dedicated scientist, Vandana Shiva (2016) investigated the impacts of the GR in India by analyzing the destructive effects of monoculture and commercial agriculture and exposing the detailed association between ecological destruction and poverty. In this standard work, the influential activist and scholar also inspect latest advances in gene technology.

Prahladachar (1983) reviewed the Indian empirical evidence to study the influence of the GR on particular dimensions of income distribution in India. The Indian empirical reports suggest a wide dispersion of modern varieties (MVs) among farmers, over time, regardless of farm size and tenurial status. But, the rate of dispersal of MVs of a given crop among farms, through the regions and over the years has unveiled a propensity to be connected with the nature and level of regions development in physical and institutional organization. From the view of increased demand for labor, the GR has certainly exerted a favorable effect on the absolute income status of the landless laborer, nevertheless the fact that the owners of land and capital have got comparatively more than the laborers in the form of improved production due to MVs.

The role of GR in revolutionizing rice production was discussed by Khush (1995). The population is increasing at the rate of 2% every year, but the growth rate of rice production has reduced to 1.2%. This necessitates the development of rice cultivars with higher yield potential. GR technology, focused on varieties with high yield potential, superior grain quality, shorter growth duration, multiple resistance to diseases and insects and tolerance to problem soils. At present 70% of the World's rice, cultivated lands are planted with high yielding varieties. Most of the Asian rice belt countries have become self-sufficient in rice and some have exportable gluts, which led to reduced real price of rice in these countries. The urban poor and rural landless were benefitted from this price decline. The increased labor demand from higher intensity of cropping raised the income of the rural landless workers. The cultivation of rice varieties with multiple resistances decreased the usage of agrochemicals and enabled the implementation of integrated pest management (IPM) practices.

The sensitiveness of new technology performance to unobserved individual characteristics results in weaker information flows in a heterogeneous population. This prevents individuals from learning from neighbors' experiences. Munshi (2004) tested this feature of social learning using wheat and rice data from the Indian GR. The greater heterogeneity was noted in rice-growing environments and the new rice varieties were found to be sensitive to unnoticed farm characteristics. As expected, wheat growers responded strongly to neighbors' experiences, whereas rice growers do not. Further, to compensate for their lack of social information rice growers seemed to test new technology more on their own land.

11.4 PROGRESS OF RELEASING RICE HYBRIDS AND VARIETIES

Within the next few years, several transgenic rice varieties will be released for commercial use. These transgenic rice varieties may be characterized with higher yield potential, greater tolerance of biotic and abiotic stresses, and resistance to herbicides, enhanced nutritional quality, and new pharmaceutical proteins. Though rice is primarily self-pollinating, the pollen-mediated gene flow may disperse its transgenes to adjacent weedy and wild relatives. Mostly, sexually compatible *Oryza* species occurs with the crop, particularly in tropical countries. But, much information is not available on how rapidly fitness-enhancing transgenes will get accumulated in these populations and whether this process will have any undesirable environmental effects. For example, if weedy rice gets herbicide resistance due to the spread of some transgenes, it could produce more seeds, or occurs in a wider range of habitats and become much more difficult to control. Therefore, the rice-growing countries should assess the ecological risks and advantages of transgenic rice before releasing new varieties (Lu and Snow, 2005).

Tu et al. (2000) described the development of transgenic elite rice lines. The transgenic rice lines were developed by transferring rice *actinI* promoter controlling *Bt* fusion gene from *cryIA(b)* and *cryIA(c)*. The Minghui 63, *indica* CMS restorer line and the hybrid rice derived from it, Shanyou 63 were used in the study. In Minghui 63 (T51-1) plants, the level of *Bt* fusion protein CryIA(b)/CryIA(c) identified was 20 ng mg^{-1} soluble protein. The *Bt* Shanyou 63 was field-tested in natural and recurrent severe infestation of two lepidopteran insects, leaffolder, and yellow stem borer. The transgenic hybrid plants showed high resistance against both insect pests without decreasing yield.

11.5 GENETIC DIVERSITY OF RICE GERMPLASM COLLECTION AND ITS EFFECTIVE UTILIZATION

The major World collection of rice accessions held at IRRI which supplies germplasm to breeders. Virk et al. (1995a) analyzed the genetic diversity of these rice accessions by applying RAPD technology. Methodologies for the rapid extraction of DNA representative of a rice accession and its amplification by PCR, and numerical techniques for the analysis of the banding data have been established. Further, with regard to earlier work on classification and cross ability, the biological importance of RAPD data has been verified.

The genetic diversity and cultivar identity can be resolved accurately with the help of molecular markers. Ni et al. (2002) carried out an investigation with an objective to study the genetic diversity within a various assemblage of rice accessions, and to determine differences in the forms of diversity within the two rice subspecies *indica* and *japonica*. Thirty-eight rice cultivars and two wild species accessions (*O. rufipogon* and *O. nivara*) were screened with 111 microsatellite markers scattered over the whole rice genome. Total 753 alleles were found, with 1 to 17 alleles per marker. The number of alleles per locus was found to be positively correlated with the maximum number of repeats within a microsatellite marker. All rice cultivars and lines could be exclusively differentiated, and the resultant groups resembled exactly to the *indica* and *japonica* subspecies, with *japonica* separated into temperate and tropical types. In comparison, *japonica* group exhibited significantly higher genetic diversity on chromosomes 6 and 7, and lower diversity on chromosome 2. Further, with stepwise discrimination, two subsets of nearly 30 markers were detected that produced genetic distance matrice and dendrograms similar to those produced by 111 markers. The study findings suggested that the genetic diversity and rice cultivar detection could be carried out using a relatively small number of microsatellite markers.

The subdivision of a whole germplasm collection is known as core collection, it represents most of the variation present in the whole collection. The core collection allows more efficient assessment of variability and easy to manage due to its small size. The United States Department of Agriculture (USDA) rice collected the rice core subset (RCS) using stratified random sampling; this collection includes 1,790 entries from 114 countries and denotes about 10% of the 18,412 accessions in the rice whole collection (RWC). Yan et al. (2007) conducted a study on RCS, by collecting RCS data from an evaluation performed in 2002 and RWC data from the USDA germplasm system at www.ars-grin.gov. The strong correlation between the RCS and RWC was revealed from the comparative analysis of 14 descriptors frequency distributions. Therefore, information acquired from the RCS could be efficiently utilized to evaluate the RWC with 88% certainty. Further, the correlation coefficients between the RCS and the RWC for eight descriptors were ≥ 0.9, which showed that the RCS extensively represents the RWC.

The crop improvement approaches are mostly based on the availability of genetic diversity and genetic associations among breeding materials. Mohammadi and Prasanna (2003) undertook a review on the utilization of statistical tools and methods in assessment of genetic diversity at the intra-specific level in crop plants. At present, numerous methods are available for the assessment of genetic diversity in germplasm accessions, breeding lines, and populations. These statistical methods depend on pedigree data, morphological data, agronomic performance data, biochemical data, and currently on molecular (DNA-based) data. For moderately precise and unbiased assessments of genetic diversity, sufficient concentration has to be assigned to (i) sampling methods; (ii) use of different data sets based on the understanding of their strengths and limitations; (iii) selection of genetic distance measure(s), clustering techniques, and other multivariate systems in analyses of data; and (iv) objective determination of genetic associations. Judicious combination and use of statistical tools and methods, like bootstrapping, is needed to deal with complicated issues associated with data analysis and interpretation of results from various types of data sets, especially through clustering techniques.

Virk et al. (1995b) screened a group of *rice* accessions from the IRRI (Philippines) using RAPD markers. The accessions include known and suspected duplicates along with closely related germplasms. To differentiate these categories of accessions and to identify true and suspected duplicates, the number of primers, the number of polymorphic bands, and the total number of bands were applied. Two methods have been defined that could be utilized on a more general basis for the detection of duplicates in genetic resources collections, and additional discussion on the values of such activities is presented.

11.6 METHODS OF RICE BREEDING

Breeding rice for maximum yield is a global hot topic and major challenge. For China, which is the most populated country in the World and where rice is a staple food, increasing rice production is of vital importance. As we look for higher and higher yield, we meet more and more difficulties. As there is no ready theory and technique for us to follow, we have to search through trial and error. Shouren (1996) made a broad and systematic analysis of the significant experiences acquired by rice scientists and growers through decades of painstaking labor, in an attempt to further improve the theories and techniques in rice breeding for maximum yield and the standards for hybrid progeny selection.

Genomic selection (GS) is a novel breeding method which utilizes genome-wide markers to envisage the breeding value of individuals in a breeding population. Its efficiency in improving the breeding efficacy of dairy cattle and some crop plant species has been proved. In this study, Spindel et al. (2015) evaluated its efficiency for breeding inbred lines of rice. On a population of 363 elite breeding lines from the IRRI's irrigated rice breeding program a genome-wide association study (GWAS) in combination with five-fold GS cross-validation was conducted. For GY, plant height, and flowering (FL) time, genomic prediction models surpassed prediction based on pedigree records alone. The marker subsets analysis showed that only one marker for every 0.2 cM is enough for GS in this group of rice breeding materials. RR-BLUP was found to be the best statistical method for GY where large effects QTL were not detected by GWAS. But, for FL time, in which single very large effect QTL was noticed, the non-GS multiple linear regression method performed better than the GS models. For plant height, where four mid-sized QTL were detected by GWAS, random forest made the most consistently precise GS models. The study findings suggested that GS, informed by GWAS analyses of genetic architecture and population structure, could develop into a valuable tool for increasing the efficacy of rice breeding.

The linkage disequilibrium (LD)-mapping involving the use of nonrandom associations of loci in haplotypes has been emerged as a powerful high-resolution mapping tool for complicated quantitative traits. The success of this method as a "powerful gene tagging" tool for crops in the plant genomics era of 21st century has been demonstrated from the current advancements made in the development of unbiased association mapping strategies and their effective use in analyzing a number of simple to complex traits in many crop species. In this review, Abdurakhmonov and Abdukarimov (2008) provided the information on basic concept, merits, and simple description of current approaches for an association mapping along with the latest improvements for plant populations. Further, to determine the practicality, success, complications, and future prospects of the association mapping in plants, the data of some pioneer and recent studies on association mapping in various crop species has been provided. This information can be applied in future breeding programs based on association mapping.

In the identification or breeding of improved crop cultivars, the farmer participatory methods can be categorized into participatory varietal selection (PVS) and participatory plant breeding (PPB). Here, Witcombe et al. (1996) reviewed different PVS and PPB methods. If an appropriate choice of cultivars exists, the identification of cultivars preferred by farmers can be made rapidly and in a cost effective manner by using PVS. If this is not possible, then the more resource-expending PPB is needed. PPB can use, as parents, cultivars that were recognized in effective PVS programs. PPB is more probable to provide farmer-acceptable products, especially for marginal environments, when compared with conventional plant breeding. The effect of farmer participatory research on biodiversity is measured. The enduring influence of PVS is to raise biodiversity, however wherever indigenous variability is high it can also decrease it.

11.7 CURRENT AND FUTURE OBJECTIVES AND TRENDS OF RICE BREEDING AND CROP IMPROVEMENT

Tilman et al. (2002) discussed that the projected global food demand for the next 50 years is a major challenge for the sustainability of both food production and terrestrial and aquatic ecosystems and the services they offer to society. Agriculturalists are the primary supervisors of Worldwide operational lands and will shape, perhaps irrevocably, the surface of the Earth in the upcoming years. To meet the needs of increasing yields without compromising environmental integrity or public health, new incentives and strategies for safeguarding the sustainability of agriculture and ecosystem facilities will be critical.

At the turn of the 20th century, the scientific foundation for plant breeding and genetics was established with the central discoveries of Darwin and Mendel. Likewise, the current integration of developments in biotechnology, genomic research, and molecular marker applications with conventional plant breeding methods has made the basis for molecular plant breeding, interdisciplinary science that is transforming 21st century crop improvement. Even though the molecular plant breeding approaches will continue to progress and are a subject of deep interest among plant breeders and crop scientists, the majority of the plant biologists involved in basic scientific research paid relatively little interest to these approaches. In this paper, Moose and Mumm (2008) discussed the significant historic advances in molecular plant breeding, important principles affecting the present molecular plant breeding practices, and aspects that affect the implementation of molecular plant breeding in crop improvement programs. Moreover, they emphasized how the use of molecular plant breeding is now promoting the detections of genes and their functions, which provides new opportunities for basic plant biology research.

To increase the efficiency and accuracy of crop improvement new tools such as functional molecular markers and informatics have been generated through genomics research. These tools helped in understanding the inheritance phenomenon of yield-related traits. Especially, the clarification of the basic mechanisms of heterosis and epigenetics, and their manipulation, has great potential. Ultimately, information of the relative values of alleles at all loci segregating in a population could permit the breeder to design a genotype *in silico* and to exercise whole genome selection. Presently, the application of genomics-assisted in crop improvement, specifically for inbreeding and/or minor crops is limited due to its high costs. However, marker-aided breeding and selection will slowly advance into 'genomics-assisted breeding' for crop improvement (Varshney et al., 2005).

A review has been undertaken by Hay (1995) on the harvest index (HI) concept in crop improvement and physiology. Evidence from the past 20 years literature showed that the term HI has been mostly used for small grain cereal crops and pulses in India, Western Europe, and the USA, while its use for maize and tuber crops is limited. In this context, the author reviewed the standard methods of quantifying HI, the related problems of measurement and interpretation, and characteristic values for a range of World species. For the MVs of most intensively-cultivated grain crops, the HI values ranged from 0.4 to 0.6. The old, outclassed, and MVs of temperate and Mediterranean wheat and barley showed trends in the harvest indices. This exhibits a progressive improvement in HI all through the present century, though progress has been much slower in Australia and Canada than in the UK. Generally, the enhancement in HI has been an outcome of improved grain population density and constant individual grain weight. The analysis of HI response to variation in environmental factors (fertilization, population density, application of growth regulators) in the absence of severe stress revealed the high heritability of HI. A comprehensive view of HI of rice, maize, and tropical pulses was gained by reviewing the literature. In rice, the significant interactions were noted among biomass, GY and season length. Further, it was observed that adhering lemma and palea which are not, primary part of economic yield can make up 20% of grain weight; therefore, they should be considered while breeding for higher yield in rice. Expansion of the HI concept to direct the partitioning of mineral nutrients along with dry matter has offered a series of responses whose inferences for production and breeding has to be investigated.

The efficiency of selection could be improved with the use of DNA markers in plant breeding, which accelerates the development of new crop varieties. The success of marker assisted selection in rice has driven the establishment of molecular breeding labs by many breeding institutes. In recent years, immense genomics research has been performed in rice, which includes the detection of thousands of QTLs for agronomically significant traits, detection of single nucleotide polymorphisms (SNP), the generation of huge amounts of gene expression data, and cloning and characterization of new genes. The highest achievement of the genomics research is the accomplishment and annotation of genome sequences for *indica* and *japonica* rice. This information together with the improvement of new genotyping methods and platforms, and the advancement of bioinformatics databases and software tools— offers even more exciting opportunities for rice molecular breeding in the 21st century. But, the major challenge for molecular breeders is to use this genomics data in real breeding programs. Here, Collard et al. (2008) reviewed the present status of MAS in rice, recent genomics projects, and promising new genotyping methods. They also assessed the possible effect of genomics research. So as to realize the full potential of MAS, the key research areas to "bridge the application gap" between QTL detection and practical breeding were identified. Further, they have suggested ideas and strategies to establish rice molecular breeding labs in the post-genome sequence era and to combine molecular breeding approaches with the rice breeding and research programs.

The development of crops tolerant to abiotic stresses is crucial for increasing the food production. Predicted fluctuations in climate and its variability, mainly extreme temperatures and variations in rainfall, are expected to make crop improvement. Varshney et al. (2011) undertook a review of two important biotechnology approaches, molecular breeding and genetic engineering, and their integration with conventional breeding to develop abiotic stress-tolerant crops. Besides a multidisciplinary approach, they also examined some limitations that need to be overwhelmed to realize the full potential of agricultural biotechnology for sustainable crop production and to meet the needs of an estimated World population of 9 billion in 2050.

China is the major producer and user of rice in the World and a developer of hybrid rice technology. In the past decades, though hybrid rice has contributed greatly to Chinese agriculture, its potential to further improve the grain quality is being doubted. More elite rice cultivars are needed to encounter the problems caused by serious crop damage due to pests and diseases, the widespread usage of pesticides and chemical fertilizers, and a scarcity of water and energy. In recent times, China experienced sustained advances in rice genetics, driven by functional genomics to maintain its rice production. In this study, Wang et al. (2005) reviewed the current position of rice breeding in China with the integration of new approaches such as hybrid rice technology, molecular marker-assisted breeding, functional genomics, and genetically modified rice.

11.8 CONCLUSION

The application of breeding and new biotechnology techniques to rice has advanced the conventional breeding methods to get rice cultivars with desired traits such as high yield, tolerance to pest and diseases and improved grain quality. The modern biotechnology massively increases the accuracy and reduces the time with which required changes in plant characteristics can be made and significantly increases the potential source from which required traits can be gained. In this chapter, the history of rice cultivar's improvement along with the current and future objectives of rice breeding are discussed.

KEYWORDS

- **bacterial blight resistance**
- **brown planthopper resistance**
- **hybrids**
- **improved cultivars**
- **introgression lines**
- **linkage disequilibrium**

REFERENCES

Abdurakhmonov, I. Y., & Abdukarimov, A., (2008). Application of association mapping to understanding the genetic diversity of plant germplasm resources. *International Journal of Plant Genomics*, *18*, doi: 10.1155/2008/574927

Akino, M., & Hayami, Y., (1975). Efficiency and equity in public research: Rice breeding in Japan's economic development. *American Journal of Agricultural Economics, 57*, 1–10.

Ashikari, M., & Matsuoka, M., (2006). Identification, isolation, and pyramiding of quantitative trait loci for rice breeding. *Trends in Plant Science, 11*, 344–350.

Cai, D. T., Yuan, L. P., & Lu, X. G., (2001). A new strategy of rice breeding in the 21st century II. Searching a new pathway of rice breeding by utilization of double heterosis of wide cross and polyploidization. *Acta Agronomica Sinica, 27*(1), 110–116.

Cheng, S. H., Cao, L. Y., & Zhuang, J. Y., (2007). Hybrid rice breeding in China: Achievements and prospects. *Journal of Integrated Plant Biology, 48*, 805–810.

Cheng, S. H., Zhuang, J. Y., Fan, Y. Y., Du, J. H., & Cao, L. Y., (2007). Progress in research and development on hybrid rice: A super-domesticate in China. *Annals of Botany, 100*, 959–966.

Collard, B. C. Y., Vera, C. C. M. C., Nally, K. L., Virk, P. S., & Mackill, D. J., (2008). Rice molecular breeding laboratories in the genomics era: Current status and future considerations. *International Journal of Plant Genomics, 25*, doi:10.1155/2008/524847

Fukai, S., (2005). Molecular breeding for rain fed lowland rice in the Mekong region, grain quality, marker-assisted selection. *Rice, 8*, 330–333.

Hay, R. K. M., (1995). Harvest index: A review of its use in plant breeding and crop physiology. *Annals of Applied Biology, 126*, 197–216.

Hazell, P. B. R., & Ramasamy, C., (1991). The green revolution reconsidered: The impact of high-yielding rice varieties in South India. Johns Hopkins University Press, ISBN: 0801841852.

Khush, G. S, Coffman, W. R., & Beachell, H. M., (2001). The history of rice breeding: IRRI's contribution. In: Rice Research and Production in the 21st Century. W.G. Rockwood (ed.). International Rice Research Institute. Los Baños, Philippines. Pp. 117–135.

Khush, G. S., (1995). Modern varieties—their real contribution to food supply and equity. GeoJournal, 35, 275–284.

Lu, B. R., & Snow, A. A., (2005). Gene flow from genetically modified rice and its environmental consequences. *Bio Science, 55*, 669–678.

Maurya, D. M., Bottrall, A., & Farrington, J., (1988). Improved livelihood, genetic diversity, and farmer participation: A strategy for rice breeding in rain fed areas of India. *Experimental Agriculture, 24*, 311–320.

Mohammadi, S. A., & Prasanna, B. M., (2003). Analysis of genetic diversity in crop plants—Salient statistical tools and considerations. *Crop Science: Review and Interpretation, 43, 1235–1248.*

Moose, S. P., & Mumm, R. H., (2008). Molecular plant breeding as the foundation for 21st century crop improvement. *Plant Physiology, 147*, 969–977.

Munshi, K., (2004). Social learning in a heterogeneous population: Technology diffusion in the Indian Green Revolution. *Journal of Development Economics, 73*, 185–213.

Ni, J., Colowit, P. M., & Mackill, D. J., (2002). Evaluation of genetic diversity in rice subspecies using microsatellite markers. *Crop Science: Plant Genetic Resources, 42, 601–607.*

Peng, S., Gurdev, S. K., Virk, P., Tang, Q., & Zou, Y., (2008). Progress in ideotype breeding to increase rice yield potential. *Field Crops Research, 108*, 32–38.

Prahladachar, M., (1983). Income distribution effects of the green revolution in India: A review of empirical evidence. *World Development, 11*, 927–944.

Spindel, J., Begum, H., Akdemir, D., Virk, P., & Collard, B., (2015). Correction: Genomic selection and association mapping in rice (*Oryza sativa*): Effect of trait genetic architecture, training population composition, marker number and statistical model on accuracy of rice genomic selection in elite, tropical rice breeding lines. *PLoS Genetics, 11*(2), e1004982.

Shiva, V., (2016). *The violence of the green revolution.* University Press of Kentucky.

Shouren, Y., Longbu, Z., Wenfu, C., Zhenjin, X., & Jinmin, W., (1995). Theories and methods of rice breeding for maximum yield. Zuo Wu Xue Bao, *22*, 295–304.

Tilman, D., Cassman, K. G., Matson, P. A., Naylor, R., & Polasky, S., (2002). Agricultural sustainability and intensive production practices. *Nature, 418*, 671–677.

Tu, J., Zhang, G., Datta, K., Xu, C., Yuqing, H. Y., Zhang, Q., Singh, G. K., & Kumar, D. S., (2000). Field performance of transgenic elite commercial hybrid rice expressing *Bacillus thuringiensis* δ-endotoxin. *Nature Biotechnology, 18*, 1101–1104.

Varshney, R. K., Bansal, K. C., Aggarwal, P. K., Datta, S. K., & Craufurd, P. Q., (2011). Agricultural biotechnology for crop improvement in a variable climate: Hope or hype? *Trends in Plant Science, 16*, 363–371.

Varshney, R. K., Graner, A., & Sorrells, M. E., (2005a). Genomics-assisted breeding for crop improvement. *Trends in Plant Science, 10*, 621–630.

Virk, P. S., Lloyd, B. V. L., Jackson, M. T., & Newbury, H. J., (1995b). Use of RAPD for the study of diversity within plant germplasm collections. *Heredity, 74*, 170–179.

Virk, P. S., Newbury, H. J., Jackson, M. T., & Ford, L. B. V., (1995). The identification of duplicate accessions within a rice germplasm collection using RAPD analysis. *Theoretical and Applied Genetics, 90*, 1049–1055.

Wang, Y., Xue, Y., & Li, J., (2005). Towards molecular breeding and improvement of rice in China. *Trends in Plant Science, 10*, 610–614.

Witcombe, J. R., Joshi, A., Joshi, K. D., & Sthapit, B. R., (1996). Farmer participatory crop improvement: I. varietal selection and breeding methods and their impact on biodiversity. *Experimental Agriculture, 32*, 445–460.

Wünn, J., Klöti, A., Burkhardt, P. K., Ghosh, G. C. B., Launis, K., Iglesias, V. A., & Potrykus, I., (1966). Transgenic *indica* rice breeding line IR58 expressing a synthetic *crylA(b)* gene from *Bacillus thuringiensis* provides effective insect pest control. *Biotechnology, 14*, 171–176.

Yan, W. G., Rutger, J. N., & Bryant, R. J., (2007). Development and evaluation of a core subset of the USDA rice germplasm collection. *Crop Science: Plant Genetic Resources, 47*, 869–876.

Zhao, K., Wright, M., Kimball, J., Eizenga, G., Clung, A., Kovach, M., Tyagi, W., & Ali, L., (2010). Genomic diversity and introgression in *O. sativa* reveal the impact of domestication and breeding on the rice genome. *PLoS ONE, 5*, e10780.

CHAPTER 12

Research Advances in Breeding and Biotechnology of Rice

ABSTRACT

As rice is an economically and socially important crop, many advances have been made in its research. Increased use of new biotechnological techniques could result in an increased rate of improved rice production. In this chapter, the significant advances made in rice breeding and biotechnology are discussed.

12.1 HIGH YIELDING RICE VARIETIES

Presently, in Colombia, nearly total irrigated rice area is occupied with high yielding varieties which increased the average yields from 1.8 to 4.4 tons/ha while the production of upland rice was declined from 50 to 9% of Colombian rice output during 1965 to 1975. Mostly large irrigated holdings were focused and rice output of 70% was got from irrigated farms of more than 50 ha. The nonexistence of high-yielding varieties (HYVs) would have lowered the rice price and the low-income consumers had the greatest net benefit but the upland rice producers would have affected severely (Scobie and Posada 1977).

Different high yielding rice genotypes were selected to evaluate the genetic variability of 10 quality parameters by Vanaja and Babu (2006). The presence of high variability in alkali-spreading value (ASV) and moderate variability in L/B ratio of grain, milling percentage, amylose content (AC), water uptake and volume expansion was unveiled from phenotypic and genotypic coefficients of variation. Further, all quality traits showed high broad sense heritability. For the quality parameters such as ASV, grain L/B ratio, milling percentage, AC, volume expansion ratio, and water uptake high heritability, high estimated genetic gain and moderate genotypic coefficient of variation were recorded signifying that these parameters can be effectively use in future breeding programs.

Gurdev (1995) discussed during the past 25 years, with the adoption of high-yielding semi-dwarf rice varieties on large scale the rice production increased significantly. But the rate of increase is much less than the rate of rice consuming population growth which may lead to severe food shortages in 20–30 years if the trend is not upturned. To overcome this problem the high yielding rice varieties are required. Under best tropical conditions the modern HYVs has a yield potential of 10 t ha^{-1} and further research program are going on to increase the yield potential up to 15 t ha^{-1}. The modern HYVs harvest index (HI) is 0.5 and to increase it to 0.6 (60% grain: 40% straw by weight) along with an increased photosynthesis ability leading to increased total biological yield new strategies were developed. The new plant type (NPT) varieties must have a 12.5–13 tons ha^{-1} yield potential which could be used to produce rice hybrid with a yield gain of 25% over best parent and would have 15 t ha^{-1} yield potential.

Research findings revealed that in hypothetical valuation questions the willingness-to-pay is overstated when compared to actual payment required. In recent times, to ignore the possible bias in hypothetical valuation questions "cheap talk" has been presented. Cheap talk is a method of clarifying the hypothetical bias to individuals before asking a valuation question. Using a conventional value elicitation technique, Lusk (2003) investigated the cheap talk effect in a mass mail survey. The findings indicated that for most of the participants of survey at decreasing willingness-to-pay cheap talk was operative but for knowledgeable consumers cheap talk did not decrease willingness-to-pay.

For the successful implementation of water, saving technology such as aerobic rice its physiological characteristics and yield performance should be compared with conventional flood-irrigated rice. Patel et al. (2010) conducted an experiment at ICAR Research Complex for NEH Region, Meghalaya with aerobic rice variety from IRRI and some recommended important HYVs from the region under both aerobic and flooded conditions. The objective of the research was to assess the frequent mid-season drainage influence as water saving technique measure along with its preconditioning effect on genotypes to tolerate water stress during the crop growth period. They had compared the crop performance of aerobic rice with the flooded rice so to detect traits accountable for the yield gap between aerobic and flooded rice. Based on varieties about 37.8% yield difference was unveiled from the results between aerobic (average yield, 1.67 t ha^{-1}) and flooded rice (average yield, 2.31 t ha^{-1}) and the rice hybrid DRRH1 exhibited the highest difference. When compared to flooded rice 27.5% less yield was recorded under aerobic condition. Among the yield components evaluated sink size (spikelets per panicle) found to be most important factor contributing to the more yield and to the yield gap between aerobic and flooded rice. In this study for both stress (aerobic) and normal (flooded) condition the variety Sahsarang1 which is characterized with moderate photosynthesis rate, transpiration rate and water use efficiency (WUE) together with higher grain yield (GY) was suggested as a suitable option. Therefore, for the effective cultivation of aerobic rice, the varieties with minimum yield gap should be selected.

For irrigated rice system, one of the proven and efficient water-saving technology water is Alternate wetting and drying (AWD) irrigation but only limited information is available on "super" hybrid rice performance under these conditions. Yao et al. (2012) compared GY and its attributing traits of “super” hybrid rice variety with a water-saving and drought-resistance rice (WDR) variety to find plant characteristics accountable for varietal difference in GY under AWD conditions. Under AWD and continuously flood-irrigated (CF) conditions, a “super” hybrid rice variety Yangliangyou 6 (YLY6) and a WDR variety Hanyou 3 (HY3) were grown by applying different nitrogen levels. The GY, yield traits, total water input, water productivity, and nitrogen use efficiency were evaluated. Compared with continuously flooded rice 24–38% irrigation water was saved under AWD. Though for AWD and CF the GY difference was insignificant, under AWD conditions YLY6 gave 21.5% higher yield than HY3. Further, YLY6 exhibited constantly higher water productivity and physiological nitrogen use efficiency leading to higher total dry weight and HI. Large sink size produces by more spikelets per panicle was found to be the important yield component contributing for higher yield of YLY6 than HY3. The findings of this study suggested that the HYVs produced for the continuously flood-irrigated rice system could give higher yield under safe AWD and the more water input is not needed for “Super” hybrid rice varieties to get high GY. For AWD conditions, the number of spikelets per panicle is a key yield component and should be taken into account while breeding high-yielding rice varieties.

During the flowering (FL) period, a high temperature in the daytime induces floret sterility. To clarify this high-temperature-induced sterility mechanism, Matsui et al. (2001) exposed nine *japonica* rice varieties to 35.0, 37.5 or 40.0°C day-temperature conditions (1000–1600) for six successive days using sun-lit phytotrons. The fertility percentage, pollination, and germination of pollen grains on the stigmas were assessed. The varieties exhibited differences pertaining to 50% sterility caused by the temperature and a difference between the most tolerant and a susceptible variety was about 3.0°C. The percentage of florets with 10 or more germinated pollen grains at 37.5°C day-temperature was approximately equivalent with the percent fertility while, at 40°C day-temperature it was higher than the percent fertility. Under both the conditions, the florets with less than 10 germinated pollen grains had less than 20 total pollen grains on their stigmas. Finally, they concluded under extremely high temperature conditions to get assured pollination and fertility, high-temperature tolerant varieties need to be developed.

In salt-affected rice paddy field on the western coast of South Korea for eco-friendly rice production, the appropriate nitrogen and phosphorus application levels, nitrogen split application ratio and application method were determined by Cho et al. (2008). 11 treatments viz. three treatments for the appropriate nitrogen and phosphorus fertilizers (A1–A3) application levels evaluation; five treatments to evaluate the nitrogen split application system (T1–T5) and three treatments for the appropriate application method determination for chemical fertilizer (M1–M3) were used. Among all the 11 treatments, significant differences were not observed in amylose and protein content. Further, there were no significant differences in GY and yield components of rice among all the treatments. The 140 kg ha^{-1} nitrogen fertilizer was determined as appropriate application level, the suitable nitrogen fertilizer split application ratio at four different periods was basal fertilization-40%, maximum tilling stage-20% and booting stage-10% and the deep layer application and whole layer application methods were recognized as best application methods.

Earlier assuming that year-to-year agricultural production variations are negligible organic agriculture was compared with conventional agriculture applying life cycle assessment (LCA). As in the conversion process, organic farming performance is unstable, it is essential to make comparisons on the basis of averages. In this paper, Hokazono and Hayashi (2012) obtained time-series data from a five-year on-farm trial and applied to an LCA of three rice production systems viz. organic, environment-friendly and conventional in Japan. The environment impact valuation was done using four categories of impact viz. global warming, acidification, eutrophication, and non-renewable energy. The findings revealed that through all impact categories only the environmental impacts related with organic farming varied extensively over the years which diminished over time. Compared with the other two modes of rice production, organic rice production had higher impact of environment in all four categories on average. But, in the last stage of conversion phase the environmental impact in both conventional and organic farming was almost same and at the initial phase the higher impacts variability in organic farming was mainly related with the instability of the organic rice yield indicating that for successful conversion to organic farming understanding of the trends in inter-temporal variability is essential.

One of the important determining factors for rice eating quality is amylose content (AC) and the rice with low AC viz., with 5% to 15% of amylose possesses different qualities such as fluffy texture, glossy appearance of the cooked rice, soft texture of cooled rice and outstanding expansibility for food processing. The rice with low AC can be utilized for cooking and as a material for instant, mixed rice and expanded foods. Zhu et al. (2004) reviewed the present status of low AC rice characterization, inheritance, molecular mechanism, and breeding. They emphasized that the forthcoming research should be directed towards screening and improving the germplasm. In addition to this, a low-AC rice breeding program should give emphasis to the development of high-quality functional rice cultivars for special purposes using genes independent of Wx, by means of clarifying the low AC mutation molecular mechanism and by pyramiding low amylose gene and other special quality genes in rice cultivars.

12.2 ADVANCES IN BIOTECHNOLOGY OF RICE

About 15 years ago, with the production of first transgenic rice plants, rapid advances have been made in rice biotechnology. Recently, field trials of transgenic rice, particularly for insect-resistant rice have been carried out and many studies have been initiated to release these transgenic crops safely. The transgenic rice system helped in understanding the transformation process, integration pattern of transgene and to modulate gene expression. These advancements resulted in the development of high frequency, routine, and reproducible genetic transformation protocols for rice. This technology has been utilized to develop rice plants resistant to abiotic stresses and tolerant to different pests and diseases. Further, to improve the quality and increase the nutritional content many traits were transferred into rice. Superior transgenic product can be produced with the new molecular inventions such as transgene inducible expression and selectable marker free technology. This will also help to alleviate public concerns associated to transgenic issues and to get regulatory approval. This acquaintance from rice can also be used for the improvement of other cereals. Further, with the completion of rice genome sequencing and assemblage of full-length cDNA resources surplus opportunities has opened up which will help in integration of data from the large-scale projects to unravel numerous biological problems.

Micronutrients that are essential for a healthy life cannot be produced by humans and should be provided through the food chain. The main food source of mankind are the staple crops like rice and as these crops are mostly deficient in one or the other micronutrient they should be supplemented by other foodstuffs. Owing to the absence of genetic variability for the preferred trait, the breeding for micronutrient-dense crops is not always a feasible option. Further, a number of crop plants are not amenable to conventional breeding because of sterility problems and their complex genetic makeup at that point only feasible option is genetic modification. Recently, with the identification of genes involved in various biochemical pathways at an extraordinary rate, many tools become available to produce several micronutrients rich staple crops (Beyer, 2010).

The *tlp* gene construct was transferred by Datta et al. (1999) into three *indica* rice cultivars viz. 'ChinsurahBoro II,' 'IR72,' and 'IR51500.' For this under the CaMV 35S promoter control rice transformation vector pGL2 was used to clone a 1.1-kb DNA fragment having the thaumatin-like protein (TLP-D34) coding region, which is a member of the PR-5 group and transferred by PEG-mediated direct gene transfer to protoplasts and using immature embryos by biolistic transformation. With Southern blot analysis T_0, T_1 and T_2 transgenic plants were detected for the chimeric 23-kDa *tlp* gene presence which was confirmed using Western blot analysis and by staining with Coomassie Brilliant Blue. The transgenic plants bioassays were tested by inoculating the sheath blight pathogen,

Rhizoctonia solani, the transgenic plants exhibited improved resistance compared to control plant indicating the over-expression of *tlp* gene.

12.2.1 RICE BASED EXPRESSION OF HUMAN PROTEIN FROM RICE

The infant formula is partly deficient in components like human lactoferrin (hLF) due to which the breast-fed infants are generally healthier than formula-fed infants. Thus, the health benefits of infant formula can be increased by fortifying it with some bioactive proteins such as recombinant HLF (rHLF). For this rice, cells were transformed with a synthetic HLF gene related to a rice glutelin 1 promoter and its signal sequence to get rHLF. The rHLF was found to express especially in the rice grain but not in other tissues with an expression level of nearly 0.5% of dehusked rice grain weight. The rHLF expressed rice grains were advanced for four generations and a stable expression levels were observed. The similarity between the rHLF and HLF in N-terminal sequence, pI, iron-binding capacity, antimicrobial activity against a human pathogen and protease digestion resistance was revealed from biochemical and physical characterization. Further, the rHLF retained the receptor binding activity which was measured by Caco-2 cells and a human small intestinal cell line. As rice may be utilized in infant formula, the incorporation of rHLF expressing rice grains into infant formula directly without purification is a viable option and provides an appropriate benefit over other expression system (Nandi et al., 2002).

Simultaneously, the stable expression of recombinant human lactoferrin (rhLF) was identified by Nandi et al. (2005) at 0.5% brown rice flour weight for nine generations. From rice flour, the efficient extraction of rhLF in 20 mM phosphate buffer (pH 7.0) with up to 0.5 M NaCl and at a ratio of 1 kg flour to 10 L buffer was obtained by process development. After separating solid/liquid, the extract was directly loaded onto an ion-exchange column then using 0.8 M NaCl the rhLF was eluted and the resultant rhLF was about 95% pure. Further, a series of biochemical and biophysical analyses were performed and the purified rhLF was found to be identical to its native human counterpart in all aspects except its glycosylation. Economic analysis revealed that to produce 1 g of rhLF of pharmaceutical-grade the cash cost is US$ 5.90 and the scale is 600 kg^{-1} year. Correspondingly, the analysis indicated that the expression level has an intense effect on costs associated to planting, milling, extraction, and purification suggesting that in plants the high level of recombinant protein expression is one of the significant factors for the success of plant-made pharmaceuticals.

In children suffering with acute diarrhea and dehydration, the rhLF and recombinant human lysozyme containing rice-based oral rehydration solution was compared with glucose and rice-based oral rehydration solution by conducting double-blind controlled trial (Zavaleta et al., 2007). To obtain low osmolarity WHO-ORS (G-ORS), rice-based ORS (R-ORS), plus lactoferrin and lysozyme (Lf/Lz-R-ORS) one hundred and forty children with age group of 5 to 33 months' were block randomized. By means of home and clinic follow-up for 14 days, intake, and output were observed continuously for 48 hours in the ORU. In diarrhea outcomes, significant difference was not shown by G-ORS and R-ORS groups and hence joined as the control group. In comparison, a significant decrease diarrhea (3.67 d vs. 5.21 d, $P = 0.05$) duration was observed in the Lf/Lz-R-ORS group from intent-to-treat analysis and in the number of children who attained 48 hours with solid stool, 85% vs. 69% ($P < 0.05$) a significant increase was observed. Further, after attaining the endpoint a decrease in diarrhea volume and the children percentage that had a new diarrhea episode was observed indicting the beneficial effects of rhLF and lysozyme addition to a rice-based oral rehydration solution on children suffering with diarrhea.

In maturing rice grains, the human lysozyme expression was studied by Huang et al. (2002). The codon-optimized structural gene for human lysozyme was delivered into the callus of rice cultivar Taipei 309 using particle bombardment-mediated transformation. The promoter and signal peptide sequence for rice storage protein Glutelin 1 controls the expression of Lysozyme. From separate transformation events, total 33 fertile plants were regenerated of which 12 plants exhibited significant lysozyme expression levels and advanced to next generations. By means of PCR and Southern blot analysis, transgenes were characterized and a typical Mendelian 3:1 segregation ratio was obtained, indicating a single locus or tightly linked loci of gene insertion. In seven transgenic breeding lines lysozyme expression levels reached 0.6% of the brown rice weight or 45% of soluble proteins and were advanced for six generations. In all generations, lysozyme expression levels were retained. The identical N-terminal sequence, molecular weight, pI, and specific activity was found between native and recombinant human lysozyme from biochemical, biophysical, and functional comparisons. Moreover, identical bactericidal activity was showed towards an *E. coli* laboratory strain. This study indicated the possibilities of using recombinant human lysozyme containing baby foods rice flour or rice extract to improve the medical and nutritional quality of infant formulas.

To produce plant-made pharmaceuticals (PMPs) the transgenic plants with heterologous protein expression levels are of crucial importance. In rice endosperm, a *puroindoline b* promoter and signal, peptide (*Tapur*) driving expression of human lysozyme was studied by Hennegan et al. (2005). The results showed that under the *Tapur* cassette control human lysozyme expression is seed-specific, readily extractable, active, and properly processed. From this cassette, the localized expression of lysozyme in protein bodies I and II in rice endosperm cells was indicated by immuno-electron microscopy signifying the use of non-storage promoter and signal peptide to target human lysozyme to rice protein bodies. They successfully employed a strategy to improve the expression of human lysozyme in transgenic rice grain by combining the *Tapur* cassette with our well established *Gt1* expression system. The results demonstrated that when the two expression cassettes were combined, the expression level of human lysozyme increased from 5.24 ± 0.34 mg g^{-1} flour for the best single cassette line to 9.24 ± 0.06 mg g^{-1} flour in the best double cassette line, indicating an additive effect on expression of human lysozyme in rice grain.

Globally, the two major health issues are iron deficiency and diarrhea. Nearly 25% of the World population was affected with iron deficiency anemia (UNICEF/WHO, 1999). In developing countries the iron deficiency prevalence is exemplified by comparing with other deficiencies; 3.5 billion people are affected with iron deficiency whereas, vitamin A affect 0.3 billion people and iodine deficiency affect 0.8 billion people. The young children and women of childbearing age (particularly pregnant women) are mostly affected with iron deficiency. It is expected that by correcting iron deficiency in developing countries national productivity levels could be improved as much as 20%. To combat iron deficiency, Ventria Bioscience evaluated the rhLF, expressed, and extracted from rice seed as a dietary supplement. Worldwide 60% of children die below age five due to pneumonia, diarrhea or measles and in the last 50 years, annually the lives of 1 to 2 million children is saved by World Health Organization oral rehydration solution (WHO-ORS) which is one of the most significant medical advances. In many studies, the decreased diarrhea frequency in breast-fed children and improvement of health due to the lactoferrin and lysozyme proteins in human milk was documented. By hLF and human lysozyme, the inhibited growth of the diarrheal related organisms such as rotavirus, ETEC, cholera, salmonella, and shigella was documented with *In vitro* data. In rice hLF and human lysozyme were expressed with Ventria's Express Tec™ system and these proteins in R-ORS formulation besides providing the benefits of decreased stool volume and increased weight gain reduce the diarrheal episodes course through antimicrobial activity against the causative agents (Bethell and Huang, 2004).

12.2.2 GOLDEN RICE

Increasing birth rate is a main cause of food security problem and a Thai businessman who is working in programs intended to reduce birth rate coined the term "golden rice." As it turned out, the term "golden rice" has confirmed to be extremely successful in upsetting the public interest. It is hard to assess how much it appeals public with its catchy name and the technological innovation it represents. We reside in a society that is strongly influenced by the media and the "catchy" titles are particularly useful in drawing the attention of media consumers. However, the "story," must contain an important message. Here, the purely noble use of genetic engineering technology has possibly resolved a crucial and formerly difficult health problem for the poor of the developing World. But after more than 10 years of the public argument about genetically modified organisms (GMOs) it is recognized that even with the aid of the media, rational disputes succeed in influencing only a small section of the public-at-large. The opposition of GMO, specifically in Europe, has been tremendously successful in guiding all negative sentiments related with the hypothetical risks of all novel technologies along with economic "globalization" onto the unproven threats presented by the release of GMOs into the food chain. This is one reason why the story of "golden rice" is so important: In the short history of GMO research, "golden rice" is exceptional in having been involved by the public-at-large. The cause for this rests in its emotive appeal: People are really worried about the fate of blind children, and they are eager to strengthen a technology that provides the opportunity to elude blindness.

In developing countries, the "food security" is one of the foremost challenges for mankind. The researchers are the fortunate group of citizens, having responsibility to advance science: They should take a higher social responsibility and wherever possible, apply science to resolve the significant problems of people. In this case, our scientific community is not in balance, and the public senses this instinctively. This, in turn, has made it easy for the GMO opposition to wage a war of propaganda against GMO technology. The common people often question if for developing countries food security is such a severe problem, and if researchers believe that GMO technology should be advanced to contribute to a solution, then why are so many researchers working on *Arabidopsis* and so few on those plants that feed the poor?. Obviously, one should know the significance of fundamental research and in what way all the information acquired from *Arabidopsis* will eventually advance the development of major crops. Press statements from the agrobiotechnology industry associated with work on food security in developing countries are dishonest and help only to raise ill will against the GMO technology. Thus, we need more examples on the "golden rice" to get better public response about the technology. Effective schemes are to be developed using public funding to address such innovative mission. Literally, there is no harmful impacts on the environment or human health, however an entire research team need 4-5 years of work for environmental risk assessment.

In developing countries, where the deficiency of Provitamin A deficiency is widespread Golden Rice is supposed to help in combating this deficiency by accumulating the Provitamin A in its genetically transformed grains. From the time when Golden Rice is originally produce, the original Golden Rice has experienced intense research to increase the provitamin A content, to found the scientific basis for its carotenoid complement and to better fulfill the regulatory obligations. Currently, the major challenge is how to effectively transfer this Golden Rice into the hands of farmers, which is a new avenue for public sector research, carried out with the support of international research consortia. Further, new research is in progress to increase the Golden Rice nutritional value (Babili and Beyer, 2005).

In developing countries the rice consumers mainly women and children suffer with the vitamin A deficiency because the endosperm of rice grain cannot produce beta-carotene. So, to improve vitamin A content in rice genetically modified Golden Rice has been produced which is still at the R&D stage. The potential impacts of this transgenic rice were analyzed by Zimmermanna and Qaim (2004) in a Philippine context. By means of disability-adjusted life years (DALYs) methodology, the health effects of golden rice are measured. As the vitamin A, deficiency problems such as blindness or increased mortality cannot be eliminated completely by Golden Rice this technology should be viewed as a complement but not as a substitute for alternate micronutrient interventions. However, significant benefits could be generated from the technology.

Rice is generally deficient in β-carotene (pro-vitamin A) and to improve its nutritional status genetically engineered (GE) variety of rice 'Golden Rice' is produced. It is estimated that its influence to reduce vitamin A deficiency would be significantly enhanced with even higher β-carotene content 2. Paine et al. (2005) assumed that the limiting step in accumulation of β-carotene is one of the two genes used in developing Golden Rice that is daffodil gene encoding phytoene synthase (*psy*). By testing the other *psy* plant systematically from maize a *psy* that significantly improved accumulation of carotenoid in a model plant system was identified. Then introducing this *psy* along with the *Erwinia uredovora* carotene desaturase (*crtI*) which is used to produce the original Golden Rice 1, a 'Golden Rice 2' was developed. Compared with the original Golden Rice 23-fold (maximum 37 μg g^{-1}) increased total carotenoids and preferential β-carotene accumulation was observed in Golden Rice 2.

The cDNA coding for phytoene synthase (*psy*) and lycopene β-cyclase (β-*lcy*) from *Narcissus pseudo narcissus* which is under the control of endosperm-specific glutelin promoter is combined with *Erwinia uredovora* bacterial phytoene desaturase controlled by constitutive 35S promoter were introduced into rice endosperm in a combined transformation by Beyer et al. (2002) to get a functioning biosynthetic pathway of provitamin A (β-carotene). This combination meets the β-carotene synthesis requirements and as expected, rice endosperm with yellow β-carotene was observed in the T_0-generation. The experiment unveiled that to guide synthesis of β-carotene and for the additional downstream xanthophylls formation only *psy* and *crt*I are required but the presence of β-*lc*y is not needed. Reasonable descriptions for this finding are that in rice endosperm the constitutive expression of these downstream enzymes are induced by the transformation, e.g., by enzymatically formed products. Finally, the assumption that present working model, trans-lycopene, or a *trans*-lycopene derivative works as an inductor in a type of feedback mechanism stimulating endogenous carotenogenic genes was developed from *N. pseudonarcissus* model system results.

In "Golden Rice" about 35 μg β-carotene per gram of rice is present and to assess the possible effect of this biofortified grain in rice-consuming vitamin A deficient population determination of vitamin A equivalency of Golden Rice β-carotene is important. So, Tang et al. (2009) determined the vitamin A equivalency of Golden Rice β-carotene by growing Golden Rice plants hydroponically with heavy water (deuterium oxide) to produce deuterium-labeled [^{2}H] β-carotene in the rice grains. Then 5 healthy adult volunteers (3 women and 2 men) were fed with 65–98 g of Golden Rice having 0.99–1.53 mg β-carotene with 10 g butter and 1 week before ingestion of the Golden Rice dose each volunteer was fed with a reference dose of [$^{13}C_{10}$] retinyl acetate (0.4–1.0 mg). Blood samples were collected over 36 d. The outcomes revealed that Golden Rice β-carotene supplied 0.24–0.94 mg retinol when compared with the reference dose [$^{13}C_{10}$] retinyl acetate (84.7 ± 34.6 μg·d) and the conversion factor of Golden Rice β-carotene to retinol is 3.8 ± 1.7 to 1 with a range of 1.9–6.4 to 1 by weight, or 2.0 ± 0.9 to 1 with a range of 1.0–3.4 to 1 by moles. This indicated the effective conversion of β-Carotene from Golden Rice to vitamin A in humans.

12.2.3 RNA INTERFERENCE (RNAI)

The rice plant height is controlled by OsGA20ox2 gene and to regulate it Qiao and Zhao (2011) suppressed the OsGA20ox2 expression using double-stranded RNA interference (RNAi) method. For these two different RNAi vectors, pCH1CK, and pCH12CK were made and transferred into rice varieties QX1 and Zhongzuo0201 using Agrobacterium-mediated transformation. When compared with the control plants a clear height reduction was seen in RNAi transgenic lines obtained from pCH1CK and pCH12CK. From pCH1CK and pCH12CK the semi-dwarf lines transcripts were also reduced when compared with the control plants. From both RNAi semi-dwarf lines and wild-type plants seeds per panicle, 1,000-grain weight, and seed yield per plant were analyzed. The results unveiled that the GY production in RNAi semi-dwarf lines was more stable than wild-type plants. Further, the results showed that using different fragments of the same RNAi target the semi-dwarf of rice plants with different magnitude of height can be produced and by applying exogenous GA3 on these plants the normal height of the plant could be restored.

Lipoxygenases (LOXs) are involved in oxidative rancidity in rice which makes it unfit for human consumption. Chowdhury et al. (2016) analyzed the functions of the LOXs specifically present in bran/seed by inhibiting LOXs expression with RNAi. The bran/seed-specific LOX were r9-LOX1 and L-2 (9-LOXcategory), whereas plastid-specific LOX was RCI-1 (13-LOX category). A tissue/cultivar specific expression of three LOXs was indicated by real-time PCR. Compared to the in stabilized bran, mature seed, and regenerated plant higher expression of r9-LOX1 and L-2 was observed in active bran/seed. RCI-1 was hardly expressed in seed. Compared to the non-transgenic controls the expression of both r9-LOX1 and L-2 in transgenic lines was severely down-regulated. The r9-LOX1RNAi transgenic lines SPME/GC-MS analysis revealed 74.33% reduction in non-anal content formed during oxidation of linoleic acid by lipoxygenase. However, the acetic acid and hexanal (direct products of 13-LOX) contents were increased by 388.24% and 184.84%, respectively. This indicated the positive regulation of non-anal amount and negative regulation of acetic acid and hexanal by r9-LOX1. In this study, a comprehensive understanding of the bran/seed-specific LOXs, r9-LOX1 was provided which can be applied in future breeding programs to improve the storage quality of rice.

The rice bran oil (RBO) is produced from the bran of polished rice grains. Earlier using RNAi down-regulation of OsFAD2-1 high oleic (HO) RBO has been produced. Owing to its high oxidative stability, RBO could be directly utilized in the food industry without hydrogenation, and it is free of trans fatty acids. However, when compared to common oilseeds, a lipid metabolism in rice grains is poorly studied. In this study, Tiwari et al. (2016) further analyzed the OsFAD2-1 role in the developing rice grain. In developing rice grain the Illumina-based NGS transcriptomics analysis revealed that in the HO rice line the knockdown of Os-FAD2-1 gene expression go together with the downregulation of the key genes expression in the lipid biosynthesis pathway. The HO-RBO also showed a slightly higher level of oil accumulation. Further, by modifying the lipid biosynthetic genes expression present in the HO line the oleic acid content in rice oil can be increased.

12.2.4 CRISPR/CAS9 SYSTEM

Rice being a staple food and important source of income for a large portion of the World's population, therefore improvement of rice quality is a focal point of new research approaches such as CRISPR genome editing. Shamik et al. (2016) indicated that genome editing is a successful and viable tool in rice which increased the ability of researchers to alter more rice genes so as to develop improved varieties. The CRISPR system generates transgene-free genome-edited plants and do not involve regulatory issues. The successful application of the CRISPR genome editing tool in rice will enable the site-specific integration of genes together with regulation of gene expression, gene discovery and rice functional genomics.

A most important source of dietary cadmium (Cd) intake is rice grain with excessive Cd and a serious threat to health of the rice consuming population. It is tough for conventional breeding methods to develop rice cultivars with constantly low Cd content and there is urgent need for the development of new strategies. Li Tang et al. (2017) reported the production of new *indica* rice lines with low Cd accumulation by knocking out the OsNramp5 metal transporter gene with CRISPR/Cas9 system. For this purpose, the rice plants were grown hydroponically and in shoots and roots of osnramp5 mutant plants reduced Cd accumulation was observed which protected the plants from impaired growth in high Cd condition. The results from Cd-contaminated paddy field trials revealed the consistently less than 0.05 mg kg^{-1} of Cd accumulation in OsNramp5 protein sequences containing grains, contrary to 2.90 mg kg^{-1} Cd accumulation in Huazhan (the wild-type *indica* rice) grains. Moreover, promising hybrid rice lines with extremely low Cd concentration in grains were developed. This research provides a practical method for developing Cd pollution-safe *indica* rice cultivars that reduces Cd contamination risk in grains.

12.2.5 BIOFORTIFICATION FOR NUTRITION

Deficiency of nutrient elements like iron, zinc in human health can create short and long term health problems. Dietary diversity, use of industrially fortified foods, and medical interventions are all effective solutions to this suite of related problems. Biofortification, or the nutritional enhancement of staple and specialty crops, represents a low cost, sustainable, and potentially effective solution to addressing dietary deficiency and malnutrition in the rural poor (Hoekenga, 2014). Gibson (2006) estimated that 60–70% of population in Asia and sub-Saharan Africa could be at risk of low Zn deficiency intake and bring threat to almost 2 billion people in Asia and 400 million people in sub-Saharan Africa (IRRI, 2006). Further

Zn deficiency has been estimated to be responsible for approximately 4% of the worldwide burden of morbidity and mortality in under-5-year children and a loss of nearly 16 million global disability-adjusted life years (Black et al., 2008; Walker et al., 2009). Genetic biofortification through both traditional breeding and biotechnological tools has led to programs such as HarvestPlus aimed to enrich Zn and other micronutrients in grain (Stein, 2010). IRRI, Philippines had screened of close to 1000 rice genotypes at the grain Zn concentrations ranged from 15.9 to 58.4 mg kg1 (Graham et al., 1999). Interestingly, in most cases, grain yield has inverse relation with grain Zn concentration (Garvin et al., 2006; McDonald et al., 2008). Swamy et al. (2016) reviewed that several QTLs and gene specific markers had been identified for grain Zn with its huge potentiality in marker-assisted Breeding. A holistic breeding approach involving high Zn trait development, high Zn product development, product testing and release, including bioefficacy and bioavailability studies is essential for successful Zn biofortification. Sathya et al. (2016) revealed that even microbes, especially actinomycetes, with metal-mobilizing and PGP traits for biofortification as this strategy may act as a complementary sustainable tool for the existing biofortification strategies. Kumari et al. (2019) studied on eighteen entries selected from two extremes of grain Zn distribution range. There were then subjected to molecular profiling using a panel of 14 candidate genes specific 12 reported and 14 designed primer pairs. Only eight (OsZIP1-1, OsZIP3a, OsZIP4a, OsZIP5-3, OsZIP7-2, OsZIP8b, OsNRAMP7 and OsNAAT1) reported and eight (OsZIP3K, OsZIP4K, OsZIP5K, OsZIP7K, OsNRAMP7K, OsNAAT1K, OsNACK and OsYSL14K) designed primers generated polymorphic amplified products showing sequence length variation due to targeted amplification of candidate genes specific genomic regions. They found that molecular analysis based on candidate genes specific primers appeared to be an efficient approach for the elucidation of genetic differentiation and divergence in relation to variation of grain Zn concentration among entries. Thus they conclude that these markers can be effectively and efficiently utilized for grain Zn concentration related discrimination of rice genotypes and selection of parental genotypes for grain Zn biofortification.

Masuda et al. (2013) studied seven transgenic approaches, and combinations thereof, that can be used to increase the concentration of Fe in rice seeds. The first approach is to through expression of the Fe storage protein ferritin under the control of endosperm-specific promoters. Seed

Fe concentration of transformants was increased by approximately 2-fold in polished seeds. In the second approach using Fe translocation by over-producing the natural metal chelator nicotianamine, polished seeds had 3-fold increase. Another, by expressing the Fe(II)-nicotianamine transporter gene *OsYSL2* under the control of an endosperm-specific promoter and sucrose transporter promoter for enhancing Fe influx to the endosperm proved to increase Fe content up to 4-fold. The introduction of the barley mugineic acid synthesis gene *IDS3* to enhance Fe uptake and translocation within plants resulted in a 1.4-fold increase during field cultivation. In addition to these, Fe-biofortified rice was produced using a combination of the first, second, and third approaches. Most importantly, without any yield reduction, polished seeds contained 4.4 and 6 times higher than non-transgenic seeds under field and greenhouse-grown condition. Lee et al. (2012) found that by increasing the expression of *nicotianamine synthase* (*NAS*), the level of bioavailable iron could be fortified in rice seeds. Activation of iron deficiency-inducible *OsNAS2* resulted in a rise in Fe content (3.0-fold) in mature seeds. Enhanced expression led to higher tolerance of Fe deficiency and better growth under elevated pH. Mice fed with *OsNAS2-D1* seeds recovered more rapidly from anemia, indicating that bioavailable Fe contents were improved by this increase in *OsNAS2* expression. Interesting in a study, it has been found that germination rates were increased after iron biofortification process. Accumulation of total iron in brown rice germinated in 2 g^{L-1} solution of ferrous sulfate was increased by 12.05–16.50 times compared with non-biofortified germinated brown rice. Total phenolic and flavonoid contents were significantly ($p<0.05$) increased 1.29 and 3.91 times compared with brown rice seed. The antioxidant activities were notably enhanced, especially at high iron concentrations (Li et al., 2018).

12.3 ENVIRONMENTAL FRIENDLY RICE

Owing to wide use of silica nanoparticles (SNPs) in different industrial products many methods to extract silica from various waste products were developed. The agricultural waste, rice husk contain high amorphous silica but the color and purity of silica may get affected with the metal ion impurities and unburned carbon present in the rice husk. So, to produce SNPs with high quality and purity from rice husk Ghorbani et al.

(2015) optimized silica extraction procedure by eco-friendly technique. To evaluate the various rice husk pretreatments effect nitric, sulfuric and hydrochloric acids were used. Under controlled calcination conditions at 600°C the process of silica extraction was performed in an electric furnace. The XRF and XRD analyses results revealed the highest performance of hydrochloric acid treatment compared with the other acid treatments. The amorphous silica with 95.55% of purity and K_2O, CaO, and P_2O_5 as major impurities was obtained which indicated the removal of significant portion of amorphous silica by acid treatment. Then to get SNPs the amorphous silica was subjected to chemical precipitation and slow gelation. Finally, the successful SNPs production was indicated with the 409 $m^2.g^{-1}$ of BET specific surface area, spherical particle shape, the average particle size of 200±20 nm and 97% purity of SNPs.

Simultaneously, using K_2CO_3 both silica and activated carbon can be produced from rice husk as shown by Liu et al. (2012). The obtained activated carbon had 1,713 m^2 g^{-1} of surface area and 4 nm of average pore size. For methylene blue, the activated carbon maximum adsorption capacity was 210 mg g^{-1} and the capacitance value was 190 F/g. The silica yield was 96.84% with 40–50 nm of particle size. Further, the potassium carbonate used in this process can be recycled. This complete synthetic procedure was simple, ecofriendly, and economically effective.

Before fermentation to ease, the release of sugars from a lignocellulosic biomass pretreatment technology is a prerequisite. The current pretreatment methods which use ionic liquids were expensive and unpractical. To recover the bio-digestible cellulose from a lignocellulosic byproduct, rice straw and increase ionic liquid use efficient pretreatment method was developed by Nguyen et al. (2010) using ammonia and ionic liquid. A synergistic effect was seen between the combined use of ammonia and ionic liquid ([Emim] Ac) treatment and rice straw which recovered 82% of the cellulose and converted 97% of the enzymatic glucose. With a lowered enzyme usage and incubation time, also the glucose conversion was over 90% by this cooperative effect. The ionic liquid was recycled more than 20 times successfully with the cellulose recovery of 74% and 78% of the glucose conversion to rice straw at 20th-recycle ([Emim] Ac). For pretreatment of lignocellulosic biomass, the combined method was found to be more economical and eco-friendly than conventional pretreatment.

Rice husk is cheap and highly available agricultural waste which has metal absorption capacity. Rocha et al. (2013) investigated the mercury absorption of rice husk using representative concentrations of mercury. Two initial concentrations of Hg(II), one with maximum Hg value for discharges from industrial sectors (0.05 mg L^{-1}) and the other with 10 times higher concentration were taken to evaluate the effectiveness of rice husk. More than 80% reduction was observed for initial concentration 0.05 mg L^{-1} and more than 90% concentration was reduced for 0.50 mg L^{-1} with small amount of rice husk viz., 0.25 and 0.50 g L^{-1} and the residual values of these concentrations were 0.048 and 0.009 mg L^{-1}, respectively. This suggested that rice husk is a good bio absorbent which can be reused for cleaning treatments, maintaining the efficiency and high performance. This Hg-rice husk system absorption kinetics is fitted with Elovich model and the diffusion models which indicated that by using intraparticle diffusion or both film and intraparticle diffusion the rice husk absorption process can be controlled based on the initial concentrations of Hg (II).

12.4 CONCLUSION

The modern breeding techniques and biotechnology tools have advanced the traditional breeding methods to get plants with desirable traits such as increased yield, disease resistance, and improved quality. The modern biotechnology techniques such as genetic engineering and genome editing enormously increased the precision and reduced the time with which desirable changes in plant characteristics can be made. In this chapter, the recent developments in breeding and biotechnology methods to improve rice are summarized.

KEYWORDS

- **alternate wetting and drying**
- **continuously flood-irrigated**
- **human lactoferrin**
- **life cycle assessment**
- **lipoxygenases**
- **recombinant human lactoferrin**

REFERENCES

Babili, S. A., & Beyer, P., (2005). Golden rice—five years on the road—five years to go. *Trends in Plant Science*, *10*, 565–573.

Bethell, D. R., & Huang, J., (2004). Recombinant human lactoferrin treatment for global health issues: Iron deficiency and acute diarrhea. *Biometals*, *17*, 337–342.

Beyer, P., (2010). Olden rice and 'golden' crops for human nutrition. *New Biotechnology*, *27*, 478–481.

Beyer, P., Babili, S. A., Ye, X., Lucca, P., Schaub, P., Welsch, R., & Potrykus, I., (2002). Golden Rice: Introducing the β-Carotene biosynthesis pathway into rice endosperm by genetic engineering to defeat vitamin A deficiency. *The Journal of Nutrition*, *132*, 506–510.

Black, R.E., Allen,L.H., Bhutta, Z.A., Caulfield, L.E., de Onis, M., Ezzati,M., Mathers, C., & Riviera, J., (2008). Maternal and child health consequences. *The Lancet*, *371*, 243–260.

Cho, J. Y., Son, J. G., Song, C. H., Hwang, S. A., Lee, Y. M., Jeong, S. Y., & Chung, B. Y., (2008). Integrated nutrient management for environmental-friendly rice production in salt-affected rice paddy fields of Saemangeum reclaimed land of South Korea. *Paddy and Water Environment*, *6*, 263–273.

Chowdhury, M. R., Xiaobai, L., Hangying, Q., Wenxu, L., Jian, S., Cheng, H., & Dianxing, W., (2016). Functional characterization of 9-/13-LOXs in rice and silencing their expressions to improve grain qualities. *BioMed Research International*, *2*, 1–8.

Datta, K., Velazhahan, R., Oliva, N., Ona, I., Mew, T., Khush, G. S., Muthukrishnan, S., & Datta, S. K., (1999). Over-expression of the cloned rice thaumatin-like protein (PR-5) gene in transgenic rice plants enhances environmental friendly resistance to *Rhizoctonia solani* causing sheath blight disease. *Theoretical and Applied Genetics*, *98*, 1138–1145.

Garvin, D.F., Welch, R.M., & Finley, J.W., (2006). Historical shifts in the seed mineral micronutrient concentration of US hard red winter wheat germplasm. *Journal of the Science of Food and Agriculture*, *86*, 2213–2220.

Graham, R., Senadhira, D., Beebe, S., Iglesias, C., & Monasterio, I., (1999). Breeding for micronutrient density in edible portions of staple food crops: conventional approaches. *Field Crops Research*, *60*, 57–80.

Ghorbani, F., Sanati, A. M., & Maleki, M., (2015). Production of silica nanoparticles from rice husk as agricultural waste by environmental friendly technique. *Environmental Studies of Persian Gulf*, *2*, 56–65.

Gibson, R.S., (2006). Zinc: the missimg link in combating micronutrient nutrition in developing countries. *Proceedings of the Nutrition Society*, *65*, 51–60.

Gurdev, S. K., (1995). Breaking the yield frontier of rice. GeoJournal, *35*, 329–332.

Hazell, P. B. R., & Ramasamy, C., (1991). The green revolution reconsidered: The impact of high-yielding rice varieties in South India. *CABI Book, 301*.

Hennegan, K., Yang, D., & Nguyen, D., (2005). Improvement of human lysozyme expression in transgenic rice grain by combining wheat (*Triticum aestivum*) puroindoline b and rice (*Oryza sativa*) Gt1 promoters and signal peptides. *Transgenic Research*, *14*, 583–592.

Hoekenga, O., (2014) Genomics of Mineral Nutrient Biofortification: Calcium, Iron and Zinc. In: Tuberosa R., Graner A., Frison E. (eds) Genomics of Plant Genetic Resources. Springer, Dordrecht.

Hokazono, S., & Hayashi, K., (2012). Variability in environmental impacts during conversion from conventional to organic farming: A comparison among three rice production systems in Japan. *Journal of Cleaner Production*, *28*, 101–112.

Huang, F., Sun, Z., Hu, P., & Tang, S., (1998). Present situations and prospects for the research on rice grain quality forming. *Chinese Journal of Rice Science*, *3*, 1–7.

Huang, J., Nandi, S., Wu, L., Yalda, D., Bartley, G., Rodriguez, R., Lonnerdal, B., & Huang, N., (2002). Expression of natural antimicrobial human lysozyme in rice grains. *Molecular Breeding*, 10, 83–94.

IRRI, (2006). Bringing Hope, Improving Lives: Strategic Plan 2007-2015. International Rice Research Institute, Manila, Philippines, p. 61.

Kumari, K., Kumar, P., Sharma, V.K., & Singh, S.K., (2019). Genomic marker assisted identification of genetic loci and genes associated with variation of grain zinc concentration in rice. *Journal of Genetics*, *98*, 111. <https://doi.org/10.1007/s12041-019-1144-8>.

Lee, S., Kim, Y., Jeon, U.S., Kim, Y.K., Schjoerring, J.K., & An, G., (2012). Activation of Rice nicotianamine synthase 2 (OsNAS2) enhances iron availability for biofortification. *Molecules and Cells*, *33*, 269-275. <https://doi.org/10.1007/s10059-012-2231-3>.

Li, K., Hu, G., Yu, S., Tang, Q., & Liu, J., (2018). Effect of the iron biofortification on enzymes activities and antioxidant properties in germinated brown rice. *Food Measure*, *12*, 789–799.

Liu, Y., Guo, Y., Gao, W., Wang, Z., Ma, Y., & Wang, Z., (2012). Simultaneous preparation of silica and activated carbon from rice husk ash. *Journal of Cleaner Production*, *32*, 204–209.

Lusk, J. L., (2003). Effects of cheap talk on consumer willingness-to-pay for golden rice. *American Journal of Agricultural Economics*, *85*, 840–856.

Matsui, T., Omasa, K., & Horie, T., (2001). The difference in sterility due to high temperatures during the flowering period among *japonica*-rice varieties. *Plant Production Science*, *4*, 90–93.

Masuda, H., Aung, M.S. & Nishizawa, N.K., (2013). Iron biofortification of rice using different transgenic approaches. *Rice*, *6*, 40. <https://doi.org/10.1186/1939-8433-6-40>.

McDonald, G.K., Genc, Y., & Graham, R.D., (2008). A simple method to evaluate genetic variation in zinc grain concentration by correcting for differences in grain yield. *Plant and Soil*, *306*, 49–55.

Nandi, S., Suzuki, Y. A., Huang, J., Yalda, D., Pham, P., Wu, L., Bartley, G., Huang, N., & Lönnerdal, B., (2002). Expression of human lactoferrin in transgenic rice grains for the application in infant formula. *Plant Science*, *163*, 713–722.

Nandi, S., Yalda, D., Lu, S., Nikolov, Z., Misaki, R., Fujiyama, K., & Huang, N., (2005). Process development and economic evaluation of recombinant human lactoferrin expressed in rice grain. *Transgenic Research*, *14*, 237–249.

Nguyen, T. A. D., Kyoung, R. K., Se, J. H., Hwa, Y. C., Jin, W. K., Sung, M. P., Jae, C. P., & Sang, J. S., (2010). Pretreatment of rice straw with ammonia and ionic liquid for lignocelluloses conversion to fermentable sugars. *Bioresource Technology*, *101*, 7432–7438.

Paine, J. A., Shipton, C. A., & Chaggar, S., (2005). Improving the nutritional value of golden rice through increased pro-vitamin content. *Nature Biotechnology*, *23*, 482–487.

Patel, D. P., Das, A., Munda, G. C., Ghosh, P. K., Bordoloi, J. S., & Kumar, M., (2010). Evaluation of yield and physiological attributes of high-yielding rice varieties under aerobic and flood-irrigated management practices in mid-hills ecosystem. *Agricultural Water Management*, *97*, 1269–1276.

Potrykus, I., (2001). Golden rice and beyond. *Plant Physiology, 125*, 1157–1161.

Qiao, F., & Zhao, K. -J., (2011). The influence of RNAi targeting of OsGA20ox2 gene on plant height in rice. *Plant Molecular Biology Reporter*, *29*, 952–960.

Rocha, L. S., Lopes, C. B., Borges, J. A., Duarte, A. C., & Pereira, E., (2013). Valuation of unmodified rice husk waste as an eco-friendly sorbent to remove mercury: A study using environmental realistic concentrations. *Water, Air, and Soil Pollution*, *224*, (1599).

Sathya A., Vijayabharathi R., & Gopalakrishnan S., (2016). Exploration of Plant Growth-Promoting Actinomycetes for Biofortification of Mineral Nutrients. In: Subramaniam G., Arumugam S., Rajendran V. (eds) *Plant Growth Promoting Actinobacteria*. Springer, Singapore.

Scobie, G. M., & Posada, T. R., (1977). The impact of high-yielding rice varieties in Latin America, with special emphasis on Colombia. Centro Internacional de Agricultura Tropical (CIAT), Cali, Colombia. CAB International. 171 p.

Shamik, M., Paul, Q. W., & Anindya, B., (2016). CRISPR-Cas9 mediated genome editing in rice, advancements and future possibilities. *Indian Journal of Plant Physiology*, *21*(4), doi: 10.1007/s40502-016-0252-1.

Stein, A.J., (2010). Global impacts of human mineral nutrition. *Plant and Soil*, *335*, 133–154.

Swamy, B.P.M., Rahman, M.A., Inabangan-Asilo, M.A. Amparado, A., Manito, C., Mohanty, P.C., Reinke, R., & Slamet-Loedin, I.H., (2016). Advances in breeding for high grain Zinc in Rice. *Rice*, *9*, 49. <https://doi.org/10.1186/s12284-016-0122-5>.

Tang, L., Mao, B., Li, Y., et al., (2017). Knockout of OsNramp5 using the CRISPR/Cas9 system produces low Cd-accumulating *indica* rice without compromising yield. *Scientific Reports*, *7*, 1–12.

Tang, G., Qin, J., Dolnikowski, G. G., Russell, R. M., & Grusak, M. A., (2009). Golden rice is an effective source of vitamin A. *The American Journal of Clinical Nutrition*, *89*, 1776–1783.

Tiwari, G. J., Liu, Q., Shreshtha, P., Li, Z., & Rahman, S., (2016). RNAi-mediated down-regulation of the expression of OsFAD2-1: Effect on lipid accumulation and expression of lipid biosynthetic genes in the rice grain. *BMC Plant Biology*, *16*, 1–13.

Vanaja, T., & Babu, L. C., (2006). Variability in grain quality attributes of high yielding rice varieties (*Oryza sativa* L.) of diverse origin. *Tropical Agriculture*, *44*, 61–63.

Walker, C.L.F., Ezzati, M., & Black, R.E., (2009). Global and regional child mortality and burden of disease attributable to zinc deficiency. *European Journal of Clinical Nutrition*, *63*, 591–597.

Yao, F., Huang, J., Cui, K., Nie, L., Xiang, J., Liu, X., Wua, W., Chen, M., & Peng, S., (2012). Agronomic performance of high-yielding rice variety grown under alternate wetting and drying irrigation. *Field Crops Research*, *126*, 16–22.

Zavaleta, N., Figueroa, D., Rivera, J., Sánchez, J., Alfaro, S., & Lönnerdal, B., (2007). Efficacy of rice-based oral rehydration solution containing recombinant human lactoferrin and lysozyme in Peruvian children with acute diarrhea. *Journal of Pediatric Gastroenterology and Nutrition*, *44*, 258–264.

Zhu, C., Shen, W., Zhao, H., Wan, J., & Zhongguo, N. K., (2004). Advances in researches of the application of low-amylose content rice gene for breeding. *Europe PMC*, *37*, 157–162.

Zimmermann, R., & Qaim, M., (2004). Potential health benefits of golden rice: A Philippine case study. *Food Policy*, *29*, 147–168.

Index

B

C

H

I

J

K

Q

R

S

Z